# Primate Perspectives
on Behavior and Cognition

# Primate Perspectives on Behavior and Cognition

EDITED BY
David A. Washburn

AMERICAN PSYCHOLOGICAL ASSOCIATION
WASHINGTON, DC

Copyright © 2007 by the American Psychological Association. All rights reserved. Except as permitted under the United States Copyright Act of 1976, no part of this publication may be reproduced or distributed in any form or by any means, including, but not limited to, the process of scanning and digitization, or stored in a database or retrieval system, without the prior written permission of the publisher.

Published by
American Psychological Association
750 First Street, NE
Washington, DC 20002
www.apa.org

To order
APA Order Department
P.O. Box 92984
Washington, DC 20090-2984
Tel: (800) 374-2721
Direct: (202) 336-5510
Fax: (202) 336-5502
TDD/TTY: (202) 336-6123
Online: www.apa.org/books/
E-mail: order@apa.org

In the U.K., Europe, Africa, and the Middle East, copies may be ordered from
American Psychological Association
3 Henrietta Street
Covent Garden, London
WC2E 8LU England

Typeset in New Century Schoolbook by World Composition Services, Inc., Sterling, VA

Printer: United Book Press, Baltimore, MD
Cover Designer: Berg Design, Albany, NY
Technical/Production Editor: Harriet Kaplan

The opinions and statements published are the responsibility of the authors, and such opinions and statements do not necessarily represent the policies of the American Psychological Association.

**Library of Congress Cataloging-in-Publication Data**

Primate perspectives on behavior and cognition / edited by David A. Washburn.
    p. cm.— (Decade of behavior)
    Festschrift for Duane M. Rumbaugh.
    Includes bibliographical references and index.
    ISBN-13: 978-1-59147-422-7
    ISBN-10: 1-59147-422-1
    1. Apes—Psychology. 2. Apes—Behavior. 3. Learning in animals. 4. Psychology, Comparative. I. Washburn, David A., 1961-    II. Rumbaugh, Duane M., 1929- III. Series.

QL737.P96P738 2006
599.8'15—dc22
    2006002458

**British Library Cataloguing-in-Publication Data**
A CIP record is available from the British Library.

*Printed in the United States of America*
*First Edition*

## APA Science Volumes

*Attribution and Social Interaction: The Legacy of Edward E. Jones*

*Best Methods for the Analysis of Change: Recent Advances, Unanswered Questions, Future Directions*

*Cardiovascular Reactivity to Psychological Stress and Disease*

*The Challenge in Mathematics and Science Education: Psychology's Response*

*Changing Employment Relations: Behavioral and Social Perspectives*

*Children Exposed to Marital Violence: Theory, Research, and Applied Issues*

*Cognition: Conceptual and Methodological Issues*

*Cognitive Bases of Musical Communication*

*Cognitive Dissonance: Progress on a Pivotal Theory in Social Psychology*

*Conceptualization and Measurement of Organism–Environment Interaction*

*Converging Operations in the Study of Visual Selective Attention*

*Creative Thought: An Investigation of Conceptual Structures and Processes*

*Developmental Psychoacoustics*

*Diversity in Work Teams: Research Paradigms for a Changing Workplace*

*Emotion and Culture: Empirical Studies of Mutual Influence*

*Emotion, Disclosure, and Health*

*Evolving Explanations of Development: Ecological Approaches to Organism–Environment Systems*

*Examining Lives in Context: Perspectives on the Ecology of Human Development*

*Global Prospects for Education: Development, Culture, and Schooling*

*Hostility, Coping, and Health*

*Measuring Patient Changes in Mood, Anxiety, and Personality Disorders: Toward a Core Battery*

*Occasion Setting: Associative Learning and Cognition in Animals*

*Organ Donation and Transplantation: Psychological and Behavioral Factors*

*Origins and Development of Schizophrenia: Advances in Experimental Psychopathology*

*The Perception of Structure*

*Perspectives on Socially Shared Cognition*

*Psychological Testing of Hispanics*

*Psychology of Women's Health: Progress and Challenges in Research and Application*

*Researching Community Psychology: Issues of Theory and Methods*

*The Rising Curve: Long-Term Gains in IQ and Related Measures*

*Sexism and Stereotypes in Modern Society: The Gender Science of Janet Taylor Spence*

*Sleep and Cognition*

*Sleep Onset: Normal and Abnormal Processes*

*Stereotype Accuracy: Toward Appreciating Group Differences*

*Stereotyped Movements: Brain and Behavior Relationships*

*Studying Lives Through Time: Personality and Development*

*The Suggestibility of Children's Recollections: Implications for Eyewitness Testimony*

*Taste, Experience, and Feeding: Development and Learning*

*Temperament: Individual Differences at the Interface of Biology and Behavior*

*Through the Looking Glass: Issues of Psychological Well-Being in Captive Nonhuman Primates*

*Uniting Psychology and Biology: Integrative Perspectives on Human Development*

*Viewing Psychology as a Whole: The Integrative Science of William N. Dember*

## APA Decade of Behavior Volumes

*Acculturation: Advances in Theory, Measurement, and Applied Research*

*Animal Research and Human Health: Advancing Human Welfare Through Behavioral Science*

*Behavior Genetics Principles: Perspectives in Development, Personality, and Psychopathology*

*Categorization Inside and Outside the Laboratory: Essays in Honor of Douglas L. Medin*

*Child Development and Social Policy: Knowledge for Action*

*Children's Peer Relations: From Development to Intervention*

*Computational Modeling of Behavior in Organizations: The Third Scientific Discipline*

*Developing Individuality in the Human Brain: A Tribute to Michael I. Posner*

*Emerging Adulthood: Coming of Age in the 21st Century*

*Experimental Cognitive Psychology and Its Applications*

*Family Psychology: Science-Based Interventions*

*Memory Consolidation: Essays in Honor of James L. McGaugh*

*Models of Intelligence: International Perspectives*

*The Nature of Remembering: Essays in Honor of Robert G. Crowder*

*New Methods for the Analysis of Change*

*On the Consequences of Meaning Selection: Perspectives on Resolving Lexical Ambiguity*

*Participatory Community Research: Theories and Methods in Action*

*Personality Psychology in the Workplace*

*Perspectivism in Social Psychology: The Yin and Yang of Scientific Progress*

*Primate Perspectives on Behavior and Cognition*

*Principles of Experimental Psychopathology: Essays in Honor of Brendan A. Maher*

*Psychosocial Interventions for Cancer*

*Racial Identity in Context: The Legacy of Kenneth B. Clark*

*Strengthening Research Methodology: Psychological Measurement and Evaluation*

*The Social Psychology of Group Identity and Social Conflict: Theory, Application, and Practice*

*Unraveling the Complexities of Social Life: A Festschrift in Honor of Robert B. Zajonc*

*Visual Perception: The Influence of H. W. Leibowitz*

# Contents

Contributors .................................................................... xiii

Series Foreword ............................................................... xv

Volume Foreword ............................................................. xvii
*William A. Mason*

Preface ........................................................................... xix

Introduction ..................................................................... 3
*David A. Washburn*

**Part I. Studying Primate Behavior** ................................... 5

    1. The Comparative Psychology of Duane Rumbaugh
       and His Influence on Zoo Biology ............................... 7
       *Terry L. Maple and Christopher W. Kuhar*

    2. Apes, Intelligent Science, and Conservation ................ 17
       *Russell H. Tuttle*

    3. Studies at the Great Ape Research Institute, Hayashibara ..... 29
       *Gen'ichi Idani and Satoshi Hirata*

    4. Continuity of Cognition Across Species:
       Darwin in Cyberspace ............................................... 37
       *Katherine A. Leighty, Dorothy M. Fragaszy, and*
       *James M. Brown*

    5. Dimensions of the Ape Mind: Adding Personality to
       Behavior and Cognition ............................................. 47
       *James E. King*

**Part II. Interpreting Primate Behavior** ............................ 61

    6. Species of Parsimony in Comparative Studies of Cognition ..... 63
       *J. David Smith*

    7. The Significance of the Concept of Emergence for
       Comparative Psychology ............................................ 81
       *Gary Greenberg, Ty Partridge, and Elizabeth Ablah*

8. The Emergence of Emergents: One Behaviorist's Perspective ..... 99
   M. Jackson Marr

9. The Perception of Emergents ..... 109
   David A. Washburn

10. New Models of Ability Are Needed:
    New Methods of Assessment Will Be Required ..... 125
    H. Carl Haywood

## Part III. Learning and Cognition ..... 135

11. The Transfer Index as a Precursor of
    Nonhuman Language Research and Emergents ..... 137
    James L. Pate

12. Monkeys Making a List: Checking It Twice? ..... 143
    F. Robert Treichler

13. Animals Count: What's Next? Contributions From
    the Language Research Center to Nonhuman Animal
    Numerical Cognition Research ..... 161
    Michael J. Beran, Jonathan P. Gulledge, and
    David A. Washburn

14. Do Primates Plan Routes? Simple Detour
    Problems Reconsidered ..... 175
    Emil W. Menzel Jr. and Charles R. Menzel

15. Willful Apes Revisited: The Concept of Prospective Control ..... 207
    R. Thompson Putney

## Part IV. Language and Tools ..... 221

16. The Past, Present, and Possible Futures of Animal
    Language Research ..... 223
    William A. Hillix

17. A Comparative Psychologist Looks at Language ..... 235
    Herbert L. Roitblat

18. Evolution of Language and Speech From a
    Neuropsychological Perspective ..... 243
    William D. Hopkins

19. Symbol Combination in *Pan*: Language, Action,
    and Culture ..... 255
    Patricia Greenfield and Heidi Lyn

20. Epigenesis, Mental Construction, and the Emergence
    of Language and Toolmaking ..... 269
    Kathleen R. Gibson

21. Kanzi Learns to Knap Stone Tools .................................... 279
     E. Sue Savage-Rumbaugh, Nicholas Toth, and Kathy Schick

An Afterword—and Words of Thanks .................................... 293
Duane M. Rumbaugh

Author Index ............................................................. 299

Subject Index ............................................................ 309

About the Editor ........................................................ 319

# Contributors

**Elizabeth Ablah,** Wichita State University, Wichita, KS
**Michael J. Beran,** Georgia State University, Atlanta
**James M. Brown,** University of Georgia, Athens
**Dorothy M. Fragaszy,** University of Georgia, Athens
**Kathleen R. Gibson,** Houston Medical School, University of Texas
**Gary Greenberg,** Wichita State University, Wichita, KS
**Patricia Greenfield,** University of California, Los Angeles
**Jonathan P. Gulledge,** Lee University, Cleveland, TN
**H. Carl Haywood,** Vanderbilt University, Nashville, TN
**William A. Hillix,** San Diego State University, San Diego, CA
**Satoshi Hirata,** Great Ape Research Institute, Hayashibara, Japan
**William D. Hopkins,** Yerkes National Primate Research Center, Emory University, Atlanta, GA, and Berry College, Mount Berry, GA
**Gen'ichi Idani,** Great Ape Research Institute, Hayashibara, Japan
**James E. King,** University of Arizona, Tucson
**Christopher W. Kuhar,** TECHlab/Zoo, Atlanta, GA, and Georgia Institute of Technology, Atlanta
**Katherine A. Leighty,** University of Georgia, Athens
**Heidi Lyn,** University of St. Andrews, St. Andrews, Fife, Scotland
**Terry L. Maple,** TECHlab/Zoo Atlanta, GA, and Georgia Institute of Technology, Atlanta
**M. Jackson Marr,** Georgia Institute of Technology, Atlanta
**William A. Mason,** University of California, Davis
**Charles R. Menzel,** Georgia State University, Atlanta
**Emil W. Menzel Jr.,** State University of New York, Stony Brook
**Ty Partridge,** Wayne State University, Detroit, MI
**James L. Pate,** Georgia State University, Atlanta
**R. Thompson Putney,** Georgia State University, Atlanta
**Herbert L. Roitblat,** OrcaTec LLC, Ojai, CA
**Duane M. Rumbaugh,** Georgia State University, Atlanta, and Great Ape Trust, Des Moines, IA
**E. Sue Savage-Rumbaugh,** Georgia State University, Atlanta, and Great Ape Trust, Des Moines, IA
**Kathy Schick,** Stone Age Institute, Gosport, IN, and Indiana University, Bloomington
**J. David Smith,** University at Buffalo, State University of New York
**Nicholas Toth,** Stone Age Institute, Gosport, IN, and Indiana University, Bloomington
**F. Robert Treichler,** Kent State University, Kent, OH
**Russell H. Tuttle,** The University of Chicago, Chicago, IL
**David A. Washburn,** Georgia State University, Atlanta

# Series Foreword

In early 1988, the American Psychological Association (APA) Science Directorate began its sponsorship of what would become an exceptionally successful activity in support of psychological science—the APA Scientific Conferences program. This program has showcased some of the most important topics in psychological science and has provided a forum for collaboration among many leading figures in the field.

The program has inspired a series of books that have presented cutting-edge work in all areas of psychology. At the turn of the millennium, the series was renamed the Decade of Behavior Series to help advance the goals of this important initiative. The Decade of Behavior is a major interdisciplinary campaign designed to promote the contributions of the behavioral and social sciences to our most important societal challenges in the decade leading up to 2010. Although a key goal has been to inform the public about these scientific contributions, other activities have been designed to encourage and further collaboration among scientists. Hence, the series that was the "APA Science Series" has continued as the "Decade of Behavior Series." This represents one element in APA's efforts to promote the Decade of Behavior initiative as one of its endorsing organizations. For additional information about the Decade of Behavior, please visit http://www.decadeofbehavior.org.

Over the course of the past years, the Science Conference and Decade of Behavior Series has allowed psychological scientists to share and explore cutting-edge findings in psychology. The APA Science Directorate looks forward to continuing this successful program and to sponsoring other conferences and books in the years ahead. This series has been so successful that we have chosen to extend it to include books that, although they do not arise from conferences, report with the same high quality of scholarship on the latest research.

We are pleased that this important contribution to the literature was supported in part by the Decade of Behavior program. Congratulations to the editors and contributors of this volume on their sterling effort.

Steven J. Breckler, PhD  
*Executive Director for Science*

Virginia E. Holt  
*Assistant Executive Director for Science*

# Volume Foreword

## *William A. Mason*

Duane M. Rumbaugh is a person of many accomplishments. His contributions range over a broad and varied spectrum, including major methodological innovations, groundbreaking research, and seminal theories of learning and intelligence. He must also be credited with the creation and nurture of a unique, world-class language research center. The contributors to this volume are people whose lives he has touched. His visionary leadership, his kindness, and his encouragement and support are qualities that are acknowledged throughout these pages. The volume is also a rich and provocative store of original research, ideas, essays, and theories revolving around fundamental issues in primate psychology. These stand as a fitting tribute to Rumbaugh's achievements as a sponsor, colleague, and friend. I am sure this will please him.

Nevertheless, I suspect that nothing could give him quite the same pleasure or professional satisfaction as what he has discovered while exploring the mind of the chimpanzee. Although much has been revealed about the remarkable abilities of this fascinating animal, thanks to the efforts of Rumbaugh and those who share his commitment and zeal, I have no doubt that he will be among the first to say that more is to come. Who can say what the future will bring? As he said recently about the prospects of research on the abilities of animals to acquire human language, "History will tell." Whatever the record may contain, we can be sure that it will include Duane M. Rumbaugh in a prominent place.

# Preface

This book represents the efforts of an accomplished group of scientists from around the world to celebrate the ongoing career and colleagueship of Duane M. Rumbaugh. Comparative psychologists, cognitive psychologists, neuropsychologists, biologists, primatologists, and anthropologists were among those gathered not to mark the end of Rumbaugh's contributions but rather to commemorate the areas of science that continue to drive his research interests after 5 decades of productivity. The 2-day Festschrift was hosted by Georgia State University and was cosponsored by the American Psychological Association and by Georgia State University's Language Research Center, Department of Psychology (particularly the Social/Cognitive and Neuropsychology/Behavioral Neuroscience graduate programs), and College of Arts and Sciences. A follow-up Festschrift symposium was held as part of the program at the annual meeting of the Southern Society for Philosophy and Psychology (an organization for which Rumbaugh served as president in 1996). The theme for both Festschrift sessions was "Emergents and Rational Behaviorism."

The present book reflects some of the outstanding contributions to those sessions. The authors included here bring a diverse range of expertise and perspectives. Within the field of psychology, there are authors who identify themselves as comparative psychologists, cognitive psychologists, behaviorists, neuropsychologists or biopsychologists, and developmental psychologists. Beyond this discipline, there are contributions from anthropologists and primatologists who share Rumbaugh's passion and understanding that behavioral science should be studying behavior as it is manifest across species.

In addition to the groups listed above who generously supported the Festschrifts, I am grateful for the support, encouragement, and assistance of Robin Morris, Judith Sizemore, Charlene Weiters, Bill Hopkins, Lauren Adamson, and many others who helped make the sessions a success. Michael J. Beran organized a student poster session for the October Festschrift, and many students contributed to the overall strength and benefits of the conference by participating with their posters. As the plans for the Festschrift became increasingly ambitious, Kimberly MacQueen assumed the incredible burden of coordinating the details. Without her talents and tireless effort, the Festschrift would never have succeeded. Support from the National Institute of Child Health and Human Development (particularly through Grants HD 38051 and 06016) has been critical for the organization of the conference, the editing of this book, and much of the scientific accomplishment recorded therein. Finally, I gratefully acknowledge the generosity of Steve Woodruff, who established the Rumbaugh Fellowship at Georgia State University to support the education of new generations of students who are interested in emergents and rational behaviorism.

Throughout this process, Rumbaugh has been uncomfortable with the attention and unwilling to accept superlatives, accolades, or anything at all resembling eulogies. He has correctly noted that his career is far from over, and he has characteristically deflected the focus to the science and to the nonhuman primates that have enriched his life. Although he has resisted veneration, his praises hardly need be spoken. The enthusiasm with which some of the most prominent scholars in the world accepted the invitation to participate stands as a testament to the regard with which Rumbaugh is held among his peers. Although some of those participants could not contribute to the present volume, each was steadfast in support of the project. Many other distinguished scientists, former students or colleagues, and friends attended the sessions or expressed their regrets at being unable to participate in a fitting tribute to this accomplished and influential man. All of these factors attest to our admiration and appreciation of Rumbaugh's many talents, not the least of which is his capacity for investing in others and in the animals he loves. So often in his career, Rumbaugh has identified the important research questions to be addressed, has developed the innovative methods by which these questions might be investigated, and has provided the theoretical framework for integrating a wide range of data. He serves as our colleague, our mentor, our inspiration, our critic, our administrator, and our friend. It is thus with anticipation for the future and appreciation for the past that we dedicate this volume to Duane M. Rumbaugh.

# Primate Perspectives on Behavior and Cognition

# Introduction

## David A. Washburn

Comparative psychologist. Anthropologist. Biologist. Cognitive psychologist. Psycholinguist. Developmental psychologist. Primatologist. Neuroscientist. Learning theorist. This could certainly be an incomplete list of professional identifications appropriate to the career of Duane M. Rumbaugh, whose contributions are honored in this book. It also reflects a partial list of professional identifications of the contributors to this volume and, we hope, of the readers who might be interested in these chapters. This compilation of essays provides a broad perspective on animal behavior and, more generally, on the nature of learning, behavior, and science itself.

The volume is divided into four parts, the first two dealing with methods and perspectives on animal behavior and the latter two focused on the content of studies of animal behavior. Part I: Studying Primate Behavior provides a unique source of background information for students of animal behavior. Terry L. Maple and Christopher W. Kuhar provide a brief biography of the man who serves as the inspiration for these chapters and for many of the careers that are reflected in the chapters. This biography is provided so that Rumbaugh's influence on zoo biology can be evaluated. Russell H. Tuttle provides a similarly useful introduction, in this case to portray the nature of the primate order and the implications for current political debates. The next two chapters illustrate specific paradigmatic contributions Rumbaugh has made to the field and serve as guideposts for new generations of researchers seeking to study behavior as it is manifest across primate species. Gen'ichi Idani and Satoshi Hirata accomplish this in their overview of the Great Ape Research Institute in Japan. Katherine A. Leighty, Dorothy M. Fragaszy, and James M. Brown focus on cyberspace rather than a physical environment in their review of the revolutionary ways that computer-based testing has changed comparative study of behavior. The continuity in psychological performance across primate species that is studied so easily with this apparatus is illustrated very well in James E. King's contribution showing parallels in personality between humans and great apes.

Part II: Interpreting Primate Behavior turns from method to theory. J. David Smith provides a critical perspective on this section, defending the continuity of process view against the species-centric bias to adopt different explanations for comparable behaviors in other species. Gary Greenberg, Ty

Partridge, and Elizabeth Ablah then provide a critical review of Rumbaugh's notion of emergent behaviors that are manifest (to varying degrees) across the primate order and, indeed, across the animal kingdom. An alternative perspective is provided by M. Jackson Marr, whose experimental analysis of behavior denies the necessity or utility of emergent forms of learning. My chapter attempts a reconciliation of the behaviorist and cognitive perspectives, offering a neo-Gibsonian view of learning across species. Finally, H. Carl Haywood summarizes the theoretical and practical contributions of comparative research, with a particular eye toward implications for translations for children with disabilities.

The second half of the book could serve as a textbook for courses in comparative cognition. In Part III: Learning and Cognition, primate research on most of the major higher order cognitive constructs is discussed: intelligence (James L. Pate), learning and memory (F. Robert Treichler), numerical cognition (Michael J. Beran, Jonathan P. Gulledge, and David A. Washburn), spatial cognition (Emil W. Menzel Jr. and Charles R. Menzel), and intentionality/executive functioning (R. Thompson Putney).

The constructs reflecting the topic of study for which Rumbaugh is most renowned receive particularly comprehensive treatment in Part IV: Language and Tools. William A. Hillix provides a historical perspective to the comparative study of language. Herbert L. Roitblat defends ape-language research with a comparative analysis of human language. William D. Hopkins and Kathleen R. Gibson contribute complementary analyses of the evolution of language from neuroscientific and anthropological perspectives. Sandwiched between those chapters, a developmental framework is used by Patricia Greenfield and Heidi Lyn in their analysis of linguistic utterances by apes. These three chapters include a consideration of the role of culture in cognitive competence. The volume concludes with E. Sue Savage-Rumbaugh, Nicholas Toth, and Kathy Schick's chapter describing the tool-use and toolmaking behavior of a bonobo named Kanzi.

In addition to serving as a comprehensive and integrative review of a wide range of research areas, the chapters in this volume establish a research agenda for years (and careers) to come. How does one bridge the gap between observable behavior as primary data on the one hand and the cognitive potential for behavior on the other? Is the notion of emergents a useful one, and if so, what are the processes that support emergent and relational forms of learning? Why do animals (including humans) occasionally fail to learn relationally? How can reinforcement, salience, attention, or memory be defined and studied in noncircular ways? Having established that nonhuman primates can learn and use humanlike language, why should we care? That is, what is it that language allows a nonhuman animal to do (or to do better)? How can we translate a history of studying behavior and cognition as they are manifest across species into interventions and applications with adults and children of our own species? Almost 100 years of radical behaviorism have failed to answer these and other important questions. The future will reveal whether Rumbaugh's more rational behaviorism continues to make progress in these areas.

# Part I

# Studying Primate Behavior

# 1

# The Comparative Psychology of Duane Rumbaugh and His Influence on Zoo Biology

*Terry L. Maple and Christopher W. Kuhar*

Duane Rumbaugh, who spent 2 decades studying primates at the San Diego Zoo, also significantly contributed to the strong scientific foundation of Zoo Atlanta, and by his successful research in laboratories, primate centers, and zoos, he should be recognized as a leader in the emergence of scientific zoo biology. Throughout his career, he has demonstrated an ability to see ahead of the curve; he anticipated opportunities despite organizational impediments and the obstructions of those with less developed vision and creativity.

In the early 1970s, Rumbaugh collaborated with Geoffrey Bourne to develop "Articles of Incorporation" for the newly formed Zoological Society of Atlanta. The documents, filed with the State of Georgia in October 1970 and coauthored by prominent Atlanta attorney (and published zoo historian) Richard Reynolds, anticipated future private management of Atlanta's municipal zoo, much like the organizational structure of the esteemed San Diego Zoo. Rumbaugh was familiar with the history of the San Diego Zoo, where he had successfully conducted benchmark behavioral research during his service in the Psychology Department at San Diego State University. Collaborating together on behalf of an enlightened constituency, Rumbaugh and Bourne codified a scientific foundation for the new society and its zoo. They proclaimed that the society was fundamentally organized to

1. advance the sciences of zoology and natural history by emphasizing research, education, and conservation;
2. assist and cooperate with any and all zoological gardens and parks; and
3. sponsor, own, and direct zoological gardens and collections of living animal forms in a manner consistent with the above-stated goals and which would encourage their breeding and the conservation of the species.

In these directives, Rumbaugh and Bourne envisioned the importance of the scientific management of endangered species and the need for zoos worldwide to cooperate in advancing the cause of global conservation. They also identified the urgent priority to formulate a pragmatic model for private, nongovernmental zoos, operated with sufficient autonomy, authority, and entrepreneurial spirit to upgrade standards. In short, Duane Rumbaugh and his close collaborators accurately forecast a vision for a new kind of zoo. Zoo Atlanta embodies all of these attributes and more, inspired by a local vision fashioned well in advance of the first opportunity to engage significant organizational change. A management crisis in the early 1980s provided the spark to invoke and expand on the "Rumbaugh–Bourne vision" (Desiderio, 2000).

## Duane Rumbaugh—Zoo Biologist

Zoo Atlanta hosted the fifth "Geoffrey Bourne Lecture" in 2005 in honor of the first president of the Zoological Society of Atlanta, generously supported by an endowment fund established by Nelly Bourne. Duane Rumbaugh, a founding member of the Zoological Society of Atlanta, delivered the Bourne lecture in 2002, revealing some of the inside details of how he and Bourne envisioned the zoo. Zoo Atlanta historians are still discovering their history, so a better understanding of Rumbaugh's leadership in promulgating and implementing the scientific standards for Zoo Atlanta is anticipated.

Rumbaugh actually started work at the San Diego Zoo on their rodent collection—not the exhibited rodents, but those destined to become prey for reptiles, birds, and other carnivorous zoo animals. Once at the zoo, he clearly recognized the opportunity presented by the large and diverse collection of nonhuman primates (Figure 1.1). From conversations with Rumbaugh, we learned that research support was modest by current standards, and he had to overcome formidable bureaucratic obstacles to carry out his ambitious program of research. As an example, his acclaimed and pioneering film *Survey of the Primates* (Rumbaugh, Riesen, & Lee, 1970) was resisted by zoo administrators who did not want to fund it. In the end, the film brought widespread recognition and respect to the zoo as a center for disseminating knowledge. Given the public nature of zoological parks, research is always a logistical challenge in the zoo, and Rumbaugh reflected on this in two publications in the journal *BioScience* (Rumbaugh, 1971, 1972b). For example, he provided this advice to those who seek to use the zoo as a classroom:

> Whereas zoo staff are willing and prepared to accommodate occasional requests, their long-term and continuing cooperation can be achieved only by careful planning. Avoid having your instructional needs appear to zoo personnel as costly and problematic activities that interfere with the zoo's prime function—the exhibiting of animal forms. (Rumbaugh, 1971, p. 806)

Rumbaugh's access to the San Diego Zoo collection, then and now the world's largest captive collection of living animals, gave him an opportunity to conduct truly comparative research. Only Robert Yerkes before him had been able to evaluate the learning ability of chimpanzees, gorillas, and orangutans. In 1916, Yerkes pursued his research interests in nonhuman primate

**Figure 1.1.** Although Duane Rumbaugh's early research in zoos involved rodents, he soon realized the potential of working with the zoos' vast diversity of primate species, including this white-cheeked gibbon.

*ideation* (a type of insight learning) at a private estate in Montecito, California, where he tested a privately owned orangutan named Julius (Yerkes & Yerkes, 1929), and with the cooperation of the Ringling Brothers Circus, he conducted experiments with their young gorilla Congo (Yerkes, 1928). To further his knowledge of chimpanzees, Yerkes traveled to Cuba to study the animals in Rosalia Abreu's unique breeding colony at Quinta Palatino (Yerkes, 1925), but he never attained the depth of experience with great ape subjects that Rumbaugh achieved in his career.

We have taken a special interest in the relationship between San Diego State University and the San Diego Zoo. (The first author of this chapter, Terry Maple, grew up in San Diego County, and his two brothers graduated from San Diego State University.) The San Diego Zoo has exerted a powerful influence on the children of San Diego. Indeed, from its inception in 1916, the zoo has been dedicated to the children of San Diego. Duane Rumbaugh provided educational leadership at San Diego State from 1954 until he was recruited to a leadership

**Figure 1.2.** Rumbaugh (right) followed the model established by Robert Yerkes and Harry Harlow (here pictured at left) using primate centers as a cognitive research lab.

position at the Yerkes Primate Research Center in 1969. His research lab at the San Diego Zoo provided many opportunities for students to observe nonhuman primates as they learned the principles of biology and psychology. His legacy in San Diego is his record of productivity. Some 30 publications resulted from his research at the zoo. No one has utilized a zoo collection for behavioral research more successfully than Rumbaugh did from 1954 until he arrived in Atlanta in 1969. Unfortunately, the Atlanta Zoo was not yet ready for serious scientific endeavors; otherwise Duane Rumbaugh might have continued his zoo-based studies of primate learning. However, with Yerkes's superior collection of great apes (the largest in the world at the time), Rumbaugh's focus shifted from the zoo to the primate center (Figure 1.2).

Rumbaugh and Harry F. Harlow of the University of Wisconsin (who is also depicted in Figure 1.2) both used zoos for their earliest behavioral research on primates. Starting somewhat earlier in 1930 at a smaller and more specialized zoo in Madison, Wisconsin, Harlow, like Rumbaugh, had no lab in the psychology department and used the nearby zoo out of necessity. Both men made contributions to the psychology of learning and generated normative data illuminating the psychological well-being of captive primates.

In San Diego, Rumbaugh studied more than a dozen different species of primate, from squirrel monkeys to orangutans. At a time when comparative psychologists were mesmerized by rats and pigeons, he tapped into the massive inventory of biodiversity available only in the zoo (Rumbaugh, 1965; Rumbaugh, 1972a; Rumbaugh & Arnold, 1971; Rumbaugh & Rice, 1962). He clearly grasped the idea that a zoo could become a living laboratory for comparative psychologists, physical anthropologists, and ethologists. He also noted that observations of zoo animals could prove inspirational and fun for aspiring young scientists (e.g., Rumbaugh, 1971). Certainly, his work fits the definition of science conducted under the banner of "zoo biology" as Heini Hediger characterized it in his influential book *Man and Animal in the Zoo* (1969).

Hediger, who in addition to his duties as professor of ethology at the University of Zurich was a director of three Swiss zoos (Basel, Bern, and Zurich), proclaimed once that "science is always last in the zoological garden" (Hediger, 1969, p. 47). He lamented this conclusion, but it was ultimately based on his observation that issues that affected the bottom line would always be more important to a zoo's board of directors. Zoo priorities have not changed much since Hediger's time, but the San Diego Zoo and Zoo Atlanta have benefited from a historical focus on science, education, and conservation. The elite zoos of the United States (Bronx, Brookfield, National, and San Diego) and the aspiring second tier (Atlanta, Lincoln Park, New Orleans, and Seattle) have written science into their strategic vision. Scientific zoo biology is beginning to take hold in the United States (e.g., Finlay & Maple, 1986; Maple, 1999; Stoinski, Lukas, & Maple, 1998), and Duane Rumbaugh has contributed to the depth and breadth of this foundation for nearly 50 years.

## The Zoo as Scientific Resource

The scientific zoo is an important development, because animal management is a complex and challenging field. Good science leads to effective medicine; better husbandry practices; superior (and more appropriate) facilities and enclosures; and more accurate predictions, evaluations, and diagnostics. By its very nature, zoo biology is more science than art, although it remains a highly creative and sometimes intuitive discipline. In examining Duane Rumbaugh's contributions to zoo biology, we have found him to be nimble and creative in the design of experiments, apparatus, and experimental methodology for a diversity of unique and intriguing creatures (e.g., see Figure 1.3). Because comparative studies of cognition are challenging to experimenters in the zoo, what works for psychologists is helpful to veterinarians. No wonder that animal training (an absolute requirement for effective animal management, husbandry, and medicine) is enjoying a renaissance in zoos and providing new professional opportunities for applied behavior analysts (Lukas, Marr, & Maple, 1998; Pryor, 1995).

Rumbaugh's research has also contributed significantly to a new emphasis on animal welfare and psychological well-being. Entire textbooks have been written on the subjects (e.g., Shepherdson, Mellon, & Hutchins, 1998). We believe that Rumbaugh's findings (Rumbaugh, 1970), along with those of other relevant behavioral primatologists, have elevated the public's interest in and appreciation for the cognitive potential of apes and other animals. Indeed, studies have shown that zoo visitors value the cognitive ability of exotic animals (Burghardt & Herzog, 1989; Kellert, 1989), and it is only through zoo-based

**Figure 1.3.** A modified version of the Wisconsin General Test Apparatus supplied by Rumbaugh and National Science Foundation funding and species diversity supplied by the San Diego Zoo were the raw materials for the development of the transfer index and eventually emergents and rational behaviorism.

psychological research that this information can be gathered and disseminated to the public at large.

Certainly, the tremendous improvement in living standards for zoo animals is testimony to our belief that the animals need challenges in their lives as well as environments that are designed to facilitate activity, curiosity, interaction, and mentality. The applied field of environmental enrichment pioneered by California comparative psychologist Hal Markowitz (1982) is testimony to the renewed commitment of zoo professionals to provide a better quality of life for animals in captivity. As applications have proliferated, scientific journals dedicated to the study of animal welfare applications, methods, and systems have been founded in North America and in Europe. As we endeavor to understand animals, our growing awareness of their innate talent ensures that their captors and collaborators will improve their standards of living. Rumbaugh has opened a window into the animal mind, but it is also a window into the animal's emotions and personality. As a construct that can be applied to apes, personality clearly resonates with psychologists (see, e.g., Gold & Maple, 1994, one of the most cited papers on Maple's curriculum vitae). Through years of working in zoos, we have concluded that it is not only acceptable to like the animals that you study, it is indeed impossible not to like them.

The ape, as Rumbaugh has demonstrated throughout his career, is capable of much more than we ever imagined. It is not so much the ape's creativity that is challenged by our experiments as it is our own creative limitations. As Rumbaugh spent quality time with Albert the gorilla, Lana the chimpanzee, and Kanzi the bonobo, he helped us to appreciate these animals for their uniqueness and their individuality. Proponents of animal welfare (aren't we all?) prefer that zoo animals be named and personalized, establishing formal recognition and identity in the presence of human caretakers and experimenters. Rumbaugh and his associates (see Rumbaugh, Savage-Rumbaugh, & Beran, 2001) learned long ago that apes were individuals as they found that their behavior could be differentiated in many ways, including their relative "brightness" in response to formal testing. As Rumbaugh et al. (2001) further concluded, "Apes are sentient, feeling, sensitive-to-pain, intelligent creatures . . . they have symbolic thought, basic dimensions of language, elemental numeric skills, impressive memory and planning capabilities, and other cognitive skills" (pp. 246–247).

As scientists who conduct psychological research in the zoo, as Rumbaugh has done, we also recognize that the nonhuman primate's similarity to humankind has stimulated objections about their subservience. Critics have suggested that apes, and indeed many other taxa, should be regarded as "world citizens," and their freedom guaranteed by law (Wise, 2001). If this idea gains traction with the public, zoos may cease to exist just as we are beginning the most enlightened period of exhibition the world has ever known. To avoid this or some other extreme scenario, the zoo must evolve as a setting where the animal's individuality, autonomy, privacy, and opportunity are respected, protected, and extended.

There is much work to do if the zoo is to fulfill its potential as a venue for education, inspiration, and good science. Rumbaugh's (1969) *transfer index*, the foundation for the development of his theory of emergent behaviors (Rumbaugh, 2002) and rational behaviorism (Rumbaugh, Savage-Rumbaugh, & Washburn, 1996), was possible because of the availability of a variety of species in zoological parks. Where will the Duane Rumbaughs of the future find the diversity for such broad-reaching theories? Primate centers and similar institutions are resources for the few. The San Diego Zoo was a resource of necessity. A young, unfunded professor needed subjects, and the zoo was the answer. Not all zoological parks have openly embraced research, but those that have carved out their niche have found a bounty of resources (see Figure 1.4). To this end, behavioral

**Figure 1.4.** Although noted for his cognitive work, Rumbaugh's work at the San Diego Zoo covered a breadth of topics. Experience rearing young apes provided insight into the development of species-typical behaviors (Rumbaugh, 1967).

scientists with the talent, creativity, and tenacity to tap into the zoo's potential will always be necessary and welcome.

The research of Duane Rumbaugh continues to inform, inspire, and influence comparative psychologists and zoo biologists. Zoo administrators for their part have recognized that the public is intrigued by animal behavior, a subject that the zoo is well positioned to teach to visitors of all ages. The synergy between university and zoo produces many benefits for both institutions. Hediger once told us (personal communication, 1988) that he could not understand why more European universities did not establish working relationships with local zoos as they had routinely done with botanical gardens. Duane Rumbaugh built a zoo research empire in San Diego 50 years ago, and he followed up by designing a mechanism that would one day inspire another research empire in a smaller but fully committed scientific enterprise known as Zoo Atlanta.

It is always difficult to measure the contributions of a scientist with over 6 decades of published work, but we acknowledge Duane Rumbaugh's proper place as a founder of American zoo biology. His work also firmly established comparative psychology and behavioral primatology as major fields of inquiry within the domain of zoo biology.

This discourse provides some perspective on the impact of Duane Rumbaugh's explorations in zoo biology. We think that his many contributions have not been fully appreciated nor acknowledged in this domain. With his cooperation, we will surely learn more about the daily challenge of conducting serious research in one of the world's most visited and valued zoological parks.

## References

Burghardt, G. M., & Herzog, H. A., Jr. (1989). Animals, evolution, and ethics. In R. J. Hoage Jr. (Ed.), *Perceptions of animals in American culture* (pp. 129–151). Washington, DC: Smithsonian Institution Press.

Desiderio, F. (2000). Raising the bars: The transformation of Atlanta's zoo (1889–2000). *Atlanta History, 18*(4), 8–64.

Finlay, T. W., & Maple, T. L. (1986). A survey of research in American zoos and aquariums. *Zoo Biology, 5,* 261–268.

Gold, K. C., & Maple, T. L. (1994). Personality assessment in the gorilla and its utility as a management tool. *Zoo Biology, 13,* 509–522.

Hediger, H. (1969). *Man and animal in the zoo*. New York: Seymour Lawrence/Delacorte Press.

Kellert, S. R. (1989). Perceptions of animals in America. In A. J. Hoage Jr. (Ed.), *Perceptions of animals in American culture* (pp. 5–24). Washington, DC: Smithsonian Institution Press.

Lukas, K. E., Marr, M. J., & Maple, T. L. (1998). Teaching operant conditioning at the zoo. *Teaching of Psychology, 25,* 112–116.

Maple, T. L. (1999). Zoo Atlanta's scientific vision. *Georgia Journal of Science, 57,* 159–179.

Markowitz, H. (1982). *Behavioral enrichment in the zoo*. New York: Van Nostrand Reinhold.

Pryor, K. (1995). *On behavior*. North Bend, WA: Sunshine Books.

Rumbaugh, D. M. (1965). Maternal care in relation to infant behavior in the squirrel monkey. *Psychological Reports, 16,* 171–176.

Rumbaugh, D. M. (1967). Alvila—San Diego Zoo's captive-born gorilla. In C. Jarvis (Ed.), *International zoo yearbook* (Vol. 7, pp. 98–107). London: Zoological Society of London.

Rumbaugh, D. M. (1969). The transfer index. In C. R. Carpenter (Ed.), *Proceedings of the Second International Congress of Primatology: Vol. 1. Behavior* (pp. 267–273). Basel, Switzerland: Karger.

Rumbaugh, D. M. (1970). Learning skills of anthropoids. In L. A. Rosenblum (Ed.), *Primate behavior: Vol. 1. Developments in field and laboratory research* (pp. 1–70). New York: Academic Press.

Rumbaugh, D. M. (1971). Zoos: Valuable adjuncts for the instruction of animal behavior. *BioScience, 21,* 806–809.

Rumbaugh, D. M. (Ed.). (1972a). *Gibbon and Siamang: A series of volumes on the lesser apes: Vol. 1. Evolution, ecology, behavior and captive maintenance*. Basel, Switzerland: Karger.

Rumbaugh, D. M. (1972b). Zoos: Valuable adjuncts for the instruction and research in primate behavior. *BioScience, 22,* 26–29.

Rumbaugh, D. M. (2002). Emergents and rational behaviorism. *Eye on Psy Chi, 6,* 8–14.

Rumbaugh, D. M., & Arnold, R. C. (1971). Learning: A comparative study of *Lemur* and *Cercopithecus*. *Folia Primatologica, 14,* 154–160.

Rumbaugh, D. M., & Rice, C. P. (1962). Learning-set formation in young great apes. *Journal of Comparative and Physiological Psychology, 55,* 866–868.

Rumbaugh, D. M., Riesen, A. H., & Lee, R. E. (1970). *Survey of the primates*. New York: Appleton-Century-Crofts.

Rumbaugh, D. M., Savage-Rumbaugh, E. S., & Beran, M. J. (2001). The grand apes. In B. B. Beck, T. S. Stoinski, M. Hutchins, T. L. Maple, B. Norton, A. Rowan, et al. (Eds.), *Great apes and humans: The ethics of coexistence* (pp. 245–260). Washington, DC: Smithsonian Institution Press.

Rumbaugh, D. M., Savage-Rumbaugh, E. S., & Washburn, D. A. (1996). Toward a new outlook on primate learning and behavior: Complex learning and emergent processes in comparative perspective. *Japanese Psychological Research, 38,* 113–125.

Shepherdson, D. J., Mellon, J., & Hutchins, M. (Eds.). (1998). *Second nature: Environmental enrichment for captive animals.* Washington, DC: Smithsonian Institution Press.

Stoinski, T. S., Lukas, K. E., & Maple, T. L. (1998). Research in American zoos and aquariums. *Zoo Biology, 17,* 167–180.

Wise, S. M. (2001). A great shout: Legal rights for great apes. In B. B. Beck, T. S. Stoinski, M. Hutchins, T. L. Maple, B. Norton, A. Rowan, et al. (Eds.), *Great apes and humans: The ethics of coexistence* (pp. 274–294). Washington, DC: Smithsonian Institution Press.

Yerkes, R. M. (1925). *Almost human.* New York: Century.

Yerkes, R. M. (1928). The mind of a gorilla. *Comparative Psychology Monographs, 5,* 1–92.

Yerkes, R. M., & Yerkes, A. W. (1929). *The great apes.* New Haven, CT: Yale University Press.

# 2

# Apes, Intelligent Science, and Conservation

*Russell H. Tuttle*

During the past half century, great apes have advanced notably up Darwin's scale of evolutionary continuity, and some scientists and humanists have granted chimpanzees (*Pan troglodytes*) and various other apes, culture, theory of mind, consciousness, personhood, bellicosity, claims to full human rights, and other characteristics once thought to be limited to *Homo sapiens* and perhaps select precedent *Pleistocene* species of *Homo* (Byrne & Whiten, 1988; Cavalieri & Singer, 1994; de Waal, 1982; Goodall, 1986; Matsuzawa, 1999; McGrew, 1983, 1992, 2004; Nishida, 1990; Parker & McKinney, 1999; van Lawick-Goodall, 1973; van Schaik, 2004; van Schaik et al., 2003; Whiten et al., 1999; Wrangham, McGrew, de Waal, & Heltne, 1994; Wrangham & Peterson, 1996). Cladistically inclined systematists, who heavily weight molecular genetic findings, have lumped extant great apes and humans in ever-lower taxonomic categories to the point that bonobos, chimpanzees, and people are congenerically *Pan* or *Homo,* and, in common parlance, people ought to be other great apes (Diamond, 1988; Goodman et al., 1998). One of the more remarkable examples of terminological abuse in an effort to assimilate chimpanzees with humans is that of McGrew (1998), who referred to *Pan troglodytes* and *Homo sapiens* as "two sibling species of hominids" (p. 607). Canonically, sibling species are "morphologically similar or identical natural populations that are reproductively isolated" (Mayr, 1963, p. 34). Although naive individuals

---

I thank David Washburn for inviting me to participate in this much-merited celebration and the staff at Georgia State University for assistance with travel arrangements to the conference that inspired this book. Our family counts Duane Rumbaugh among our dearest friends, and I cherish the many intellectual and social interactions that I have experienced with him over 35 years of colleagueship. The essay is partly drawn from a quartet of lessons, under the general title "Apes and Human Evolution," that I presented at the Collège de France in Paris in November 1995. I am profoundly grateful to Yves Coppens for this challenge, and for sponsoring me, and to him, Pascal Picq, and James W. Fernandez for their good colleagueship and generosity during the visit in Paris. I also happily recall the many critical, attentive undergraduate and graduate students who have helped to test my ideas over the past 3 decades of teaching the course "Apes and Human Evolution" at the University of Chicago.

commonly confuse chimpanzees, bonobos, and even gorillas with one another, I doubt that anyone could misidentify a naked human for one of them.

Certainly, there are genetic, physiological, morphological, and behavioral continuities among humans, apes, and many other organisms that grace or plague Earth, the most fundamental of which is life itself. For instance, were the war hawks to succeed in arresting the genetic chain of life via panglobal devastation, it probably would not begin again. Further, Darwin (1872/1998) was probably correct to link some human facial expressions of emotion with counterparts in other mammals; for example, one must be careful not to mistake a grin for a smile in chimpanzee or human.

My more modest complaint is that people who are humanely and politically motivated to save apes from extinction and abuse need not—indeed, should not—cite Darwin's genre of continuity to achieve their important goals. Further, if behavioral, ethological, genetic, evolutionary biological, and other scientific findings are to be used to support arguments for ape dignity and rights, one had better be certain that they are rock solid and necessary.

My reading of Darwin's landmark trilogy, particularly the second book *The Descent of Man and Selection in Relation to Sex* (1871), indicates that he roundly exemplified the classist, racist, sexist, privileged Victorian English gentleman, who at times was quite naive about human biological unity and behavior. In the introduction to the above-named book, Darwin praised Ernst Haeckel's *Natürliche Schöpfungsgeschichte* (The History of Creation; 1868), "in which he fully discusses the genealogy of man" as having confirmed "almost all the conclusions at which I have arrived" (Darwin, 1871, p. 390).

The diabolical use of Haeckel's (1868) pseudoscientific racist and classist human phylogeny by the National Socialists and eugenicists in the 20th century is well known, as is the fact that today many people suffer in the shadow of its folkish, morally invidious legacy (Gasman, 1971; Stein, 1998). However, one seldom sees cited Darwin's descendant echo on the nature of links among monkeys and apes and different people, whom he erroneously considered to embody biological races. After ominously predicting that "the civilized races of man will . . . replace the savage races" and that "the anthropomorphous apes . . . will . . . be exterminated" (Darwin, 1871, p. 521), Darwin went on to say that

> the break between man and his nearest allies will then be wider, for it will intervene between man in a more civilized state . . . even than the Caucasian, and some ape as low as a baboon, instead of as now between the negro or Australian and the gorilla. (Darwin, 1871, p. 521)

Is this really the sort of archaic rubbish that Fouts (1997), Bekoff (2002), Sheets-Johnstone (1996), and other advocates of ape rights and dignity should associate with a worthy cause? Should we be telling people whose support is needed to save apes and their habitats that they are actually apes, albeit in a different sense from that of classic and current racism?

As one who has not had people who are racist compare my physical features with those of apes—though my thin lips, frontal baldness, noncoiled hair, light tannable skin, and chimpanzoid pinnae could be earmarked for this purpose (Tuttle, 1986, pp. 14–16)—I would have no problem viewing myself as just another ape. Moreover, I am not ashamed to have evolved from creatures that lack many features of *Homo sapiens*. However, as a show-me scientist, I am not convinced that *Homo sapiens* are simply brainy, bipedal, sparsely hirsute, articulate apes (Tuttle, 2001b).

Extant apes are focal subjects in many arguments over the human career and condition, and they are commonly cited in attempts to highlight our special features or contrarily to show how close we are to other animals. When assessing the various arguments, it is important to distinguish among reconstructions, models, and scenarios. Failure to do so can be misleading and smack of hubris (Tuttle, 2001b).

Evolutionary biologists and anthropologists who purport to provide "phylogenetic reconstructions" are at best suggesting models of what might have unfolded over many generations given that they have only fragmentary empirical data sets with which to document phylogenic events. Models of ecological and behavioral evolution are more properly termed *scenarios*, because habitats and behavior are transitory and leave only tantalizing traces. Indeed, paleoanthropological theorists are basically reduced to writing scientifically informed stories (Tuttle, 2001b).

## Genomics and Adaptive Complexes

Molecular biologists have no trouble identifying the chromosomes and DNA of humans versus those of nonhuman primates (Marks, 2002). Moreover, the much-touted genomic closeness of humans and chimpanzees has dipped quantitatively from 98% to 99% overall similarity down to 95% (Britten, 2002). This uncertainty combined with ignorance of the actual numbers of genes in humans and chimpanzees (Claverie, 2001; Cohen, 1997; Fields, Adams, White, & Vernier, 1994; Hattori et al., 2000; O'Brien et al., 1999; Shouse, 2002; Venter et al., 2001), let alone how they are translated into traits and trait complexes (Carroll, 2003), make comparative genomics a shaky platform from which to argue for panhominoid parity. Were future genomic studies to reveal discrete, profound differences between humans and the African apes, and if such factors were weighted heavily in decisions about the relative status of apes versus people, efforts to conserve the apes and their natural habitats could be jeopardized (Corbey, 2005; Tuttle, 2001b).

Fine caricaturists that they can be when beaten or bribed and stuffed into inane human costumes, no ape could truly emulate a human stride, sprint, or long-distance jogging gait because they lack the requisite complex of locomotor and physiological traits (Tuttle, 1994). Whereas the spinal, pelvic, and hind limb anatomy of great apes predispose them to compliant postures and gaits when they engage in facultative arboreal or terrestrial bipedalism, humans have numerous distinctive adaptations of the spine, pelvis, hips, knees, and feet that underpin obligate terrestrial bipedalism. Indeed, the peculiar terrestrial bipedal adaptive complex is a sine qua non for paleontologists to track the hominid lineage in the Late Miocene, Pliocene, and Pleistocene epochs (Tuttle, 1988).

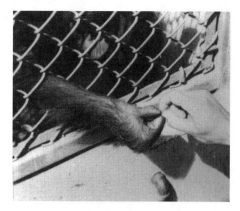 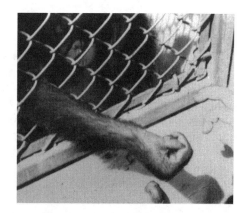

**Figure 2.1.** Left: Young orangutan and human each holding half a grain of rice. Right: The human has released the rice. Note that whereas the human grips the tiny object via pulp-to-pulp opposition of the thumb and index finger, the orangutan uses the tip of the thumb against the lateral side of the first interphalangeal joint of the index finger.

Humans have sparsely haired bodies and a superfluity of eccrine sweat glands, especially on anterior surfaces of the torso and limbs, which contact the rush of air and cool one's blood, thereby preventing damage to the highly heat-sensitive brain (Bramble & Lieberman, 2004; Tuttle, 1994). Although there probably is little difference in tactile sensitivity and basic motor control between human and great ape hands, contrasts are apparent when they grip objects of various shapes and dimensions, especially small ones. The proportions of human and great ape hands are markedly different, and human thumbs are unique in having a broad distal phalanx, to which a unique muscle— the flexor pollicis longus—attaches (Tuttle, 1992). Accordingly, compared with great apes, humans have a powerful thumb-to-index fingertip pinch (Figure 2.1) and can oppose the tip of the thumb to each of the other three fingers.

There are dramatic contrasts in the genitalia between humans and great apes, particularly chimpanzees. The human corpus penis is cylindrical and terminates in a prominent glans penis, whereas chimpanzee penes taper to a narrow point. Chimpanzee testes are notably larger than those of humans. Female humans lack the perineal estrous swellings that characterize estrous chimpanzees.

Although humans and chimpanzees are comparable in degree of sexual dimorphism in overall mass, humans are more distinctive in epigamic features, for example, the distribution of fat deposits as exemplified by the curvaceous figure, enhanced by a wide pelvis, and voluptuous breasts of human females. However, the last feature might represent a transspecific continuity because some multiparous female great apes also sport prominent breasts, albeit in the lower range of human expression.

Human speech depends on an adaptive complex that demarcates *Homo sapiens* from all other extant species on Earth. The underpinnings for this ability are largely neurological (Gannon, Kheck, & Hof, 2001; Gibson, Rumbaugh, & Beran, 2001; Preuss, 2001), but it is also effected by peculiar structures in the mouth, pharynx, and larynx, none of which lend it to being traced into the fossil record (Tuttle, 2001b).

Although human speech in myriad varieties is reasonably linked to our relatively capacious pharynx and mobile tongue, there is no compelling anatomical reason to deny some form of vocal language in ancestral hominids. Several authors argue that without a lowered larynx and expanded supralaryngeal region, articulate human speech is fairly excluded (Laitman, Heimbuch, & Crelin, 1979; Lieberman, 1994). Accordingly, even Neanderthals, who lived between 130,000 and 35,000 years ago, were inept vocally and probably also were challenged cognitively vis-à-vis *Homo sapiens* of the Upper Paleolithic. It is simplistic to expect that were its brain and cognitive capacity developed, and were its vocal cords, pharynx, and mouth innervated so that concepts as well as emotions could be expressed voluntarily, an ape or an *Australopithecus* could not express itself verbally.

Decades ago, Jones (1940) and DuBrul (1958) noted that gibbons and great apes have lowered larynges and concomitant gaps between the epiglottis and soft palate, albeit less than the human condition (Nishimura 2004; Wind 1992). The calls of gibbons are wonderfully varied in pitch and pattern. If their air columns were broken into discrete bits by consonantal sounds, they could emulate words. The same may be said for great apes. The calls of bonobos are quite different from those of chimpanzees (Taglialatela, Savage-Rumbaugh, & Baker, 2003; Tuttle, 1986). Orangutans, chimpanzees, and bonobos have notoriously mobile lips and tongues, surely transcending the human condition. All that they lack is wiring to recruit them for speech.

More basically, I am not convinced that one can veridically model the vocal tracts of fossil hominids based on degree of basicranial flexion or inferred mandibular mechanisms. There simply are too few bony landmarks related to the lips, tongue, and soft structures of the throat to anchor models of phonemic production in our macerated ancestors.

## Cognition and the Question of Culture

Six and a half decades of living with companion animals, a decade of working intimately with captive great apes (Nakatsukasa, Nakano, Kunimatsu, Ogihara, & Tuttle, 2006; Tuttle, Hallgrimsson, & Basmanian, 1999), and many films of free-ranging apes and other animals have persuaded me that humans are not the only thinking beings. Perhaps Rumbaugh's (2002; Rumbaugh & Washburn, 2003) concept of emergents will intensify research on this complex problem, along with research on self-awareness and naturalistic symboling capabilities. The last is much on my mind lately as one reads declarations that chimpanzees are cultural beings because of social learning, but no one has demonstrated unequivocally that symboling is part of the process (McGrew, 2004; Tuttle, 2001a; van Schaik, 2004; van Schaik et al., 2003; Whiten et al., 1999). This is a dead-end path toward understanding the evolution of the capacity for culture in *Homo sapiens* and the extent to which homologous structures that underpin culture are present in other animals.

Although my heart is wholly with the plight of chimpanzees and against the holocaust that we have arrogantly unleashed on other organisms with which we must share Earth, I sense that opportunities to understand the minds of chimpanzees (and probably of other vertebrates) will be missed if too many behavioral scientists accept recent declarations of chimpanzee culture (Whiten et al., 1999) and a coming of age of cultural primatology (de Waal, 1999, p. 635) without evidence that they truly are cultural beings. To date, little more has been demonstrated than that chimpanzees have local or demic behavioral traditions that are learned somehow from conspecifics.

No one has shown that naturalistically chimpanzees have symbolically mediated ideas, beliefs, and values, the sine qua non of culture as understood by most students of culture (Tuttle, 2001a, 2006). Indeed, one rarely encounters mention, let alone detailed discussion, of symbols in the arguments for naturalistic chimpanzee culture. Instead, there is an emphasis on how behavioral traditions are learned and passed on socially and that the variations in or tangible products of chimpanzee behavior are not a consequence of physical environmental (i.e., ecological) influences or genetic transmission (Boesch & Tomasello, 1998). The actual nature of culture itself, and especially the mechanism(s) by which meanings are encoded for the chimpanzees, are missing from the discussion.

Although I agree that spoken language—a cultural category—need not be invoked as the criterion for culture in other animals, nonetheless the challenge remains to discern behaviors that are influenced by symbolically encoded meanings in wild chimpanzees and other nonhuman animals. Combined with growing understanding of human cultural cognition, particularly from studies of developmental and cultural psychologists (Cole, 1996; Tomasello, 1999), we would have a better base for modeling the evolution of human cultural capacities and for appreciating actual similarities with unique features of chimpanzee minds.

Boesch and Tomasello (1998) argued that "culture is not monolithic but a set of processes" (p. 591) and attempted to devise a concept of culture to embrace the chimpanzee case. There had been and continues to be some confusion between product and process, with the assumptive former taken to indicate presence of the latter. First and foremost, chimpanzees should be revealed as cultural beings before labeling their demic traditions cultures (Tuttle, 2001a). Process is indeed preeminent in this exercise. Accordingly, one would expect much more thorough digestion of the information that emerged from the half century of intensive anthropological research, particularly that of the American school (Kuper, 1999; Shore, 1996), on the nature of culture since Kroeber and Kluckhohn (1952) compiled their classic catalog of definitions.

Although Kroeber and Kluckhohn (1952) confessed that they had no full theory of culture, they proposed as the central idea of most contemporary social scientists that

> culture consists of patterns, explicit and implicit, of and for behavior acquired and transmitted by symbols, constituting the distinctive achievement of human groups, including their embodiments in artifacts; the essential core of culture consists of traditional (i.e., historically derived and selected) ideas

and especially their attached values; culture systems may, on the one hand, be considered as products of action, on the other as conditioning elements of further action. (p. 357)

Today, I doubt that there is much disagreement among anthropologists and sociologists (Turner, 2000) that key to the concept of culture are symbols and symbolically mediated ideas, values, and beliefs, however difficult it might be to explicate the precise psychological, neurophysiologic, and social processes that underpin them or to discern them from behavior, narratives, or texts.

Boesch and Tomasello (1998, p. 610) sought to bridge the gap between the views of culture typical in biology and psychology and to find common ground between them. They concluded,

> There seems to be enough common ground concerning processes of culture and cultural evolution that investigators from many different disciplines can begin to make their voices heard in a way that results in an accumulation of modifications to the concept of culture that will facilitate everyone's empirical work. (Boesch & Tomasello, 1998, p. 611)

Until the new cultural primatologists engage the scholarly corpus of 5 decades of research on the concept of culture by a notable roster of cultural anthropologists (Geertz, 1973; Harris, 1999; Sahlins, 1976; Schneider, 1968; White & Dillingham, 1973), I doubt that much empirical progress can be made toward discerning naturalistic humanoid cultural capacities in chimpanzees and other nonhuman species, especially the phylogeny of human cultural capacities. Although it is important to understand how animals, including people, learn and transmit behavior spatiotemporally, we need greater focus on whether, and if so, how, symbolic mediation might be involved in naturalistic behaviors of other animals. Then, we might begin to construct refined models on hominid behavioral evolution over the past 5 million years (Tuttle, 2001a).

It is indeed unfortunate that beginning with Tyler (1871), definitions of culture restricted the phenomenon to *Homo sapiens* and that many sociocultural anthropologists have believed that to search for culture in other animals is futile. However, this should not dissuade others from searching for symbolically mediated, shared systems of meaning among chimpanzees and other animals (Tuttle, 2001a).

## Conservation

Sadly, over the past millennium, Earth has become the planet of people, where far too many individuals and societies behave like the omnipotent beings in Pierre Boule's (1963) *Planet of the Apes*. The corrective is to deal effectively with human hubris, poverty, greed, political corruption, and other all-too-human failings and to find a way to conserve biodiversity and natural ecological communities without privileging some nonhuman beings vis-à-vis others and while upholding the dignity and rights of humans who live most proximately with them (Tuttle, 1998).

Exhibit 2.1. The World's Top 25 Most Endangered Primates, 2002

| Asia | Africa | Neotropics | Madagascar |
|---|---|---|---|
| **White-headed langur** | Miss Waldron's red colobus | **Black-faced lion tamarin** | **Greater bamboo lemur** |
| Yunnan snub-nosed langur | Roloway guenon | **Yellow-breasted capuchin** | Perrier's sifaka |
| Guizhou snub-nosed monkey | White-naped mangabey | **Northern muriqui** | Silky sifaka |
| Delacour's langur | **Tana River mangabey** | | |
| Golden-headed langur | **Tana River red colobus** | | |
| Gray-shanked douc | **Sanje mangabey** | | |
| Tonkin snub-nosed monkey | Cross River gorilla | | |
| Eastern Black crested gibbon | Mountain gorilla | | |
| Natuna banded leaf monkey | | | |
| Pagai pig-tailed snub-nosed monkey | | | |
| Sumatran orangutan | | | |

*Note.* Taxa left unprotected by sympatry with great ape umbrella species are noted in bold font. Data from van Hoof (2002).

Exhibit 2.2. The World's Top 25 Most Endangered Primates, 2005–2006

| Asia | Africa | Neotropics | Madagascar |
|---|---|---|---|
| Sumatran orangutan | Eastern gorilla | **Black-faced lion tamarin** | **Greater bamboo lemur** |
| **Horton Plains slender loris** | Cross River gorilla | **Yellow-breasted capuchin** | **White-collared lemur** |
| **Western purple-faced langur** | **Mt. Rungwe galago** | **Brown spider monkey** | Perrier's sifaka |
| Delacour's langur | **Sanje mangabey** | **Northern muriqui** | Silky sifaka |
| Golden-headed langur | **Tana River red colobus** | | |
| Gray-shanked douc | **White-naped mangabey** | | |
| Tonkin snub-nosed monkey | **Bioko red colobus** | | |
| Hainan black-crested gibbon | | | |
| **Miller's grizzled surili** | | | |
| **Pagai pig-tailed snub-nosed monkey** | | | |

*Note.* Taxa left unprotected by sympatry with great ape umbrella species are noted in bold font. Data from Conservation International (2005).

Accordingly, I am uneasy about aspects of the Great Ape Project, the Great Ape Survival Plan, and Great Ape World Heritage Species Project, which would privilege a few—albeit marvelous—species over others in efforts to save them from mistreatment and extermination. These programs will also privilege researchers whose careers depend on access to localities with great apes. Accordingly, dubbing the great apes as flagship species may be close to an unwelcome truth, because in colonial times, a flagship carried elite opportunists to and precious commodities from the lands and peoples that they exploited. One must hope that World Heritage specific status will protect great apes better than some World Heritage Sites, like Laetoli and Olduvai Gorge, Tanzania, have fared under a similar designation, and that indigenous expert scientists can freely determine who conducts research in their nations by what they contribute to conservation efforts at their work sites (Karoma, 1996; Tuttle, 1998, 2002).

The argument that great apes will be umbrella species, protecting not only them but also their habitats and fellow denizens, is an improvement over the flagship moniker, but many more taxa of primates and other mammals that do not live with great apes should be targeted for special protection. Among the world's top 25 most endangered primates listed in 2002 (Exhibit 2.1) and 2004–2006 (Exhibit 2.2), 3 are great apes, which would be protected as World Heritage species, whereas the vast majority of other imperiled primate taxa would be left unprotected because they do not live with them (Lazaroff, 2002; Rylands, 2005; van Hoof, 2002, pp. 1367–1368).

I agree with ethicists, philosophers, and fellow scientists who believe that our conservative efforts should be based on broader moral and ethical considerations that are not dependent on presumed biological closeness to humans. And frankly, I would not want to visit African and Indonesian leaders who are of precolonial heritage and try to persuade them to help us save great apes because they too are apes.

## References

Bekoff, M. (2002). *Minding animals*. Oxford, England: Oxford University Press.
Boesch, C., & Tomasello, M. (1998). Chimpanzee and human cultures. *Current Anthropology, 39*, 591–604, 610–614.
Boule, P. (1963). *Planet of the apes*. New York: Vanguard Press.
Bramble, D. M., & Lieberman, D. E. (2004, November 18). Endurance running and the evolution of *Homo. Nature, 432*, 345–352.
Britten, R. J. (2002). Divergence between samples of chimpanzee and human DNA sequences is 5%, counting indels. *Proceedings of the National Academy of Sciences USA, 99*, 13633–13635.
Byrne, R., & Whiten, A. (1988). *Machiavellian intelligence*. Oxford, England: Clarendon Press.
Carroll, S. B. (2003, April 24). Genetics and the making of *Homo sapiens. Nature, 422*, 849–857.
Cavalieri, P., & Singer, P. (1994). *The great ape project*. New York: St. Martin's Press.
Claverie, J. M. (2001, February 16). What if there are only 30,000 human genes? *Science, 291*, 1255–1257.
Cohen, J. (1997, February 7). How many genes are there? *Science, 275*, 769.
Cole, M. (1996). *Cultural psychology*. Cambridge, MA: Harvard University Press.
Conservation International. (2005, April 26). *Primates on the brink: Mankind's closest living relatives under threat around the world*. Retrieved April 13, 2006, from http://www.primates.com/primate/endangered.html

Corbey, R. (2005). *The metaphysics of apes: Negotiating the animal–human boundary*. Cambridge, England: Cambridge University Press.

Darwin, C. (1871). *The descent of man and selection in relation to sex*. New York: Modern Library.

Darwin, C. (1998). *The expression of the emotions in man and animals* (3rd ed.). Oxford, England: Oxford University Press. (Original work published 1872)

de Waal, F. (1982). *Chimpanzee politics*. New York: Harper & Rowe.

de Waal, F. B. M. (1999, June 17). Cultural primatology comes of age. *Nature, 399,* 635–636.

Diamond, J. M. (1988, April 21). DNA-based phylogenies of the three chimpanzees. *Nature, 332,* 685–686.

DuBrul, E. L. (1958). *Evolution of the speech apparatus*. Springfield, IL: Charles C Thomas.

Fields, C., Adams, M. D., White, O., & Vernier, J. (1994). How many genes in the human genome? *Nature Genetics, 7,* 345–346.

Fouts, R. (1997). *Next of kin*. New York: William Morrow.

Gannon, P. J., Kheck, N. M., & Hof, P. R. (2001). Language areas of the hominoid brain: A dynamic communicative shift on the upper east side planum. In D. Falk & K. R. Gibson (Eds.), *Evolutionary anatomy of the primate cerebral cortex* (pp. 216–240). Cambridge, England: Cambridge University Press.

Gasman, D. (1971). *The scientific origins of national socialism*. London: Macdonald.

Geertz, C. (1973). *The interpretation of cultures*. New York: Basic Books.

Gibson, K. R., Rumbaugh, D., & Beran, M. (2001). Bigger is better: Primate brain size in relationship to cognition. In D. Falk & K. R. Gibson (Eds.), *Evolutionary anatomy of the primate cerebral cortex* (pp. 79–97). Cambridge, England: Cambridge University Press.

Goodall, J. (1986). *The chimpanzees of Gombe*. Cambridge, MA: Harvard University Press.

Goodman, M., Porter, C. A., Czelusniak, J., Page, S. L., Schneider, H., Shoshani, J., et al. (1998). Toward a phylogenetic classification of primates based on DNA evidence complemented by fossil evidence. *Molecular Phylogenetic Evolution, 9,* 585–598.

Haeckel, E. (1868). *Natürliche Schöpfungsgeschichte* [The history of creation]. Berlin: Wilhelm Engelmann, G. Reimer.

Harris, M. (1999). *Theories of culture in postmodern times*. Walnut Creek, CA: AltaMira Press.

Hattori, M., Fujiyama, A., Taylor, T. D., Watanabe, H., Yada, T., Park, H. S., et al. (2000, May 18). The DNA sequence of human chromosome 21. *Nature, 405,* 311–319.

Jones, F. W. (1940). The nature of the soft palate. *Journal of Anatomy, 74,* 147–170.

Karoma, N. J. (1996). The deterioration and destruction of archaeological and historical sites in Tanzania. In P. R. Schmidt & R. J. MacIntosh (Eds.), *Plundering Africa's past* (pp. 191–201). Bloomington: Indiana University Press.

Kroeber, A. L., & Kluckhohn, C. (1952). *Culture: A critical review of concepts and definitions*. New York: Random House.

Kuper, A. (1999). *Culture: The anthropologists' account*. Cambridge, MA: Harvard University Press.

Laitman, J. T., Heimbuch, R. C., & Crelin, E. S. (1979). The basicranium of fossil hominids as an indicator of their upper respiratory systems. *American Journal of Physical Anthropology, 5l,* 15–34.

Lazaroff, C. (2002). Threats to primates are escalating. *Environment News Service*. Retrieved April 3, 2006, from http://ens-news.com/ens/oct2002/2002-10-08-06.asp

Lieberman, P. (1994). Human language and human uniqueness. *Language & Communication, 14,* 87–95.

Marks, J. (2002). *What it means to be 98% chimpanzee*. Berkeley: University of California Press.

Matsuzawa, T. (1999). Communication and tool use in chimpanzees: Cultural and social contexts. In M. Hauser & M. Konishi (Eds.), *The design of animal communication* (pp. 645–671). Cambridge, England: Cambridge University Press.

Mayr, E. (1963). *Animal species and evolution*. Cambridge, MA: Harvard University Press.

McGrew, W. C. (1983). Animal foods in the diets of wild chimpanzees (*Pan troglodytes*): Why cross-cultural variation? *Journal of Ethology, 1,* 46–61.

McGrew, W. C. (1992). *Chimpanzee material culture*. Cambridge, England: Cambridge University Press.

McGrew, W. C. (1998). Comments. *Current Anthropology, 39,* 607–608.

McGrew, W. C. (2004). *The cultured chimpanzee: Reflections on cultural primatology*. Cambridge, England: Cambridge University Press.

Nakatsukasa, M., Nakano, Y., Kunimatsu, Y., Ogihara, N., & Tuttle, R. H. (2006). Hidemi Ishida: 40 years of footprints in Japanese primatology and paleoanthropology. In H. Ishida, R. H. Tuttle, M. Pickford, M. Nakatsukasa, & N. Ogihara (Eds.), *Human origins and environmental backgrounds* (pp. 1–14). New York: Springer Publishing Company.

Nishida, T. (1990). *The chimpanzees of the Mahale Mountains*. Tokyo: University of Tokyo Press.

Nishimura, T. (2004). Developmental changes in the shape of the vocal tract in chimpanzees. *Folia Primatologica, 75*, 401–402.

O'Brien, S. J., Menotti-Raymond, M., Murphy, W. J., Nash, W. G., Wienberg, J., Stanyon, R., et al. (1999, October 15). The promise of comparative genomics in mammals. *Science, 286*, 458–481.

Parker, S. T., & McKinney, M. L. (1999). *Origins of intelligence*. Baltimore: Johns Hopkins University Press.

Preuss, T. M. (2001). The discovery of cerebral diversity: An unwelcome scientific revolution. In D. Falk & K. R. Gibson (Eds.), *Evolutionary anatomy of the primate cerebral cortex* (pp. 138–164). Cambridge, England: Cambridge University Press.

Rumbaugh, D. M. (2002). Emergents and rational behaviorism. *Eye on Psi Chi, 7*, 8–14.

Rumbaugh, D. M., & Washburn, D. A. (2003). *Intelligence of apes and other rational beings*. New Haven, CT: Yale University Press.

Rylands, A. (2005). Report from the IUCN/SSC Primate Specialist Group. *International Journal of Primatology, 25*, 253–256.

Sahlins, M. (1976). *Culture and practical reason*. Chicago: University of Chicago Press.

Schneider, D. M. (1968). *American kinship: A cultural account*. Englewood Cliffs, NJ: Prentice-Hall.

Sheets-Johnstone, M. (1996). Taking evolution seriously: A matter of primate intelligence [Special issue]. *Etica & Animali, 8/96*, 115–130.

Shore, B. (1996). *Culture in mind*. Oxford, England: Oxford University Press.

Shouse, B. (2002, February 22). Revisiting the numbers: Human genes and whales. *Science, 295*, 1457.

Stein, G. J. (1998). Biological science and the roots of Nazism. *American Scientist, 76*, 50–58.

Taglialatela, J. P., Savage-Rumbaugh, S., & Baker, L. A. (2003). Vocal production by a language-competent *Pan paniscus*. *International Journal of Primatology, 24*, 1–17.

Tomasello, M. (1999). *The cultural origins of human cognition*. Cambridge, MA: Harvard University Press.

Turner, J. H. (2000). *On the origins of human emotions*. Stanford, CA: Stanford University Press.

Tuttle, R. H. (1986). *Apes of the world*. Park Ridge, NJ: Noyes.

Tuttle, R. H. (1988). What's new in African paleoanthropology? *Annual Review of Anthropology, 17*, 391–426.

Tuttle, R. H. (1992). Hands from newt to Napier. In S. Matano, R. H. Tuttle, H. Ishida, & M. Goodman (Eds.), *Topics in primatology: Vol. 3. Evolutionary biology, reproductive endocrinology and virology* (pp. 3–20). Tokyo: University of Tokyo Press.

Tuttle, R. H. (1994). Up from electromyography: Primate energetics and the evolution of human bipedalism. In R. S. Corruccini & R. L. Ciochon (Eds.), *Integrative paths to the past: Paleoanthropological advances in honor of F. Clark Howell* (pp. 269–284). Englewood Cliffs, NJ: Prentice Hall.

Tuttle, R. H. (1998). Global primatology in a new millennium. *International Journal of Primatology, 19*, 1–12.

Tuttle, R. H. (2001a). On culture and traditional chimpanzees. *Current Anthropology, 42*, 407–408.

Tuttle, R. H. (2001b). Phylogenies, fossils and feelings. In B. B. Beck, T. S. Stoinski, M. Hutchins, T. L. Maple, B. Norton, A. Rowan, et al. (Eds.), *Great apes and humans* (pp. 178–190). Washington, DC: Smithsonian Institution Press.

Tuttle, R. H. (2002). Paleoanthropology read in tooth and nail. *Review of Anthropology, 31*, 103–128.

Tuttle, R. H. (2006). Are human beings apes, or are apes people too? In H. Ishida, R. H. Tuttle, M. Pickford, M. Nakatsukasa, & N. Ogihara (Eds.), *Human origins and environmental backgrounds* (pp. 245–254). New York: Springer Publishing Company.

Tuttle, R. H., Hallgrímsson, B., & Basmanian, J. V. (1999). Electromyography, elastic energy and knuckle-walking: A lesson in experimental anthropology. In D. Lindburg & S. C. Strum (Eds.), *The new physical anthropology* (pp. 32–41). Englewood Cliffs, NJ: Prentice-Hall.

Tyler, E. B. (1871). *Primitive culture*. London: John Murray.

van Hoof, J. A. R. A. M. (2002). Proceedings of IXth Congress of the International Primatological Society, Beijing, People's Republic of China, August 4–9, 2002. *International Journal of Primatology, 23*, 1357–1371.

van Lawick-Goodall, J. (1973). Cultural elements in a chimpanzee community. In E. W. Menzel Jr. (Ed.), *Symposium of the fourth international congress of primatology: Vol. 1. Precultural primate behavior* (pp. 144–184). Basel, Switzerland: Karger.

van Schaik, C. P. (2004). *Among orangutans*. Cambridge, MA: Harvard University Press.

van Schaik, C. P., Ancrenaz, M., Borgen, G., Galdikas, B., Knott, C. D., Singleton, I., et al. (2003, January 3). Orangutan cultures and the evolution of material culture. *Science, 299*, 102–105.

Venter, J. C., Adams, M. D., Myers, E. W., Li, P. W., Mural, R. J., Sutton G. G., et al. (2001, February 16). The sequence of the human genome. *Science, 291*, 1304–1351.

White, L. A., & Dillingham, B. (1973). *The concept of culture*. Minneapolis, MN: Burgess.

Whiten, A., Goodall, J., McGrew, W. C., Nishida, T., Reynolds, V., Sugiyama, Y., et al. (1999, June 17). Cultures in chimpanzees. *Nature, 399*, 682–685.

Wind, J. (1992). Speech origin: A review. In J. Wind, B. Chiarelli, B. Bichakjian, & A. Nocentini (Eds.), *Language origin: A multidisciplinary approach* (pp. 421–448). Dordrecht, the Netherlands: Kluwer Academic Publishers.

Wrangham, R. W., McGrew, W. C., de Waal, F. B. M., & Heltne, P. G. (1994). *Chimpanzee cultures*. Cambridge, MA: Harvard University Press.

Wrangham, R., & Peterson, D. (1996). *Demonic males*. Boston: Houghton Mifflin.

# 3

# Studies at the Great Ape Research Institute, Hayashibara

### Gen'ichi Idani and Satoshi Hirata

In 2001, an institute for studying great apes was built in Japan that offers the largest enclosure in the world for conducting such study. Known as the Great Ape Research Institute (GARI), it is affiliated with Hayashibara Biochemical Laboratories, Inc., a private company in Okayama Prefecture. GARI was established in 1998; studies of 4 infant chimpanzees started in a temporary facility in January 1999 (Idani, Morimura, & Fuwa, 1999). The GARI facility was completed in August 2001, and another infant chimpanzee was added to the group at the beginning of 2002.

GARI was established to study the evolution, life, behavior, society, intelligence, and culture of apes and humans. The facility was designed to create a complex environment for maintaining captive apes. The environment includes natural vegetation, a pond, and a 13-meter tower in an open enclosure. GARI was designed to study aspects of the behavior and ecology of apes that have been observed in the wild, with the added benefit that the enclosure would allow detailed observations of behavior that cannot be seen in the apes' natural habitat. By studying our closest living relatives, the great apes, we hope to answer the questions "What is a human being?" and "What should the present and future of humans be?" Simultaneously, we aim to enhance the quality of life for the apes in captivity by considering their welfare (Fraser, Weary, Pajor, & Milligan, 1997). To this end, we have enriched their environment to improve both the physical and psychological conditions of the captive apes (Morimura, 2003; Shepherdson, 1998). This chapter explains the management policy of GARI and describes the facility; it also outlines some of the current and future studies to be conducted there.

---

We consider Duane Rumbaugh to be one of the most respected individuals in the field of ape research. It is a great honor to contribute to this book. We also thank David Washburn, who made possible this connection with Rumbaugh.

Figure 3.1. Activities in the Great Ape Research Institute: academic study and practice.

## The Basic Policy

The basic concept and policy of GARI has a dual focus: an academic emphasis, focused on basic science, and a practical side (see Figure 3.1). With respect to our academic interests, research will be undertaken to examine intelligence, behavior, and social evolution in the *Hominoidea*. What kind of society did early hominoids have? What kinds of behavior did they show? How were they related to their environment? How did the human mind evolve? Behavior and matters pertaining to the mind are not preserved in fossils, but they can be examined by means of comparative studies involving humans and other living animals (Matsuzawa, 2001; Shumaker & Beck, 2003). Chimpanzees are the closest living relatives of humans. The difference between the DNA of these two species is only 1.23% (Fujiyama et al., 2002). In other words, humans and chimpanzees have shared the longest period of time, as relatives, in terms of evolutionary history. Comparative studies of the behavior, intelligence, life, and society of these two species will increase our understanding of human beings (Matsuzawa, 1998; Savage-Rumbaugh, 1986; Rumbaugh & Washburn, 2003). It should be possible to take a broad view of human beings through the study of chimpanzees; this constitutes an important aim of the chimpanzee research at GARI (Hirata, 2003). In addition to the study of captive apes, GARI will also be conducting field studies of wild apes. Furthermore, we will be conducting anthropological and comparative developmental studies of humans. Simultaneous studies of captive and wild apes will increase our scientific understanding of apes and should lead to the development of new theories with respect to the process of human evolution.

Our practical focus emphasizes three areas: husbandry, education, and wildlife conservation. With respect to husbandry, GARI is equipped with "hardware" and "software" to enhance the quality of life of the captive animals. In this instance, *hardware* refers to the ways in which the animals' environment can be enriched to improve its livability, whereas *software* refers to the establishment of a better husbandry system and better relationships between chimpanzees and humans. The aim with both approaches is to support the research. It is necessary that we consider and evaluate chimpanzee husbandry scientifically and provide useful information to the public. We plan to promote discussion of bioethics and "husbandry-ology" as these topics relate to animal welfare and the management of captive individuals. The results of studies conducted at GARI will be made available to the general public at any time.

Humans are a species of animal. The dichotomy of humans versus primates or apes is no longer accepted. Long-term studies of apes have irreversibly reduced the artificial gap between apes and humans (de Waal, 1999; Itani, 1996). As a result, most of the observed differences now appear quantitative rather than qualitative in nature. It is essential that we understand "humans' place in nature" correctly, because this knowledge is crucial to our continuation as a species.

All wild ape species are now on the verge of extinction (Butynski, 2001; Rijksen, 2001). Their populations have decreased markedly over the past century as a result of human activities such as agriculture, deforestation, poaching, and war. We will be providing information to the general public with respect to the relationship between humans and the global environment, and we would like to contribute to the development of culture, education, and wildlife conservation.

## Research at the Great Ape Research Institute and Its Applications

A number of ongoing studies are being conducted at GARI. These include studies of visual cognition, the acquisition and spread of tool use, sign language, the development of play behavior with objects and object manipulation, feeding behavior and evaluation of the tastiness of artificial foods, motor ability, morphological and physiological development, hormone dynamics, and sperm viability. In addition, observations of social interactions and relationships, behavioral development, and the ecology of natural plant foods are continuing in the open enclosure. Data obtained from these studies will not only further our understanding of the life, behavior, and cognition of chimpanzees but will also be put to practical use in the development and evaluation of methods of husbandry. In addition, for educational purposes, we will be hosting a symposium to promote scientific understanding. We are also involved in giving study tours for children so that we can return the results of our research to the local community. Basic and applied research at GARI are closely intertwined.

There are two main aspects of our research. First, research at GARI is designed not only to satisfy the researchers' interests but also to promote animal welfare, husbandry, and education. No study should lower the quality of life of the captive animals or have a negative effect on their welfare. We aim to improve husbandry by developing our understanding of the animals' behavior.

Second, GARI does not split the staff into researchers and caregivers. All of the researchers are involved in daily husbandry work. A division of labor in terms of research and husbandry may appear to be more efficient, but it is our contention that the researchers increase their knowledge and understanding of the chimpanzees through their involvement in the daily care of the animals. The researchers care for the animals that are the focus of their research. Furthermore, we consider the chimpanzees as research partners and hope to develop an affinity between the chimpanzees and the researchers that we hope will lead to new research ideas and new husbandry methods. We will not be subjecting the apes to invasive studies; rather, we will promote our scientific understanding of these animals through noninvasive techniques.

## The Chimpanzees at the Great Ape Research Institute

There are 5 chimpanzees at GARI: Loi, Zamba, Tsubaki, Mizuki, and Misaki. Loi and Zamba are both 7-year-old males. Tsubaki, Mizuki, and Misaki are females and are 7, 6, and 4 years old, respectively. They were born at another facility in Kumamoto prefecture, Japan, and were reared there until they were moved to GARI. They are unrelated individuals. Four of the chimpanzees were reared by their own mothers and grew up with their siblings; Mizuki was nursed by humans because her mother refused to care for her.

Wild chimpanzees start to be weaned at about 2 years of age, when they begin to interact with other chimpanzees. The early mother–infant relationship influences their growth. Therefore, we planned to move the animals to GARI when they were around 3 years old. When they came to GARI, on January 13, 1999, Loi and Zamba were 3 years 6 months old, Tsubaki was 2 years 11 months old, and Mizuki was 2 years 1 month old. Because Mizuki had never lived with conspecifics, she joined the group when she was younger. Misaki joined the GARI group at the age of 3 years, on January 10, 2002. When the animals came to GARI, they already had individual personalities, and each one behaved in a different manner. Ultimately, we plan to enlarge the group to between 12 and 15 individuals.

## Facility Outline

The GARI facility is situated on a peninsula that juts out into the Setonaikai Sea (the Inland Sea of Japan). Because our company owns the entire peninsula, which is located in Setonaikai National Park, outsiders will not disturb the work undertaken at the facility. Expansion of the facility is possible. Moreover, natural vegetation thrives on the peninsula, and various wild animals are present, including foxes, raccoon dogs, hares, and wild birds.

The GARI site covers 10,000 square meters. The open enclosure is surrounded by a concrete wall 4 to 5 meters high. Natural plants and a pond (900 square meters) have been placed in the enclosure, along with a climbing structure, a jungle gym, and an observation booth. Adjacent to the enclosure is a building that contains a gymnasium; a bedroom for the chimpanzees; a medical examination and treatment room; psychology, ethology, and other laboratories; and the control room. The chimpanzees can move freely from inside to outside.

## Open Enclosure

The open enclosure covers 7,400 square meters. It contains a 13-meter-high tower that functions as a tree would in the chimpanzees' natural habitat. There are platforms on the tower at heights of 3, 5, 7, and 10 meters, with a total area of 250 square meters. Additionally, ropes and a net surround the tower. Consequently, the chimpanzees can move vertically and horizontally and traverse slopes on the tower. In the near future, the height of the tower will be increased to 30 meters.

A hill area (2,700 square meters) in the open enclosure is covered with natural vegetation; there are over 100 species of plants here, including *Pinus*, *Rhododendron*, *Juniperus*, and herbaceous plants. The chimpanzees can search the bush in this environment.

The observation booth in the enclosure can be entered via an underground passage from outside the enclosure. The booth is made of transparent polycarbonate, 10 millimeters thick, and we can perform various outdoor experiments there. Additionally, we can observe chimpanzee behavior and social interaction at close range from the booth.

## Gymnasium

The gymnasium has an area of 150 square meters. It has a 6-meter-high ceiling covered with stainless steel mesh. Inside, there are three 5-meter-high towers. The towers are equipped with flat platforms, having a total area of about 50 square meters. Pasture plants, such as clover and Italian ryegrass, grow on the ground. The chimpanzees can spend time here when they are not involved in learning or training or while work, such as restoration work or tree planting, is being done in the open enclosure. We can also observe their behaviors and interactions through windows in the surrounding passage.

## Ethological Laboratory

There are two experimental booths in the ethological laboratory. The larger of the two is 10 meters square and 6 meters high. From this booth we can observe the physical function of the chimpanzees as well as social interactions between individuals. It should be possible to observe and record the behavior of chimpanzees from all angles, because the entire booth is made of transparent polycarbonate, 10 millimeters thick.

There is also a small booth in the ethological laboratory. Here, we can study the development of object manipulation, using objects such as building blocks and seriated cups. This space can also used for education programs to benefit school children.

*Psychology Laboratory*

The psychology laboratory consists of two identical booths. When one chimpanzee is learning something in one booth, another in the next booth can observe it and thus learn the behavior from the animal in the first booth.

One chimpanzee has already mastered nut cracking. We plan to let other chimpanzees observe him cracking nuts in the near future. We are teaching chimpanzees the names, colors, numbers, and shapes of various objects in Japanese sign language and by using a touch panel computer. At present, our chimpanzees know approximately five fruits, six colors, and eight signs.

*Chimpanzee Bedroom*

We believe that group life is an essential factor for chimpanzees because they are such social animals. Therefore, the chimpanzees all sleep in the same room at GARI. Their bedroom (42 square meters) is divided into three parts, which allows the animals to sleep in their favorite places. The bedroom is 5 meters high with two entrances in each wall, in upper and lower positions. The chimpanzees can move to the adjoining sections using either entrance. There is a sliding door in each entrance, so it is possible to maneuver and separate individuals by opening and shutting the doors. The bedroom is a complex space with wooden beds, heated floors, trees, a climbing pole, metal-mesh floors, and flat platforms.

## A Day at the Great Ape Research Institute

Our staff arrive at the institute by 8 o'clock every morning. The chimpanzees are already awake by this time, and they eat breakfast while a staff member checks their health (Figure 3.2). After breakfast and the health check, two experiment sessions are conducted in the morning. At 12:30, the chimpanzees have lunch while the staff clean the bedrooms. The bedroom is not washed with water everyday. The floor is covered with straw and sawdust, and dirty straw or sawdust is replaced by new bedding.

There are two more experiment sessions in the afternoon. At 4:30 p.m., the staff members feed the chimpanzees supper in their bedroom. After the meal, the chimpanzees spend time playing and grooming with the staff. This constitutes an important time for the formation of friendly relationships between the chimpanzee and the staff. We turn the lights out at 6 o'clock in the evening. The staff lies down with the chimpanzees; when the chimpanzees fall asleep, the staff quietly leaves the bedroom. Having the staff sleep with the chimpanzees is a unique characteristic of our institute, not seen at other facilities in Japan.

| 8:00-9:30 | 9:30-12:30 | 12:30-13:30 | 13:30-16:30 | 16:30-17:30 | 18:00 |
|---|---|---|---|---|---|
| breakfast health check | experiment/ observation | lunch cleaning | experiment/ observation | supper | light out |

**Figure 3.2.** A day at the Great Ape Research Institute.

## Future Plans

We plan to explore various fields of study on an anthropological basis. In addition, we intend to practice husbandry by imagining ourselves in the place of the captive apes. We will use our results to educate the public.

We must enhance the quality of life of apes in captivity (Beck et al., 2001). In addition, we need to determine the best handling and feeding techniques for these apes. It is hoped that as a result of these efforts, the apes will remain healthy, both physically and psychologically. We should be able to develop various research plans on the basis of our partnership with the apes.

Wild chimpanzees live in different habitats, such as rain forest and savanna woodlands, in tropical Africa (Idani, 1995; Kano, 1972; McGrew, Baldwin, & Tutin, 1981; Ogawa, Idani, & Kanamori, 1999). In Africa, the utilization of different environments by our ancestors, who had the appearance of humans, constitutes an interesting topic for consideration in the study of human evolution (Coppens, 1988; Moore, 1996). Fortunately, we have two field sites in Africa where we have been conducting active fieldwork: a forest in the Congo and a savanna in Tanzania. If we develop the captive and field studies in parallel, we might be able to unfold the mysteries of human evolution.

# References

Beck, B. B., Stoinski, T. S., Hutchins, M., Maple, T. L., Norton, B., Rowan, A., et al. (2001). *Great apes and humans: The ethics of coexistence*. Washington, DC: Smithsonian Institution Press.

Butynski, T. M. (2001). Africa's great apes. In B. B. Beck, T. S. Stoinski, M. Hutchins, T. L. Maple, B. Norton, A. Rowan, et al. (Eds.), *Great apes and humans: The ethics of coexistence* (pp. 3–56). Washington, DC: Smithsonian Institution Press.

Coppens, Y. (1988). Hominid evolution and the evolution of the environment. *Ossa, 14*, 157–163.

de Waal, F. B. M. (1999, June 17). Cultural primatology comes of age. *Nature, 399*, 635–636.

Fraser, D., Weary, M. D., Pajor, A. E., & Milligan, N. B. (1997). A scientific conception of animal welfare that reflects ethical concerns. *Animal Welfare, 6*, 187–205.

Fujiyama, A., Watanabe, H., Toyoda, A., Taylor, T. D., Itoh, T., Yaspo, M. L., et al. (2002, January 4). Construction and analysis of a human–chimpanzee comparative clone map. *Science, 295*, 131–134.

Hirata, S. (2003). Exploring the mind and behavior of chimpanzees (in Japanese). In *The Annual Report of Hayashibara Group* (pp. 58–59). Okayama, Japan: Hayashibara.

Idani, G. (1995). A preliminary report on distribution of the tschego chimpanzee (*Pan troglodytes*) in the region of Lekoumou, Republic of Congo. *African Study Monographs, 15*, 77–82.

Idani, G., Morimura, N., & Fuwa, K. (1999). A new institute for the great ape research. *Primate Research, 15*, 277–280.

Itani, J. (1996). Afterword: A new milestone in great ape research. In W. C. McGrew, L. F. Marchant, & T. Nishida (Eds.), *Great ape societies* (pp. 305–308). Cambridge, England: Cambridge University Press.

Kano, T. (1972). Distribution and adaptation of the chimpanzee on the eastern shore of Lake Tanganyika. *Kyoto University African Studies, 7*, 37–129.

Matsuzawa, T. (1998). Chimpanzee behavior: Comparative cognitive perspective. In G. Greenberg & M. Haraway (Eds.), *Comparative psychology: A handbook* (pp. 360–375). New York: Garland.

Matsuzawa, T. (2001). Primate foundations of human intelligence: A view of tool use in nonhuman primates and fossil hominids. In T. Matsuzawa (Ed.), *Primate origins of human cognition and behavior* (pp. 3–25). Tokyo: Springer-Verlag Tokyo.

McGrew, W. C., Baldwin, P. J., & Tutin, C. E. G. (1981). Chimpanzees in a hot, dry, and open habitat: Mt. Assirik, Senegal, West Africa. *Journal of Human Evolution, 10*, 227–244.

Moore, J. (1996). Savanna chimpanzees, referential models and the last common ancestor. In W. C. McGrew, L. F. Marchant, & T. Nishida (Eds.), *Great ape societies* (pp. 275–292). Cambridge, England: Cambridge University Press.

Morimura, N. (2003). A note on enrichment for spontaneous tool use by chimpanzees (*Pan troglodytes*). *Applied Animal Behaviour Science, 82*, 241–247.

Ogawa, H., Idani, G., & Kanamori, M. (1999). Chimpanzee habitat in the savanna woodland, Ugalla, Tanzania. *Primate Research, 15*, 147–151.

Rijksen, H. D. (2001). The orangutan and the conservation battle in Indonesia. In B. B. Beck, T. S. Stoinski, M. Hutchins, T. L. Maple, B. Norton, A. Rowan, et al. (Eds.), *Great apes and humans: The ethics of coexistence* (pp. 57–70). Washington, DC: Smithsonian Institution Press.

Rumbaugh, D. M., & Washburn, D. A. (2003). *Intelligence of apes and other rational beings*. New Haven, CT: Yale University Press.

Savage-Rumbaugh, S. (1986). *Ape language: From conditioned response to symbol*. New York: Columbia University Press.

Shepherdson, D. J. (1998). Tracing the path of environmental enrichment in zoos. In D. J. Shepherdson, J. D. Mellen, & M. Hutchins (Eds.), *Second nature: Environmental enrichment for captive animals* (pp. 1–12). Washington, DC: Smithsonian Institution Press.

Shumaker, R. W., & Beck, B. B. (2003). *Primates in question: The Smithsonian answer book*. Washington, DC: Smithsonian Institution Press.

# 4

# Continuity of Cognition Across Species: Darwin in Cyberspace

*Katherine A. Leighty, Dorothy M. Fragaszy, and James M. Brown*

People have wondered for millennia whether other species experience and understand the world and their actions in it in the same manner as humans. As with every other aspect of understanding the world, various cultures have developed their own beliefs of the relations among humans and nonhuman animals, of animal behavior, and of animal intellect. Some cultures, such as many Native American tribes, attribute deep wisdom or supernatural powers to various animals. Current views of the mental lives of animals in North American science, and in Western science in general, reflect the particular cultural heritage of European enlightenment philosophy, from which empirical science as we know it has developed since the 1700s. The Enlightenment view of human uniqueness has been an uncomfortable bedfellow in the past century and a half with the Darwinian doctrine of phylogenetic continuity, including continuity of intellect and of experience. We are still wrestling with the disjoint notions of human uniqueness and phylogenetic continuity, a schism reflected everywhere in psychological science.

Comparative psychologists come down on the other side of this fence, regarding cognitive processes in nonhuman species as shared with humans in some way or ways. These processes are known or are to be discovered through diligent empirical inquiry of behavior, a position that has philosophical support from those more impressed with action and feeling than with introspection (e.g., Dewey, 1942; Johnson, 1987; Langer, 1967; Rorty, 1982). Comparative psychologists studying nonhuman primates have been at the forefront of this group (e.g., Robert Yerkes, Wolfgang Köhler, Harry Harlow, and David Premack), although psychologists studying other kinds of animals are also represented (e.g., Edward Tolman, Louis Herman, and Irene Pepperberg).

Through the study of behavior, comparative psychologists have adopted several empirical strategies to examine the mental prowess of various species and the prospect of mental continuity across the animal kingdom. Earliest efforts involved collection of anecdotes of apparently intelligent behavior in nonhuman species (Romanes, 1884), a method that is still occasionally adopted even in contemporary times (Whiten & Byrne, 1988). However, psychologists soon turned to experimental methods, using highly constrained (and therefore replicable) testing conditions, presenting simple problems to an individual animal that could be solved repetitively in multiple "trials" (e.g., Harlow, 1949; Thorndike, 1911). This experimental strategy still dominates contemporary comparative psychology. Often, the aim of the experiment is to identify how experience increases the probability of a particular form of solution (i.e., learning); this method can also be used to identify perceptual capabilities and preferences. Others have presented problems in a free-form context, in which the individual may demonstrate variable behaviors to achieve a goal (e.g., Köhler, 1925). This approach is more concerned with the production of flexible behavior than with the rapidity of learning a specific behavior. All of these strategies have value; we can expect that all will continue to be used.

Enter the computer in the mid-20th century; our science has changed dramatically because of this technological advance. Church (1983, 1993, 2001) highlighted the numerous ways that computers have affected our lives as researchers. At first computers changed our science in practical ways. Computers freed us from reliance on card-catalog searches and printed indices of literature and afforded new features in experimental design, such as automatization and randomization. The use of computers to run our experiments increased our control of the parameters of our tasks as well as eased the burdens of collecting, storing, and analyzing our data. After attaining our results, the computer provided us with an efficient means to present these data to the research community.

As important as these contributions are, we believe that they should be regarded as the "first generation" benefits. We are now in the "second generation" of applications of technology to our field, and we have opportunities to develop entirely new ways of examining behavior and new vistas for psychological experimentation. In this sense, just as genomics is leading biological science to new theories and new ways of studying living systems, advances in technology are leading psychological science to develop new research areas (i.e., cyberpsychology), new theories, and new paradigms of studying cognition and action (Loomis, Blascovich, & Beall, 1999; Riva & Galimberti, 2001). Technological development offers immense opportunities for comparative inquiry.

## Technology and Comparative Psychology

Beginning in the 1970s, nonhuman subjects became the "users" of technological systems (see Leighty & Fragaszy, 2003). That is, computers were no longer being used solely as response detectors or to control presentation of stimuli but were also presenting subjects with new environments in which to act. This work was pioneered with nonhuman primates by Duane Rumbaugh and his colleagues. Rumbaugh (1977) developed a visually based symbol system for the chimpanzee Lana by presenting lexigrams on a computerized keyboard. Presenting lexigrams in a computerized format allowed researchers to address symbol use in nonhumans in a new and more efficient way. The computerized presentation of lexigrams afforded Lana the opportunity to produce "words" and "phrases" when she desired, not just in response to a request by the experimenter, as was the case in alternative paradigms.

Lana later learned to manipulate a joystick to control the movement of a cursor on a computer monitor (Rumbaugh, Hopkins, Washburn, & Savage-Rumbaugh, 1989). Her ability, and that of other chimpanzees in the Rumbaugh lab, to "use" this form of interactive computer system allowed researchers to address a number of cognitive abilities such as numerosity judgments in a new way (Rumbaugh, Hopkins, Washburn, & Savage-Rumbaugh, 1993; Washburn & Rumbaugh, 1991a). Rumbaugh and colleagues also taught rhesus macaques and baboons to "use" these interactive computerized testing systems to study trajectory prediction, path planning, discrimination, categorization, mental rotation, and learning set performance (Hopkins, Fagot, & Vauclair, 1993; Rumbaugh, Richardson, Washburn, Savage-Rumbaugh, & Hopkins, 1989; Washburn, 1992; Washburn, Hopkins, & Rumbaugh, 1989; Washburn & Rumbaugh, 1991b; Wasserman, Fagot, & Young, 2001). Not only does this methodology allow subjects considerable latitude in how and when they interact with the experimental tasks, but it also provides a "level playing field" for cross-species comparisons (see Hopkins, Washburn, & Hyatt, 1996; Washburn & Rumbaugh, 1992). Because nonhumans "use" the interactive computerized testing systems as humans do, more direct comparison of nonhumans with humans is supported.

The development of this new method of cognitive inquiry has had far-reaching impacts on comparative psychology. Inspired by the approaches of Rumbaugh and colleagues, Tetsuro Matsuzawa developed an interactive computerized testing system in his laboratory at the Primate Research Institute in Inuyama, Japan. As does the Rumbaugh laboratory, Matsuzawa and colleagues address the ability of chimpanzees to work with numbers, as well as short-term memory and visual figure discrimination, and have furthered the investigation of path planning using touch-screen testing systems (Iversen & Matsuzawa, 2001, 1996; Kawai & Matsuzawa, 2001; Matsuzawa, 1985; Tomonaga & Matsuzawa, 1992).

We believe that interactive computerized testing systems have revolutionized comparative psychology, especially with nonhuman primates. They allow the subject to play an active role in the testing environment and permit a wide array of actions. These systems moved experimental approaches away from single response and choice tasks and allow us to examine cognition in nonhumans in a wide array of situations while still affording rigorous experimental control.

As we all know too well, technology advances at a rate that seems at times to exceed our ability to keep up with it. We fight to maintain the computerized testing systems of our laboratories by updating hardware and software and attempting to maintain compatibility. Many of us who examine cognition in nonhumans are wary of deviating from our established systems, knowing the difficulty of training animals on new interfaces and knowledgeable of the potential domino effect caused by changing one piece of equipment. Yet, we cannot let these attitudes hinder the development of our field. We can look to the booming progress in human cognitive research to see our future.

## Interactive Virtual Reality

We are all familiar with the astonishing advances in neural imaging capabilities. However, another less recognized technology, the creation of virtual environments, is also poised to become a powerful ally in the quest to study perception and action. This new technology is known as *virtual reality* (VR), which refers to the use of computer graphics and peripheral devices to simulate a realistic-looking and -feeling world. This simulation is dynamic in that it responds to the inputs of the user and does so in real time. A real-time interaction between user and environment is a defining characteristic of VR (Burdea & Coiffet, 1994). Contributing to the realism of the VR environment is the fact that it may be achieved through multiple senses. In its most developed form, not only is the environment seen, but it is heard, felt, and occasionally smelled and tasted. VR thus provides an interesting leap in the application of technology to the experimental environment. Prior to the introduction of the virtual environment, computerized presentations were made in two dimensions. VR technology allows us to create the perception of a three-dimensional world.

What are the defining features of VR? How is it different from other computer simulations? Burdea and Coiffet (1994) referred to the defining features as the "three *I*s" of VR. The first of these is the *interactive* nature of the VR environment. That is, the environment responds to the user's actions. The second *I* refers to the idea that the VR environment is *immersive*, meaning that it affords perception of something that otherwise would not have been perceived without this external stimulation (Sherman & Craig, 2003). The third *I* is *imagination*, the only thing that limits the applications and functions of VR.

There are a number of different methods for the presentation of virtual environments to human participants. One often-used presentation device is the head-mounted display. The device works by presenting a pair of images on screens inside a pair of goggles or glasses, such that one image is viewed by each eye. Sensors are included within this system that track the movements of the participant's eyes and head to determine where in the image the participant is looking. Using this information, the computer alters the display to immediately reflect changes in viewpoint due to eye and head movements (Sherman & Craig, 2003). A second presentation method involves the participant being positioned in a room and surrounded by computer-generated displays. One example of such a presentation is the Cave Automatic Virtual Environment (CAVE) system developed at the University of Illinois (Sherman & Craig, 2003). One benefit of the CAVE system is that it allows for mobility of the participant and continually updates the projected images to reflect these movements.

These presentation methods are each strictly visual in form but can be augmented to provide stimulation to other senses. Sensors can be placed on the body to track movement through space. Automated systems insert sounds and odors into the virtual environment. Participants can wear receivers about the body to stimulate the touch receptors. This is most commonly conducted using a DataGlove (U.S. Patent No. 4,542,291; Zimmerman, 1985), which detects movements and force and provides pressure stimulation on the hands to simulate grasping and touch (Burdea & Coiffet, 1994).

Researchers of human cognition and perception have begun to present virtual environments in their laboratories to address a wide variety of experimental questions. They have discovered that the use of these interactive three-dimensional environments, in which behavioral responses of the participants are detected and recorded automatically, offers options for research never before available (Gaggioli, 2001). Early applications of VR were conducted by human performance researchers, often in conjunction with government projects. The most well known of these applications were flight simulators for the NASA program (Gaggioli, 2001). VR systems were soon introduced into psychological laboratories in the years following these early government projects. The areas of study to which VR has been applied are numerous and include perception (visual, haptic, etc.), visuospatial navigation, sensorimotor transformations, attention, memory, cognitive performance, and mental imagery (Gaggioli, 2001).

To give a few examples of the utilization of VR systems, researchers of human visual perception investigated the influence of optical flow on the perception of walking speed (Durgin & Kearns, 2002; Kearns, Durgin, & Warren, 2002). This same laboratory has also investigated the impact of vestibular and proprioceptive variables on path integration in the virtual environment (Kearns, Warren, Duchon, & Tarr, 2002). Fajen, Beem, and Warren (2002) had participants navigate through a virtual room of obstacles to study route selection in simple and complex environments. Harrison, Warren, and Tarr (2002) examined path selection in a virtual hedge maze while manipulating path form, landmarks, and path junctions. Sandstrom, Kaufman, and Huettel (1998) studied sex differences in the use of landmarks and room geometry in navigating a virtual water maze. Similarly, Gron, Wunderlich, Spitzer, Tomczak, and Riepe (2000) used the VR system and functional magnetic resonance imaging to identify the brain areas activated in navigation tasks and how activation in these areas differs by sex.

Integrating a haptic/force-feedback system into a VR application, Triesch, Sullivan, Hayhoe, and Ballard (2002) examined the propensity for change blindness (nondetection of changes within one's environment) in a reaching task. Researchers had the participants pick up objects within the virtual environment and place them on conveyer belts. On probe trials, the size of the object was altered between the time of picking it up and placing it on the conveyer belt. Participants were more likely to detect size manipulations if size was attended to in the sorting task (i.e., the objects were to be placed on conveyer belts by size). These findings emphasized the fragmented nature of attention, in that information of object features is often not noticed unless relevant to the particular task at hand.

These experiments exemplify some of the benefits of conducting research using VR systems in human perception and cognition. The first of these benefits is improved ecological validity (Gaggioli, 2001). Immersing the participant in the three-dimensional virtual environment can increase the ecological validity of the testing scenario over two-dimensional interfaces by creating a testing environment more comparable to that of the real world. Similarly, the researcher can place the participant in any number of virtual environments without leaving behind the control of the laboratory. The virtual environment is created by the researcher and thus is enormously flexible (Gaggioli, 2001). In this way we can accommodate for historical and subject variables in a way never possible before. The virtual environment allows the researcher to collect behavioral responses of all forms. It also affords action as we see in the real world (e.g., turning the head, walking, reaching), and the technology of the virtual environment ensures that these actions are detected and recorded. Another benefit of the applications of VR systems is the ability to provide participants with sensory feedback about their actions (Gaggioli, 2001). This feature, like that of the inclusion of normal movement in three dimensions, contributes to the ecological validity of the experimental setting and therefore the generalizability of the results.

Overall, the use of VR holds great promise as a revolutionary method of examining perception and cognition in humans. If this is the future of human perceptual and cognitive research, then we should also consider its applications for comparative psychologists. We propose that the integration of VR systems into laboratories of nonhuman cognition and perception will allow us to better examine continuity of cognition across species. This method will also increase the ecological validity of our experiments with nonhuman subjects in a similar manner as it does for human participants. In this way we can "remove" subjects from the laboratory setting and "insert" them into the virtual environment of our choosing. Once the subject is immersed in the virtual environment, we are able to observe an entire repertoire of behaviors that perhaps would not otherwise have been expressed in the laboratory.

## The Promise of Virtual Reality

Imagine for a moment the possibilities for research that integrates VR into the nonhuman laboratory. We could examine kin recognition, food selection, predatory responses, response to social and behavioral cues, object recognition, locomotion, path selection, and numerous other areas of perception and cognition. We could immerse subjects in interactive environments of our own design while systematically controlling all the variables presented within the environment.

For example, consider research that has been conducted with nonhuman primates on path selection in mazes. This research was originally conducted by presenting a bird's-eye view of the maze on a two-dimensional computer screen, much like a paper-and-pencil maze. Subjects manipulated a joystick to navigate a cursor from the start point to the goal region (Fragaszy, Johnson-Pynn, Hirsh, & Brakke, 2003; Washburn & Astur, 2003). The perceptual nature of this task is far removed from that of moving oneself through a three-dimensional maze, because the actor can see the entire maze at each choice point. Washburn and Astur (2003) advanced this methodology by presenting rhesus macaques with two-dimensional mazes from the perspective of the subject (like that of contemporary video games) instead of the overhead or bird's-eye view. This alters the demands of the task by incorporating the subject into a three-dimensional environment. The study of navigation through a virtual hedge maze by human participants exemplifies the potential of the applications of VR by comparative psychologists (Harrison et al., 2002). By using the VR technology being developed by human cognition researchers, we will bridge our fields such that we can truly study the continuity of cognition using methods that address the importance of ecological validity and perception.

A number of technical and methodological problems arise when presenting virtual environments. Although in some ways VR can increase the ecological validity and generalizability of results compared with other laboratory testing methods, the degree of relatedness between perception and action in the virtual and actual environments has yet to be determined. We therefore cannot make broad generalizations from VR research. Other problems center on the fact that current sensory feedback systems are quite basic in form (especially in the case of proprioception), are still rather expensive, are not very portable, and are not easy to install or use without special training. However, with time, all of these issues will undoubtedly be mitigated if not fully resolved.

## Challenges to Overcome

A number of logistical troubles must be resolved before we can integrate VR into studies of nonhuman cognition and perception to the degree it is used in studies with humans. The first and perhaps most obvious trouble spot will be the presentation method. How does one get an ape, monkey, pigeon, or rat to wear a headset and not remove critical wires and sensors? This is an excellent question, but we do not believe that it is an unsolvable problem. Many potential solutions exist, and we will only be limited by our own imaginations. Animals may be habituated to helmets, jackets, or gloves; certain motions of animals could be restricted (to prevent wire removal); or the CAVE-style setup could be chosen over the headset. We should also consider the pace at which VR technology has advanced over the past several decades. It is probable that miniaturized VR apparatus and remote detection devices will be developed in the near future, thus reducing the impracticality of application of current VR procedures with nonhumans.

A second issue that arises with the application of VR to nonhumans is a perceptual one. Much experimentation will be required to determine the degree to which the virtual and three-dimensional environments are equated (i.e., we must ask subjects if the virtual environment looks real). Additionally, we must consider the sensory feedback systems that we use with nonhumans. We will have to closely examine the perceptual systems of our subjects so that we can design appropriate virtual environments and feedback systems. Thus, a number of basic research issues of environmental perception by nonhumans, such as rapidity of eye movements, will need to be addressed prior to exposing them to virtual environments with the goal of simulating experiences in the real world.

## Conclusion

The advances in technology witnessed over the past 50 years have provided researchers with many new tools for studying cognition across species. The innovative ways researchers have utilized these advances have had, and will continue to have, a profound impact, not only on the speed and efficiency with which research is conducted but, more important, on the questions and issues in comparative cognition that can be addressed. There are still some obstacles to be overcome before the use of leading-edge technologies like VR begin to flourish in the comparative arena, but at the rate technology is advancing we can already see their potential impact on the "virtual" horizon.

## References

Burdea, G., & Coiffet, P. (1994). *Virtual reality technology*. New York: Wiley.
Church, R. M. (1983). The influence of computers on psychological research: A case study. *Behavior Research Methods, Instruments, & Computers, 15,* 117–126.
Church, R. M. (1993). Uses of computers in psychological research. In G. Keren & C. Lewis (Eds.), *A handbook for data analysis in the behavioral sciences: Statistical issues* (pp. 459–476). Hillsdale, NJ: Erlbaum.
Church, R. M. (2001). Animal cognition: 1900–2000. *Behavioural Processes, 54,* 53–63.
Dewey, J. (1942). How is mind to be known? *Journal of Philosophy, 39,* 29–35.
Durgin, F. H., & Kearns, M. J. (2002, May). *The calibration of optic flow produced by walking: The environment matters*. Paper presented at the annual meeting of the Vision Sciences Society, Sarasota, FL.
Fajen, B. R., Beem, N. O., & Warren, W. H. (2002, May). *Route selection emerges from the dynamics of steering and obstacle avoidance*. Paper presented at the annual meeting of the Vision Sciences Society, Sarasota, FL.
Fragaszy, D. M., Johnson-Pynn, J., Hirsh, E., & Brakke, K. E. (2003). Strategic navigation of two-dimensional alley mazes: Comparing capuchin monkeys and chimpanzees. *Animal Cognition, 6,* 149–160.
Gaggioli, A. (2001). Using virtual reality in experimental psychology. In G. Riva & C. Galimberti (Eds.), *Towards cyberpsychology: Mind, cognition and society in the Internet age* (pp. 157–174). Amsterdam: IOS Press.
Gron, G., Wunderlich, A. P., Spitzer, M., Tomczak, R., & Riepe, M. W. (2000). Brain activation during human navigation: Gender-different neural networks as substrate of performance. *Nature Neuroscience, 3,* 404–408.
Harlow, H. F. (1949). The formation of learning sets. *Psychological Review 56,* 51–65.
Harrison, M. C., Warren, W. H., & Tarr, M. J. (2002, May). *Ordinal structure in route navigation*. Paper presented at the annual meeting of the Vision Sciences Society, Sarasota, FL.
Hopkins, W. D., Fagot, J., & Vauclair, J. (1993). Mirror-image matching and mental rotation problem solving in baboons (*Papio papio*): Unilateral input enhances performance. *Journal of Experimental Psychology: General, 122,* 61–72.
Hopkins, W. D., Washburn, D. A., & Hyatt, C. W. (1996). Video-task acquisition in rhesus monkeys (*Macaca mulatta*) and chimpanzees (*Pan troglodytes*): A comparative analysis. *Primates, 37,* 197–204.
Iversen, I. H., & Matsuzawa, T. (1996). Visually guided drawing in the chimpanzee (*Pan troglodytes*). *Japanese Psychological Research, 38,* 126–135.
Iversen, I. H., & Matsuzawa, T. (2001). Acquisition of navigation by chimpanzees (*Pan troglodytes*) in an automated fingermaze task. *Animal Cognition, 4,* 179–192.
Johnson, M. (1987). *The body in the mind: The bodily basis of meaning, imagination, and reason*. Chicago: University of Chicago Press.

Kawai, N., & Matsuzawa, T. (2001). Reproductive memory processes in chimpanzees: Homologous approaches to research on human working memory. In T. Matsuzawa (Ed.), *Primate origins of human cognition and behavior* (pp. 226–234). New York: Springer-Verlag.

Kearns, M. J., Durgin, F. H., & Warren, W. H. (2002, May). *Sensitivity to the gain of optic flow during walking.* Paper presented at the annual meeting of the Vision Sciences Society, Sarasota, FL.

Kearns, M. J., Warren, W. H., Duchon, A. P., & Tarr, M. (2002). Path integration from optical flow and body senses in a homing task. *Perception, 31,* 349–374.

Köhler, W. (1925). *The mentality of apes.* London: Rutledge & Kegan Paul.

Langer, S. K. (1967). *Mind: I. An essay on human feeling.* Baltimore, MD: Johns Hopkins University Press.

Leighty, K. A., & Fragaszy D. M. (2003). Primates in cyberspace: Using interactive computer tasks to study perception and action in nonhuman animals. *Animal Cognition, 6,* 137–139.

Loomis, J. M., Blascovich, J. J., & Beall, A. (1999). Immersive virtual environment technology as a basic research tool in psychology. *Behavior Research Methods, Instruments, & Computers, 31,* 557–564.

Matsuzawa, T. (1985, May 2). Use of numbers by a chimpanzee. *Nature, 315,* 57–59.

Riva, G., & Galimberti, C. (Eds.). (2001). *Towards cyberpsychology: Mind, cognition and society in the Internet age.* Amsterdam: IOS Press.

Romanes, G. J. (1884). *Animal intelligence.* New York: Appleton.

Rorty, R. (1982). *The consequences of pragmatism: Essays 1972–1980.* Minneapolis: University of Minnesota Press.

Rumbaugh, D. M. (1977). *Language learning by a chimpanzee: The Lana project.* New York: Academic Press.

Rumbaugh, D. M., Hopkins, W. D., Washburn, D. A., & Savage-Rumbaugh, E. S. (1989). Lana chimpanzee learns to count by "Numath": A summary of videotaped experimental report. *Psychological Record, 39,* 459–470.

Rumbaugh, D. M., Hopkins, W. D., Washburn, D. A., & Savage-Rumbaugh, E. S. (1993). Chimpanzee competence for counting in a video-formatted task situation. In H. L. Roitblat, L. M. Herman, & P. E. Nachtigall (Eds.), *Language and communication: Comparative perspectives* (pp. 329–346). Hillsdale, NJ: Erlbaum.

Rumbaugh, D. M., Richardson, W. K., Washburn, D. A., Savage-Rumbaugh, E. S., & Hopkins, W. D. (1989). Rhesus monkeys (*Macaca mulatta*), video tasks, and implications for stimulus–response spatial contiguity. *Journal of Comparative Psychology, 103,* 32–38.

Sandstrom, N. J., Kaufman, J., & Huettel, S. A. (1998). Males and females use different distal cues in a virtual environment navigation task. *Cognitive Brain Research, 6,* 351–360.

Sherman, W. R., & Craig, A. B. (2003). *Understanding virtual reality: Interface, application, and design.* San Francisco: Morgan Kaufmann.

Thorndike, E. L. (1911). *Animal intelligence: Experimental studies.* New York: Macmillan.

Tomonaga, M., & Matsuzawa, T. (1992). Perception of complex geometric figures in chimpanzee (*Pan troglodytes*) and human (*Homo sapiens*): Analyses of visual similarity on the basis of choice reaction time. *Journal of Comparative Psychology, 106,* 43–52.

Triesch, J., Sullivan, B. T., Hayhoe, M. M., & Ballard, D. H. (2002, May). *Transient visual representations: A change blindness approach.* Paper presented at the annual meeting of the Vision Sciences Society, Sarasota, FL.

Washburn, D. A. (1992). Analyzing the path of responding in maze-solving and other tasks. *Behavior Research Methods, Instruments, & Computers, 24,* 248–252.

Washburn, D. A., & Astur, R. S. (2003). Exploration of virtual mazes by rhesus monkeys (*Macaca mulatta*). *Animal Cognition, 6,* 161–168.

Washburn, D. A., Hopkins, W. D., & Rumbaugh, D. M. (1989). Automation of learning-set testing: The video-task paradigm. *Behavior Research Methods, Instruments, & Computers, 21,* 281–284.

Washburn, D. A., & Rumbaugh, D. M. (1991a). Ordinal judgments of numerical symbols by macaques (*Macaca mulatta*). *Psychological Science, 2,* 190–193.

Washburn, D. A., & Rumbaugh, D. M. (1991b). Rhesus monkeys' (*Macaca mulatta*) complex learning skills reassessed. *International Journal of Primatology, 12,* 377–388.

Washburn, D. A., & Rumbaugh, D. M. (1992). Testing primates with joystick-based automated apparatus: Lessons from the Language Research Center's computerized test system. *Behavior Research Methods, Instruments, & Computers, 24,* 157–164.

Wasserman, E. A., Fagot, J., & Young, M. E. (2001). Same–different conceptualization by baboons (*Papio papio*): The role of entropy. *Journal of Comparative Psychology, 115,* 42–52.

Whiten, A., & Byrne, R. W. (1988). Tactical deception in primates. *Behavioral and Brain Sciences, 11,* 233–273.

Zimmerman, T. (1985). *U.S. Patent No. 4,542,291.* Washington, DC: U.S. Patent and Trademark Office.

# 5

# Dimensions of the Ape Mind: Adding Personality to Behavior and Cognition

*James E. King*

Reports about the behavior of the great apes extending back to the beginnings of scientific scrutiny of these primates are almost inevitably enlivened by accounts of the antics and dispositions of notable individuals. My introduction to the sometimes-dramatic individuality of great apes occurred during a visit to the San Diego Zoo with Duane Rumbaugh. On our approach to the gorilla exhibit, a large male, Albert (pictured on the cover of this book), launched into an impressive emotional display directed toward Rumbaugh. Albert's outburst was occasioned by Rumbaugh's previous unsuccessful attempt to train Albert on stimulus discrimination problems. After the first two correct and rewarded responses, Albert suffered the indignity of getting no reward following his third and incorrect response. Albert retaliated by breaking the bulletproof glass enclosing the discrimination objects. That not only ended Albert's research career but also left him with an intense animosity toward Rumbaugh. Albert's humanlike response to frustration and perhaps a sense of the injustice of it all was a dramatic illustration of a particularly sensitive and irascible ape personality. Rumbaugh has continually seen the importance of individual personality differences (see Rumbaugh & Washburn, 2003) in understanding the psychology of the great apes.

Wolfgang Köhler, in his classic book, *Mentality of Apes* (1924/1925), used the word *personality* in describing individual differences among his 9 chimpanzee subjects. Robert Yerkes (1925, 1929, 1943) later wrote a series of books focused on behavior and learning of chimpanzees. Personality descriptive adjectives occurred throughout these books in descriptions of individual chimpanzees. For example, Chim was described by Yerkes as sanguine, venturesome, trustful, friendly, and energized, whereas his companion Panzee was distrustful, retiring, and lethargic. In modern psychological jargon, Chim would be described as extraverted with high subjective well-being, whereas the phlegmatic Panzee would be described as introverted with low subjective well-being. The personality descriptive adjectives in Yerkes's accounts displayed a range similar to that of adjective lists typically used in assessing human personality differences (see Goldberg, 1990). In particular, Yerkes's profiles of chimpanzee personality used the adjectival forms of the nouns that describe human personality in the five-factor model (i.e., Extraversion, Agreeableness, Dependability [Conscientiousness], Emotionality [Neuroticism], and Openness). Similar personality-based characterizations of chimpanzees occurred frequently during the 20th century in behavioral reports (e.g., Goodall, 1986). However, the personality information was typically only background for the main point of the reported research that was oriented in other directions.

A reasonably large literature now exists on animal personality in which personality differences are defined by individual variation in either objective behavior or subjective ratings of personality traits (Gosling, 2001; Gosling & John, 1999). Subjects for these personality studies encompass large numbers of vertebrate and invertebrate taxa ranging from octopuses to chimpanzees. A review of 12 vertebrate and invertebrate species by Gosling and John (1999) showed evidence for dimensions resembling the human Extraversion, Agreeableness, and Neuroticism dimensions. Nevertheless, systematic studies of nonhuman personality traits are clearly not within the mainstream of behavioral and cognitive research on nonhuman primates, including the great apes. The inattention to primate personality traits persists despite almost a century of frequent, although informal, acknowledgment of the existence of those traits. (For exceptions, see Buirski, Plutchik, & Kellerman, 1978; Gold & Maple, 1994; Stevenson-Hinde, Stillwell-Barnes, & Zunz, 1980.)

Uneasiness about characterizations of nonhumans in what might be considered anthropomorphic terms has no doubt impeded research in animal personality. *Anthropomorphism* may be defined as the attribution of human mental or psychological traits to animals. Issues about anthropomorphism have most commonly centered on description of temporary psychological states (e.g., fear or anger). However, the same issues apply to application of long-term human personality traits to animals. Severe criticism of anthropomorphic descriptions of animal behavior is still common (e.g., Davis, 1997; Heyes, 1998; Kennedy, 1992), although it has been noted that the problems involving attribution of anthropomorphic traits to animals are fundamentally no different than attributing those same traits to humans (Schilhab, 2002). In both cases, we are making inferences about subjective experiences and mental dispositions of another being based on indirect evidence. In a widely cited and influential article, "What Is It Like to Be a Bat?", Thomas Nagel (1974) argued that it would be impossible for humans to ever experience or understand the subjective experience of an animal. Indeed, imagining a subjective perceptual world of objects in three-dimensional space based on a microchiopteran sonar detection would be beyond the capabilities of even the most imaginative humans. However, imagining the subjective experience of a closely related species such as a great ape is a far more realistic proposition. The ability of humans to detect the subjective states of other humans is referred to as *empathic ability* (Ickes, 1993). Is detection of timidity, fear, curiosity, and even sympathy in an ape much more of an inductive leap than detection of those same traits in another person?

A more general impediment to wider use of personality traits in animal behavior research may be an outgrowth of the old distrust of personality-based descriptions and explanations for human behavior. A large and well-known literature exists on the fallibility of human subjective judgments in a variety of contexts (Ross & Nisbett, 1991). Consequently, to many researchers the word *subjective* connotes biased, unreliable, and unscientific. One expression of this distrust is the often expressed "fundamental attribution error," which is the presumed error of relying on personality as an explanation of human behavior that is better explained by environmental variables (Nisbett & Ross, 1980; Ross, 1977). The debate over personality versus situation as determinates of behavior explanation persisted for decades (Kenrick & Funder, 1988). However, it is now clear that humans can accurately assess the personality of others and that assessed personality is more often than not an accurate predictor of behavior (Funder, 1999). In fact, humans are remarkably adept at forming accurate impressions of others' personalities based on a highly limited amount of information (for a review, see Hall & Bernieri, 2001). People can accurately infer considerable information about someone's personality from viewing that person's office or bedroom (Gosling, Ko, Mannarelli, & Morris, 2002) or from a handshake (Chaplin, Phillips, Brown, Clanton, & Stein, 2000). Within the context of evolutionary psychology, these findings suggest that human personality perception is a highly complex adaptation that evolved during hominid evolution in response to increasingly sophisticated and competitive social environments. Selection calibrated human ability for accurate judgment of others' personalities to a high degree of accuracy. Our understanding of human psychology would have been seriously impaired if subjective personality ratings of self as well as others were not a fundamental part of psychology. A highly unlikely lack of parsimony would be evident if humans who are so perceptive of other people's personality became totally clueless when making similar assessments of great apes.

Use of the human capability for accurate personality perception in making assessments of animal personality is a new dimension for describing and understanding animal behavior. Failure to include subjective personality measures as a fundamental component of animal behavior studies would be roughly analogous to studying animal behavior exclusively by automatic data recording while scrupulously excluding direct visual observation. However, before personality variables can be fully accepted as a legitimate part of animal psychology, at least three issues need to be addressed: the interrater reliability of personality judgments, the basic dimensions or factors of personality in the researched species, and the relationship between personality and behavior. This chapter is directed toward these three issues.

## Personality Structure in Chimpanzees

My initial explorations into assessment of chimpanzee personality were conducted on a group of 100 chimpanzees residing at 12 zoos in the United States and Australia (King & Figueredo, 1997). Workers at the zoos who regularly worked with the chimpanzees used a 7-point Likert scale to rate them on 43 adjectival personality descriptors. Interrater reliabilities were high, sometimes exceeding comparable ratings when humans rated spouses or friends. Factor analysis of the data showed six clear factors. Five of the factors were similar to the five factors that are now almost universally used in the taxonomy of human personality (see Digman, 1996). The five factors and some of the items defining them on the chimpanzee scale were as follows: Extraversion (active, playful, and sociable), Agreeableness (sympathetic, gentle, protective), Dependability (predictable, not impulsive, not erratic), Emotionality (excitable, not stable, not unemotional), and Openness (inventive, inquisitive). The sixth and largest factor was related to dominance. The Dominance factor had 12 items: dominant, not submissive, not dependent, independent, decisive, not timid, not cautious, persistent, not fearful, intelligent, bullying, and stingy. A factor containing this particular combination of traits related to dominance, fearlessness, and overall social prowess has not been reported in humans. Instead, items in the chimpanzee Dominance factor are represented in all five of the human personality dimensions when humans are rated.

The presence of the Dominance factor in chimpanzees and its total absence in humans suggest interesting possibilities for the evolution of personality and what types of personality traits work together to increase fitness. In addition, the chimpanzee Dominance factor counters a potential criticism of personality ratings on nonhumans. The criticism is based on human implicit theories of personality. People have an understanding of the pattern formed by the intercorrelations of different personality traits, that is, an implicit personality theory. As a result of implicit personality theories, the basic five-factor theory of personality emerges even when people rate total strangers (Norman & Goldberg, 1966). Therefore, an obvious question is whether the zoo workers were making up an imaginary human when rating a chimpanzee and then rating the traits of that imaginary human. The rater's implicit theories of personality would then ensure that the human five factors would emerge. However, the presence of the chimpanzee-specific Dominance factor is inconsistent with the implicit theory of personality scenario but supports the assumption that the raters were rating the chimpanzees, not imaginary humans.

## Does Chimpanzee Personality Structure Generalize Across Habitats?

About 4,505 adjectives in English describe some temporally stable aspect of human personality (Allport & Odbert, 1936). If adjectives not applicable to apes (e.g., *conservative, cosmopolitan*) are eliminated, the number of remaining items (e.g., *sociable, active*) would still be enormous. This mass of potential personality variables necessitates aggregation of intercorrelated items into a smaller number of dimensions or factors. In addition to the practical benefits of reducing the number of potential dependent variables, differences in the factors and their component items across species may yield interesting information about the evolution of personality.

However, the usefulness of animal personality factors is dependent on a basic constancy of those factors across different habitats and environments. An animal personality factor that retained its internal structure only within one type of social environment would have limited usefulness; a necessity for changing factor definitions with changes in habitat or subject population would be awkward and unsatisfactory.

The issue of factor constancy in humans has been studied extensively. The generality of the five factors noted above has been assessed across several Western and non-Western cultures. Language of the questionnaires; historical events, including wars and social disorders; and the social structure varied between the cultures. The results have shown that there is reasonable, although not perfect, factor replicability across cultures (McCrae, 2001). Extraversion, Dependability, and Agreeableness seem to be most robust across cultures (Saucier & Goldberg, 2001).

My colleagues and I have begun addressing the question of the constancy of chimpanzee personality factors across different habitats. The zoo populations used for the chimpanzee personality study described in the previous section (King & Figueredo, 1997) had a mean size of about 10 chimpanzees, with 60% female. The groups were stable and underwent infrequent changes in group composition. Obviously, substantive environmental and social differences exist between chimpanzees living in a zoo and chimpanzees living in a more naturalistic setting. King, Weiss, and Farmer (2005) compared the personality factor structure of zoo chimpanzees and chimpanzees living at a chimpanzee sanctuary and reintroduction project, Habitat Ecologique et Liberté des Primates (Ecological Habitat and Freedom of Primates; HELP), located in the Conkouati Douli National Park, Republic of the Congo, Brazzaville. Most of the ratings of the Conkouati chimpanzees were done with rating forms translated into French. The Conkouati chimpanzees lived in much larger habitats with larger interacting groups and a population density over 300 times smaller than that of the zoo-housed chimpanzees. Furthermore, the Conkouati chimpanzees were wild-born and subjected to multiple early stressors and deprivations before arriving at the sanctuary.

The study was based on unit-weighted definitions of the six factors from the original 100 chimpanzees (King & Figueredo, 1997). The unit weighting assigned a weighting of 1 to items with a positive loading of .53 or more and a weighting of −1 to items with loadings of −.53 or less.

The most basic question to be addressed was whether the factor structure differed between the zoo and the Conkouati chimpanzees. Standard procedures for comparing factor structure (e.g., Procrustes rotation) between groups require large sample sizes. The small sample of only 43 chimpanzees in the Conkouati sample precluded use of these procedures. Therefore, King et al. (2005) used a simpler procedure in which internal consistency alpha values for each factor were compared across the two samples. The sample of 56 zoo chimpanzees was different from the 100 chimpanzees that had been used to obtain the original factor definitions. Use of this independent sample was desirable, because the factor extraction procedure maximized the internal consistency of factors, making the internal consistency of the factors for the original 100 chimpanzees necessarily high. The factor internal consistencies of the zoo and the Conkouati chimpanzees were therefore compared for a set of factors that had been determined from a completely independent sample of chimpanzees.

The internal consistencies of the factors were virtually identical for the two groups of chimpanzees. Furthermore, none of the factor internal consistencies for either group differed significantly from those calculated for the original group of 100 chimpanzees used to define the factors. These results indicate that personality factors of zoo chimpanzees generalize to new populations of zoo chimpanzees as well as to chimpanzees reared in vastly different social and physical environments.

In addition, the pattern of correlations between factors in the zoo and the Conkouati chimpanzees was high. In other words, the correlation of interfactor correlations (.76) between the two populations was high. Finally, age effects on factor scores were similar for the zoo and Conkouati chimpanzees. In both populations, Dominance and Agreeableness increased with age, whereas Extraversion, Emotionality, and Openness decreased. There was no interaction of chimpanzee population with any of these age effects. The direction of the age changes for chimpanzee Agreeableness, Extraversion, Emotionality, and Openness was identical to that of humans across diverse cultures (McCrae et al., 1999).

In summary, across the two disparate groups, the six chimpanzee personality factors displayed similar internal consistencies, a similar pattern of interfactor correlation, and similar age changes. I believe that these results support the conclusion that the correlational structure of chimpanzee personality, in common with human personality, is fundamentally species typical and does not vary substantially with situation or habitat. The invariance of chimpanzee and human personality structure suggests the interesting possibility that species personality structure might be a basis for inferring evolutionary changes in personality.

## Does Chimpanzee Personality Predict Behavior?

The studies described up to this point have been based on subjective ratings of chimpanzees' long-term behavior dispositions by workers and caretakers who had spent considerable time observing the chimpanzees and their social interactions. The ratings had high interrater reliability, and correlations across chimpanzees formed meaningful factors that remained invariant across habitats and rearing conditions. However impressive these results might appear, they are nevertheless based entirely on impressions within the minds of the raters. Is it possible that these ratings were based on some consistent but superficial property of chimpanzees that attracted the attention of their raters but had little substantive reality for chimpanzee psychology? Perhaps the chimpanzees radiated a type of "anthropomorphic halo" so that their overall appearance unconsciously suggested specific human personality traits to the raters. Attributes including physical attractiveness, size, common facial expressions, and clumsiness are possible contributors to anthropomorphic halos. Although this possibility of raters being completely misled by the chimpanzees' anthropomorphic halos seems unlikely, a die-hard critic of anthropomorphism could nevertheless make that argument. The burden of proof then falls on the advocate of subjective ratings to show that personality ratings reflect long-term behavioral dispositions.

Correlations between personality ratings and overt behaviors have the potential for falsifying the anthropomorphic halo hypothesis. Personality–behavior correlations fall into at least two different categories. The simplest is when a personality trait has specific meaning close to that of a specific behavior. For example, the trait "playful" is positively correlated with frequencies of play behaviors in rhesus monkeys (Stevenson-Hinde et al., 1980). Ratings of semantically restricted traits of this type are useful because they are more easily obtained than extended behavioral observations. However, any explanations of play behavior in terms of the trait "playful" would be obviously circular because the ratings would be based on past observed instances of play behaviors.

Personality–behavior correlations become more complex and interesting when the personality trait is not associated with only a single behavior but instead is associated with a broad, open-ended class of behaviors. Traits linked to potentially large numbers of behaviors are therefore similar to latent variables. Single adjectival descriptors may have this property. Examples include the descriptors *sensitive*, *sociable*, and *timid* (King & Figueredo, 1997), which imply a large number of possible behaviors. However, other even more abstract descriptors are not readily linked to specific behaviors but instead are inferred from observing patterns or sequences of behavior over extended periods of time (e.g., *persistent*, *independent*, and *predictable*; King & Figueredo, 1997). These abstract descriptors can, however, be linked to behavior through their inclusion in a cluster of intercorrelated traits (i.e., a factor or dimension), which will imply behavioral consequences. Ideally, a factor would include descriptors from each of these three types.

It is also important that research on personality–behavior correlations address both the convergent and discriminate validity of the personality measures (Cronbach & Meehl, 1955). Convergent validity occurs to the extent that the factor has significant correlations with a variety of different behavioral measures all consistent with the meaning of the factor. Positive correlations between several social behaviors and a Sociability or an Extraversion factor would be examples. Discriminant validity would be indicated when each of two or more factors was correlated with a distinct profile or pattern of behavior. Different personality factors that predicted similar groups of behaviors would indicate either a problem with the independence of the factors or a nonrepresentative sample of behaviors.

Initial studies on personality–behavior correlations in nonhuman primates have been encouraging. Capitanio (1999) rated male rhesus monkeys living in a half-acre cage on 25 adjectival personality descriptors. Four factors were identified: Sociability, Confidence, Excitability, and Equability. After a period of up to 4.5 years, the factors generally predicted the monkeys' behaviors in settings different from the original cages. Sociability was positively correlated with affiliative behavior and Confidence with aggression, both results consistent with the convergent and discriminative validity of the original personality ratings. A separate study (Capitanio, Mendoza, & Baroncelli, 1999) showed that the Sociability factor was positively correlated with the immune response to inoculation with simian immunodeficiency virus.

In a more specific context, Lilienfeld, Gershon, Duke, Marino, and de Waal (1999) constructed a scale of 23 items related to human psychopathy. The chimpanzee subjects were also scored on a wide range of behaviors. Only a small number of behaviors were significantly related to the psychopathy scale; however, the behaviors with significant correlations (e.g., agonistic, daring, and detached behaviors) were consistent with the psychopathy construct.

We (Pederson, King, & Landau, 2005) have explored the relationship between personality ratings and behavior with the 43-item scale used by King and Figuredo (1997; Pederson et al., 2005). A group of 49 chimpanzees from the ChimpanZoo program was scored on the ChimpanZoo ethogram (Jane Goodall Institute, 1991). Instantaneous sampling (Martin & Bateson, 1993) was used to score two types of information at each sample point. The first was the overall social context in which the behavior occurred (agonistic, submissive, affinitive, public orientation, and solitary). The second was the specific behaviors that occurred. Twenty-five combinations of social context and behavior occurred with sufficient frequency for meaningful statistical tests.

Although most of the correlations between personality and behaviors were not significant, the significant correlations were consistent with the personality factor definitions (convergent validity). For example, Extraversion contained one group of items implying sociability (viz. playful, sociable, friendly, and not solitary) and another group implying high activity (viz. active, not lazy). Consistent with the sociability component, Extraversion was positively correlated with affinitive context scores and with social approach. Consistent with the high activity component, Extraversion was positively correlated with gymnastic behavior.

An unexpected finding was a high negative correlation between Extraversion and public orientation behaviors, including displays, greetings, and watching directed toward the viewing public. Apparently, extraverted chimpanzees direct their attention and behavior mainly toward other chimpanzees, whereas introverted chimpanzees focus more on the public. Perhaps excessive interest in the viewing pubic is an indicator of social maladaptation in chimpanzees.

## Can Chimpanzees Be Happy?

Personality has an affective component expressed as being either in a positive, happy mood or in a negative, unhappy mood. History indicates that psychologists have had a stronger and longer lasting fascination with human unhappiness and misery than with happiness and joy. A review of the human psychological literature by Myers and Diener (1995) found that references to negative affective states exceeded those to positive states by a ratio of 17 to 1. In fact, *happiness* did not appear as a term in the *Psychological Abstracts* until 1973. The term *happiness* is now typically replaced by the more scientific-appearing term *subjective well-being* (SWB; Diener, 1984). The past 30 years of research has shown that human SWB is a temporally stable and heritable personality trait that has surprisingly small correlations with environmental and demographic variables (for a review, see Diener, Suh, Lucas, & Smith, 1999).

The much shorter history of psychological well-being in nonhuman primates shows a similar early focus on defining unhappiness but not happiness in primates and on eliminating the sources of their unhappiness. The recent proliferation of research on enhancing the psychological well-being of captive primates has led to numerous definitions of well-being, including normal cortisol levels, absence of self-directed and self-injurious behaviors, reproductive success, and species-specific behaviors (e.g., Baker & Aureli, 1997; Eaton, Kelly, Axhelm, & Iliff-Sizmore, 1994; Novak, 2003). These studies have largely occurred in the context of attempts to increase the psychological well-being of captive primates who manifest abnormal behaviors associated with suboptimal social and environmental conditions in laboratory housing. A growing literature focused on farm animals has described specific behaviors and lack of behavioral variability associated with low psychological well-being (e.g., Wemelsfelder, 1997; Wemelsfelder, Haskell, Mendl, Calvert, & Lawrence, 2000).

Measures that have been used to assess the effectiveness of various types of environmental and social enrichment have been highly objective. The emphasis on objective indices of psychological well-being was consistent with other objective indices almost universally used in primate behavior research.

Objective measures of psychological distress in nonhuman primates probably work well in assessing the lower pole of the well-being dimension because distress and boredom have profound effects on behavior and internal physiology. However, one might ask whether the highest levels of psychological well-being may entail more than simply engaging in typical species-specific behaviors without overt symptoms of abnormality. In other words, can a stable personality trait akin to human happiness occur in animals? If so, the great apes would be the most likely place to begin.

We (King & Landau, 2003) began our explorations of the higher regions of the happiness dimension by constructing a four-item scale of happiness for chimpanzees that was labeled a Subjective Well-Being scale to be consistent with the terminology in human literature. In our study, all four items were rated on a 7-point Likert scale by zoo workers. The subjects were 128 chimpanzees living at 13 zoos.

The first item asked about the amount of time that the target chimpanzee was in a positive, happy, contented mood and enjoying itself. The second item asked about the extent to which social interactions with other chimpanzees were satisfying and enjoyable. Although innumerable definitions of human happiness have been proposed by philosophers, psychologists, and others interested in the human condition, the balance of positive and negative affect has been a central component in recent human SWB measures (Diener & Emmons, 1984; Diener et al., 1999). The first two items on the chimpanzee SWB scale are consistent with the affect-related human SWB items.

The third item asked the raters to evaluate the extent to which the target chimpanzee was effective in achieving its goals or wishes. This is based on the idea that human happiness is directly related to a feeling of personal control over important events. This aspect of happiness has also been incorporated into human SWB measures (Campbell, 1981; Cantor & Sanderson, 1999).

The fourth item asked the raters to imagine that they were the target chimpanzee for a week, experiencing the world exactly as the chimpanzee experienced it. In other words, the raters were being asked to answer the question, "What is it like to be a chimpanzee?" This might be viewed by some as a blatant and foolhardy denial of the wisdom embodied in Nagel's (1974) warning that we cannot know what it is like to be a bat. However, as noted earlier, there is an untenable lack of parsimony in the assumption that there is something about subjective experiences of all animals that place them totally beyond human comprehension in contrast to the ability of humans to comprehend some subjective experiences of other humans. Study of the chimpanzee would be particularly problematic if this rigid dichotomy were rigorously enforced.

The question of whether a person can reasonably infer the long-term average affective states of chimpanzees and then judge the desirability of those states is probably not going to be illuminated by more philosophical analysis. The meaningfulness of the question can be attacked, at least indirectly, by comparing the interrater reliabilities and the interitem correlations of the suspect fourth item with those for the other more conventional items. Both types of measures were comparable across all four items.

Furthermore, interrater reliabilities for the chimpanzees on the full Subjective Well-Being scale were at least comparable and sometimes greater than comparable reliabilities for human self–peer correlations. Likewise, the internal consistency ($\alpha = .86$) was comparable to those on human SWB scales (for details, see King & Landau, 2003).

The four chimpanzee Subjective Well-Being items displayed four similarities with human SWB scales. First, they formed a single factor or latent variable in common with a similar tendency of different human SWB items to form one factor (Stones & Kozma, 1985). Second, the six chimpanzee personality factors noted previously (King & Figueredo, 1997) accounted for 52% of the SWB variance, a result again similar to the high amount of human SWB variance accountable to personality variance (see Diener et al., 1999). Third, longitudinal data showed the temporal stability of the SWB dimension over a 4-year period. Finally, multiple regression showed that the Dominance, Extraversion, and Dependability factors had significant correlations with SWB. This pattern of significance was similar for all four items. In humans, Extraversion has been frequently associated with high SWB, and Emotionality has been associated with low SWB (Costa & McCrae, 1980; Diener, 1998). Dependability and Agreeableness also have a positive correlation with human SWB (McCrae & Costa, 1991).

What behavior indicators did the chimpanzee raters use as a basis for their SWB ratings? Unfortunately, we have only an incomplete answer to this question. Significant negative correlations occurred between SWB and submissive context behaviors ($r = -.33$) as well as submissive avoidance ($r = -.35$). SWB was not significantly correlated with any other ChimpanZoo ethogram behavior.

The lack of significant correlations between SWB and most of the ethogram behaviors probably reflects the lack of SWB-relevant behaviors in the ethogram more than any inherent lack of connection between SWB and behavior. More information is needed about behaviors that are indicators of high as well as low SWB status. One potentially useful research strategy would be to make instantaneous recordings of the affective state of chimpanzees along with behaviors that accompany the SWB judgments.

Chimpanzee SWB and Dominance show high additive heritability (Weiss, King, & Enns, 2002). In addition, Dominance and SWB have a genetic correlation approaching 1.0, although the Dominance and SWB items formed two distinct factors when the combined items were analyzed. This indicates that there is an almost complete overlap between the heritable components of these two constructs. Weiss et al. speculated that the strong genetic linkage between SWB and Dominance could be interpreted as evidence that happiness or SWB could serve as a fitness indicator in sexual selection of chimpanzees (and humans!). This speculation requires the assumption that chimpanzees can perceive the happiness of other chimpanzees and that Dominance is fitness enhancing. Both assumptions seem plausible.

## Future Directions

I believe that the studies described in this chapter make a strong case for including subjective personality ratings as a legitimate part of psychological research on the great apes. Subjective ratings are reliable, predict behavior, and form factors that generalize across habitats and are similar to those in humans.

Obviously, the ultimate value of any new measure is not simply to show that it can be measured with psychometric rigor. The new measure should also suggest new questions that can be answered resulting in new insights about animal psychology. The generation of new and interesting questions involving personality and other more conventional psychological measures would provide the best evidence that personality is not a mere disembodied epiphenomenon floating above the animal mind unperturbed by any connection to reality.

One class of questions for future research is the contributions of personality to evolutionary fitness. Measurement of personality in natural populations could answer questions about the effect of personality differences on reproduction, mate attraction, and survival. These questions have been addressed in humans (e.g., Buss, 1996; Hogan, 1983), but the necessity of using present-day humans living in environments rather different from Pleistocene habits may be problematic. Other questions involving personality variables in health, stress, and formation of groups and subgroups with larger populations and cognition are obvious areas for further personality research. Finally, the role of personality variability within as well as between individuals in adaptation to a complex competitive social environment remains a tantalizing issue for future study.

## References

Allport, G. W., & Odbert, H. S. (1936). Trait names: A psycho-lexical study. *Psychological Monographs, 47*(Whole No. 211).

Baker, K. C., & Aureli, F. (1997). Behavioural indicators of anxiety: An empirical test in chimpanzees. *Behaviour, 134,* 1031–1050.

Buirski, P., Plutchik, R., & Kellerman, H. (1978). Sex differences, dominance, and personality in the chimpanzees. *Animal Behaviour, 26,* 123–129.

Buss, D. M. (1996). Social adaptation and five major factors of personality. In J. S. Wiggins (Ed.), *The five-factor model of personality: Theoretical perspectives* (pp. 180–207). New York: Guilford Press.

Campbell, A. (1981). *The sense of well-being in America: Recent patterns and trends.* New York: McGraw-Hill.

Cantor, N., & Sanderson, C. A. (1999). Life task participation and well-being: The importance of taking part in daily life. In D. Kahneman, E. Diener, & N. Schwarz (Eds.), *Well-being: The foundations of hedonic psychology* (pp. 230–243). New York: Russell Sage Foundation.

Capitanio, J. P. (1999). Personality dimensions in adult male rhesus macaques: Prediction of behaviors across time and situation. *American Journal of Primatology, 47,* 299–320.

Capitanio, J. P., Mendoza, S. P., & Baroncelli, S. (1999). The relationship of personality dimensions in adult male rhesus macaques to progression of simian immunodeficiency virus disease. *Brain, Behavior, and Immunity, 13,* 138–154.

Chaplin, W. F., Phillips, J. B., Brown, J. D., Clanton, N. R., & Stein, J. L. (2000). Handshaking, gender, personality, and first impressions. *Journal of Personality and Social Psychology, 79,* 110–117.

Costa, P. T., Jr., & McCrae, R. R. (1980). Influence of extraversion and neuroticism on subjective well-being: Happy and unhappy people. *Journal of Personality and Social Psychology, 38,* 668–678.

Cronbach, J. P., & Meehl, P. E. (1955). Construct validity in psychological tests. *Psychological Bulletin, 52,* 281–302.

Davis, H. (1997). Animal cognition versus animal thinking: The anthropomorphic error. In R. W. Mitchell, N. S. Thompson, & H. L. Miles (Eds.), *Anthropomorphism anecdotes, and animals* (pp. 335–347). Albany: State University of New York Press.

Diener, E. (1984). Subjective well-being. *Psychological Bulletin, 95,* 542–575.

Diener, E. (1998). Subjective well-being and personality. In D. F. Barone, M. Herson, & V. B. van Hasselt (Eds.), *Advanced personality* (pp. 311–334). New York: Plenum Press.

Diener, E., & Emmons, R. A. (1984). The independence of positive and negative affect. *Journal of Personality and Social Psychology, 47,* 1105–1117.

Diener, E., Suh, E. M., Lucas, R. E., & Smith, H. L. (1999). Subjective well-being: Three decades of progress. *Psychological Bulletin, 125,* 276–302.

Digman, J. M. (1996). The curious history of the five-factor model. In J. S. Wiggins (Ed.), *The five-factor model of personality: Theoretical perspectives* (pp. 1–20). New York: Guilford Press.

Eaton, G. G., Kelly, S. T., Axhelm, M. K., & Iliff-Sizmore, S. A. (1994). Psychological well-being in paired adult female rhesus (*Macaca mulatta*). *American Journal of Primatology, 33,* 89–99.

Funder, D. C. (1999). *Personality judgment: A realistic approach to person perception.* San Diego, CA: Guilford Press.

Gold, K. C., & Maple, T. L. (1994). Personality assessment in the gorilla and its utility as a measurement tool. *Zoo Biology, 13,* 509–522.

Goldberg, L. R. (1990). An alternative "description of personality": The Big-Five factor structure. *Journal of Personality and Social Psychology, 59,* 1219–1229.

Goodall, J. (1986). *The chimpanzees of Gombe: Patterns of behavior.* Cambridge, MA: Belknap Press.

Gosling, S. D. (2001). From mice to men: What can we learn about personality from animal research? *Psychological Bulletin, 127,* 45–86.

Gosling, S. D., & John, O. P. (1999). Personality dimensions in non-human animals: A cross-species review. *Current Directions in Psychological Science, 8,* 69–75.

Gosling, S. D., Ko, S. J., Mannarelli, T., & Morris, M. E. (2002). A room with a cue: Personality judgments based on offices and bedrooms. *Journal of Personality and Social Psychology, 82,* 379–398.

Hall, J. A., & Bernieri, F. J. (2001). *Interpersonal sensitivity: Theory and measurement.* Hillsdale NJ: Erlbaum.

Heyes, C. M. (1998). Theory of mind in nonhuman primates. *Behavioral and Brain Sciences, 21*, 101–148.

Hogan, R. (1983). A socioanalytic theory of personality. In M. M. Page (Ed.), *Personality: Current theory and research* (pp. 55–89). New York: Holt.

Ickes, W. (1993). Empathic accuracy. *Journal of Personality, 61,* 587–610.

Jane Goodall Institute. (1991). *ChimpanZoo observer's guide.* Silver Spring, MD: Author.

Kennedy, J. S. (1992). *The new anthropomorphism.* Cambridge, England: Cambridge University Press.

Kenrick, D. T., & Funder, D. C. (1988). Profiting from controversy: Lessons from the person–situation debate. *American Psychologist, 43,* 23–34.

King, J. E., & Figueredo, A. J. (1997). The five-factor model plus dominance in chimpanzee personality. *Journal of Research in Personality, 31,* 257–271.

King, J. E., & Landau, V. I. (2003). Can chimpanzee (*Pan troglodytes*) happiness be estimated by human raters? *Journal of Research in Personality, 37,* 1–15.

King, J. E., Weiss, A., & Farmer, K. H. (2005). A chimpanzee (*Pan troglodytes*) analogue of cross-national generalization of personality structure: Zoological parks and an African sanctuary. *Journal of Personality, 73,* 389–410.

Köhler, W. (1925). *The mentality of apes* (E. Winter, Trans.). New York: Harcourt Brace. (Original work published 1924)

Lilienfeld, S. O., Gershon, J., Duke, M., Marino, L., & de Waal, F. B. M. (1999). A preliminary investigation of the construct of psychopathic personality (psychopathy) in chimpanzees (*Pan troglodytes*). *Journal of Comparative Psychology, 113,* 365–375.

Martin, P., & Bateson, P. P. G. (1993). *Measuring behavior: An introductory guide.* New York: Cambridge University Press.

McCrae, R. R. (2001). Trait psychology and culture: Exploring intercultural comparisons. *Journal of Personality, 69,* 819–845.

McCrae, R. R., & Costa, P. T., Jr. (1991). Adding *Liebe* and *Arbeit*: The full five-factor model and well-being. *Personality and Social Psychology Bulletin, 17,* 227–232.

McCrae, R. R., Costa, P. T., Jr., Lima, M. P. D., Simões, A., Stendorf, F., Angleitner, A., et al. (1999). Age differences in personality across the adult life span: Parallels in five cultures. *Developmental Psychology, 35,* 466–477.

Myers, D. G., & Diener, E. (1995). Who is happy? *Psychological Science, 6,* 6–10.

Nagel, T. (1974). What is it like to be a bat? *Philosophical Review, 83,* 435–450.

Nisbett, R. E., & Ross, L. (1980). *Human inference: Strategies and social shortcomings of social judgment.* Englewood Cliffs, NJ: Prentice-Hall.

Norman, W. T., & Goldberg, L. R. (1966). Raters, ratees, and randomness in personality structure. *Journal of Personality and Social Psychology, 41,* 681–691.

Novak, M. A. (2003). Self-injurious behavior in rhesus monkeys: New insights into its etiology, physiology and treatment. *American Journal of Primatology, 50,* 3–19.

Pederson, A. K., King, J. E., & Landau, V. I. (2005). Chimpanzee (*Pan troglodytes*) personality predicts behavior. *Journal of Research in Personality, 39,* 534–545.

Ross, L. (1977). The intuitive psychologist and his shortcomings. In L. Berkowitz (Ed.), *Advances in experimental social psychology* (Vol. 10, pp. 174–214). New York: Academic Press.

Ross, L., & Nisbett, R. E. (1991). *The person and the situation: Perspectives of social psychology.* New York: McGraw-Hill.

Rumbaugh, D. M., & Washburn D. A. (2003). *Emergents and rational behaviorism.* New Haven, CT: Yale University Press.

Saucier, G., & Goldberg, L. R. (2001). Lexical studies of indigenous personality factors: Premises, products, and prospects. *Journal of Personality, 69,* 847–879.

Schilhab, T. S. S. (2002). Anthropomorphism and mental state attribution. *Animal Behaviour, 63,* 1021–1026.

Stevenson-Hinde, J., Stillwell-Barnes, R., & Zunz, M. (1980). Subjective assessment of rhesus monkeys over four successive years. *Primates, 21,* 66–82.

Stones, M. J., & Kozma, A. (1985). Structural relationships among happiness scales: A second order factorial study. *Social Indicators Research, 17,* 49–55.

Weiss, A., King, J. E., & Enns, R. M. (2002). Subjective well-being is heritable and genetically correlated with dominance in chimpanzees (*Pan troglodytes*). *Journal of Personality and Social Psychology, 83*, 1141–1149.

Wemelsfelder, F. (1997). The scientific validity of subjective concepts in models of animal welfare. *Applied Animal Behavioural Science, 53*, 75–88.

Wemelsfelder, F., Haskell, M., Mendl, M. T., Calvert, S., & Lawrence, A. B. (2000). Diversity of behaviour during novel object tests is reduced in pigs housed in substrate impoverished conditions. *Animal Behaviour, 60*, 385–394.

Yerkes, R. M. (1925). *Almost human.* New York: The Century Co.

Yerkes, R. M. (1929). *The great apes: A study of anthropoid life.* New Haven, CT: Yale University Press.

Yerkes, R. M. (1943). *Chimpanzees: A laboratory colony.* New Haven, CT: Yale University Press.

# Part II

# Interpreting Primate Behavior

# 6

# Species of Parsimony in Comparative Studies of Cognition

## J. David Smith

The problem of interpreting animal behavior is a classic one in American psychology (e.g., Morgan, 1906; Thorndike, 1911; Tolman, 1927). Indeed, this problem partly defined American psychology, given the early emphasis within behaviorism on low-level, associative descriptions of behavior and the later movement toward cognitivism that sometimes granted animals cognitive processes and representations. Duane Rumbaugh has joined Harlow (e.g., 1949) and a few others in the group of comparative researchers who have carefully addressed the proper explanatory framework regarding animals' performance in cognitive tasks. One can see this in his seminal research that differentiated associative and mediational levels of learning and showed the changing balance of these levels across the primate species (Rumbaugh & Pate, 1984). One can see this in his writings on animals' emergent behaviors—that is, behaviors constructed for use in novel situations without the benefit of training (Rumbaugh, 2002; Rumbaugh, Savage-Rumbaugh, & Washburn, 1996; Rumbaugh, Washburn, & Hillix, 1996).

In the light of these important contributions, I consider in this chapter the kind of behavioral description that comparative psychologists have generally found parsimonious and preferable. I make five principal points. First, I suggest that the explanatory parsimony seemingly embodied in Morgan's canon is sometimes false. Second, I argue that the canon has sometimes been used to support an inappropriate scientific stance. Third, I show that cross-species analyses of cognition may require an alternative kind of explanatory parsimony than that which suffices for single-species analyses. Fourth, I show that the same is true of cross-task analyses of cognition. Fifth, I discuss this alternative principle of parsimony by which one tries to understand the performance of humans and animals in an integrative, not separative, way, and by which one seeks inclusive psychological ideas that will bear useful explanatory weight across species and across tasks. I also summarize some inclusive psychological ideas I have found constructive.

---

The preparation of this chapter was supported by National Institutes of Health Grant HD-38051.

## Dueling Interpretative Perspectives in Comparative Studies of Metacognition

There is an important reason why the character of explanation in comparative psychology concerns me. With my colleagues, I have been exploring the capacity that animals have for metacognition and uncertainty monitoring (Beran, Smith, Redford, & Washburn, in press; Shields, Smith, Guttmannova, & Washburn, 2005; Shields, Smith, & Washburn, 1997; Smith, Beran, Redford, & Washburn, in press; Smith et al., 1995; Smith & Schull, 1989; Smith, Shields, Allendoerfer, & Washburn, 1998; Smith, Shields, Schull, & Washburn, 1997; Smith, Shields, & Washburn, 2003a, 2003b; Smith & Washburn, 2005; Washburn, Smith, & Shields, in press). As we will see, the constructs in this field are theoretically challenging—especially when applied to nonhuman animals (hereinafter I will refer to them simply as *animals*). The idea in research on metacognition is that some minds—human minds at least—have a cognitive utility that monitors ongoing perception and memory, that evaluates whether perception and memory have been successful, and that controls these processes to make them more effective (e.g., Flavell, 1979; Koriat, 1993; Nelson, 1992; Nelson & Narens, 1990; Schwartz, 1994).

Humans do monitor and control their cognitive processes. They feel uncertain and confident. They know when they don't know and when they do. They act accordingly. When humans do these things, which seem to be sophisticated cognitive capacities, scientists comfortably make some intriguing assumptions about their minds. We take metacognition to be about hierarchical levels of cognitive processing, because there is a *meta* level (the cognitive executive) that monitors and controls the *object* level (the ongoing cognitive processing). We assume that metacognition points to awareness of the processes of mind and to self-awareness as persons (because feelings of confidence and doubt are personalized—*I* know; *I* doubt). Finally, we take metacognition to be related to declarative consciousness because these feelings are so often explicit and reportable (though not always; Reder & Schunn, 1996).

Given these inferences from the phenomenon of metacognition, whether animals have a capacity like metacognition becomes an important question because it raises issues about their cognitive awareness and perhaps their self-awareness and consciousness. This explains why this area is one in which theoretical perspectives clash and interpretative sparks fly (e.g., Smith, Shields, & Washburn, 2003b).

I can use the sparks to illuminate my research area. In one of our uncertainty-monitoring tasks, my colleagues and I placed humans in a visual density-discrimination task (Smith et al., 1997). The participants used a joystick to move a cursor to one of three objects on a computer screen (Figure 6.1A). Moving the cursor to the box was correct if it contained exactly 2,950 illuminated pixels. Choosing the "S" was correct if the box contained any fewer pixels. Choosing the star allowed participants to decline the trial and move on to a new, guaranteed-win trial. Trial difficulty was adjusted on the basis of participants' performance within a session to maintain a high level of difficulty. To do so, we adjusted the density of the sparse boxes to keep participants near their threshold of distinguishing dense from sparse.

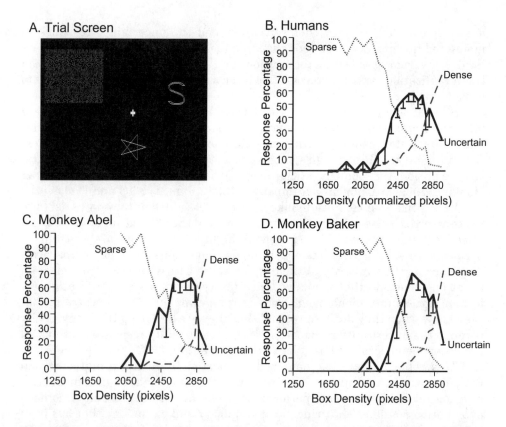

**Figure 6.1.** (A) The screen from a trial in the dense–sparse discrimination of Smith et al. (1997). (B) The performance of seven humans in the dense–sparse task. The dense response was correct for boxes with exactly 2,950 pixels (these trials are represented by the rightmost data point for each curve). All other boxes deserved the sparse response. To equate discrimination performance across participants, the data have been normalized to place each participant's discrimination crossover at a pixel density of about 2,700. The horizontal axis indicates the normalized pixel density of the box. The solid line represents the percentage of trials receiving the uncertainty response at each density level. The error bars show the lower 95% confidence limits. The percentages of trials ending with the dense response (dashed line) or sparse response (dotted line) are also shown. (C) The performance of Monkey Abel in the dense–sparse discrimination depicted in the same way. (D) The performance of Monkey Baker in the dense–sparse discrimination.

Figure 6.1B shows the human participants' performance. Box responses predominated on dense trials and the most difficult sparse trials. S responses predominated on the sparser trials. The primary discrimination was performed at chance where these two response curves cross. The critical result is that humans used the uncertainty response (the star) most in this threshold region of maximum uncertainty. They knew when they were at risk for error and declined those trials selectively. Humans even reported that their uncertainty responses were prompted by feelings of uncertainty and doubt about the correct answer on a trial, so they themselves mapped the uncertainty response to monitored uncertainty. One sees that this task has strong points as a behavioral, nonverbal assay of metacognition, and one does not strongly question the verbalizations or the metacognitive interpretation because the participants were humans.

So what about animals? Smith et al. (1997) also placed two rhesus monkeys (*Macaca mulatta*) in the dense–sparse uncertainty-monitoring task. The animals were tested using the Language Research Center's Computerized Test System, another of Rumbaugh's important contributions to the field of comparative cognition (Rumbaugh, Richardson, Washburn, Savage-Rumbaugh, & Hopkins, 1989; Washburn & Rumbaugh, 1992). Figures 6.1C and D show the monkeys' performance. Both monkeys also assessed when they were liable to make an error in the primary discrimination, and they bailed out of those trials selectively and adaptively. The graphs in Figure 6.1 show one of the strongest performance similarities between humans and animals in the comparative literature. Is the similarity in the data patterns across species strong enough to let one conclude that monkeys' uncertainty responses also indicate some form of uncertainty monitoring? Do we say that monkeys feel uncertain, that they know when they don't know, and that they are declaring this nonverbally through the uncertainty response? If so, the uncertainty response becomes an important behavioral ambassador of the animal mind.

The problem, though, is that one does question these interpretations when the participants are animals. The reason is that a core value in comparative psychology is to give animals' performance the simplest and most parsimonious explanation possible. The principle of simplicity and parsimony that has been most dominant in comparative psychology was stated famously by Morgan (1906): "In no case may we interpret an action as the outcome of the exercise of a higher psychical faculty, if it can be interpreted as the outcome of the exercise of one which stands lower in the psychological scale" (p. 53).

Accordingly, given a metacognitive-like performance by an animal, there is a 100-year-old urge to demote it to some lower associative phenomenon. One may have felt this urge as I described the result. For example, one might argue that threshold stimuli bring timeouts and aversive states that condition escape responses. Or one might argue that animals memorize a series of reinforcement histories along the density continuum and that the law of effect stamps in the best response to make at each stimulus level.

## Taking Aim at the Canon

My first principal point is to caution that in a case such as that shown in Figure 6.1, the principle of parsimony and simplicity seemingly embodied by Morgan's canon is false. The reason is that this is a cross-species comparison of cognitive processes, not a monkey study. Consequently, one cannot interpret the animals' "metacognitive" performance in a vacuum. Humans perform almost identically. Humans report that their uncertainty is conscious, and thus, by definition their uncertainty is declarative. Humans' performance is appropriately interpreted as metacognitive. Moreover, humans and monkeys share an evolutionary past that could have given them a common uncertainty-monitoring capacity for survival reasons (Griffin, 2003). They share homologous brain structures that could serve this common capacity. Thus it is unparsimonious to interpret the same graph produced by monkeys and humans in qualitatively different ways—low-level associative versus consciously metacognitive. This invokes two behavioral mechanisms even though there is only a single phenomenon. Rather, the parsimonious account would give both performances a common psychological description.

My second principal point is to caution that the dual theoretical interpretation may not even be an appropriate scientific stance. If, for example, younger and older children showed identical data patterns, one would never think to offer qualitatively different low- and high-level interpretations of them. There would be no basis for the dual interpretation. Rather, the appropriate scientific stance would be to conclude (provisionally, at least) that the two performances deserved the same psychological interpretation.

Likewise, regarding Figure 6.1, there is also no basis for making a dual interpretation of animal and human performance. It is critical to realize that the history of comparative psychology has created an environment of scientific inference in which the dual interpretation is given credence when humans and animals are involved. However, leaving aside for a moment the history of the field and the reflexive downgrading of animal minds, one sees that this is not justified. To the contrary, here too one would provisionally interpret the two data patterns as instances of the same phenomenon—a phenomenon that humans clearly describe—until other evidence required the monkey graph to be given a different and lower interpretation.

Now it remains possible that a behavioral researcher might provide evidence that the low-level, associative interpretation is justified for monkeys' performance in the dense–sparse task. In fact, one can cite a related case in point from the developmental literature. Individuals with Williams syndrome, a rare genetic disorder (Donnai & Karmiloff-Smith, 2000), display scores in the normal range on some face-processing and reading tasks. However, additional in-depth studies have revealed that these children reach their normal scores by different information-processing routes than do normal controls (Deruelle, Mancini, Livet, Casse-Perrot, & de Schonen, 1999; Karmiloff-Smith, 1998; Laing, Hulme, Grant, & Karmiloff-Smith, 2001). So in this case, the dual theoretical or process interpretation is justified despite the fact that the surface data patterns are similar. It might be justified for the animals, too, if additional in-depth studies made this case. Here the point is just that the default assumption must not be that the identical data pattern produced by two species deserves qualitatively different interpretations. Therefore, the low-level interpretation awaits the additional research. Usually, a heavy burden of proof falls on the cognitive interpretation. It is constructive for the field when—given the obvious fact in Figure 6.1 that humans and animals are performing in the same way—the burden shifts to the behavioral researcher.

This discussion raises my third principal point. The principle of parsimony embodied in Morgan's canon can seem to work well when just one animal species produces a particular data pattern, because one can then derive a simple, low-level explanation that does not conflict with anything else. However, given a cross-species study, the low-level explanation may contradict other aspects of the whole interpretative situation. Here, for example, it conflicts with how we interpret humans' performance in the identical task (and with how humans interpret their own performance). (For this reason, the inclusion of humans in a comparative study is constructive and challenging because it forces us to reckon with their experience, their verbalizations, and the continuity of their cognitive processing vis-à-vis that of animals.) As a result of the conflict, the low-level explanation may multiply levels of explanation needlessly and wrongly. Comparative studies of cognition may thus not be able to suffice with the species of parsimony that has been historically associated with Morgan's canon. They may require instead a broader, more inclusive kind of parsimony that incorporates the performances of several species and tries to explain the common psychological organization underlying their performances.

## Theoretical Tensions in Cross-Task Analyses of Cognitive Processing

The tension between species of parsimony in cross-species analyses also arises regarding cross-task analyses of cognitive processing. I illustrate this by introducing another uncertainty-monitoring phenomenon.

Researchers of human metacognition often focus on memory difficulty instead of perceptual difficulty (as created by the dense–sparse task). Thus, researchers of animal metacognition have also asked whether animals have a cognitive capacity that is analogous to humans' capacity for memory monitoring. The experimental question is whether animals, like humans, can assess the state of their memory and respond adaptively when it does not justify attempting a memory test (Hampton, 2001; Inman & Shettleworth, 1999; Smith et al., 1998).

Smith et al.'s (1998) exploration of the memory-monitoring capacity relied on the predictable changes in memory performance that occur across the serial positions of a memory list (e.g., primacy and recency effects). These predictable changes tell the experimenter which kinds of memory probe will likely be difficult and produce a state like uncertainty in participants. It lets him or her ask whether animals selectively decline memory tests when given these kinds of probes. Smith et al. adopted the influential serial probe recognition (SPR) task (Castro & Larsen, 1992; Roberts & Kraemer, 1981; Wright, Santiago, Sands, Kendrick, & Cook, 1985). In their procedure, the animal saw a "list" of four pictures and then was asked whether a subsequent probe picture had been in the list or not. To see what a memory trial was like in this task, try to encode—from left to right at about 1 second each—the pictures in Figure 6.2. Bear in mind that a test of memory might happen later in this chapter. Monkeys made a there or not-there response as they judged whether or not the probe had been a list member. We also gave the monkeys an uncertainty response with which they could decline to complete the memory test whenever they chose.

Figure 6.3A shows a monkey's performance in this task. A second monkey performed almost identically (compare Figures 7A and 7B in Smith et al.,

**Figure 6.2.** An example of the memory lists Smith et al. (1998) presented to monkeys in a serial probe recognition task. The pictures would have been presented to monkeys successively for about 1 second each, by computer and in different colors.

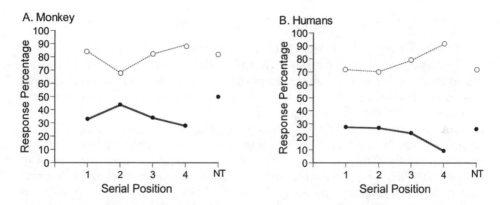

**Figure 6.3.** (A) Serial probe recognition (SPR) performance by Monkey Baker in the task of Smith et al. (1998). The serial position (1–4) of the probe in the list of pictures is also given along the x-axis for the probes on there trials. The percentage of total trials that received the uncertain response is shown (bold line). The percentage correct (of the trials on which the memory test was accepted) is also shown (dotted line). (B) Performance by 10 humans in a similar SPR task used by Smith et al. NT = not-there trials.

1998). Both animals showed classic primacy and recency effects when they elected to complete the memory test. Crucially, though, the pattern of their uncertainty responses was the mirror image of the primacy and recency effects. That is, monkeys were most likely to decline memory trials when their worst serial positions were tested. Figure 6.3B shows that 10 humans produced the same mirror-image relation between memory performance and uncertainty responding (though the experimental conditions were not right for them to show a strong primacy effect). The similarity of the animals' performance to the humans' performance is made more striking by the fact that humans were instructed to use the uncertainty response to report on and cope with memory indeterminacy. Thus, humans declined memory tests when they were uncertain that they remembered. Monkeys behaved like humans.

**Figure 6.4.** Two probes that might have followed the memory list given in Figure 6.2. Both would be there probes. The item on the left reprises the item in the list that occurred at the final, recency-advantaged serial position. The item on the right reprises the item in the list that occurred at the weakest, second serial position. The monkeys' memory probes were presented by computer in color.

The monkeys' performance is also interesting because in this case a low-level interpretation is unavailable. Stimuli could not have controlled responding in this task. Given the structure of the SPR task, every probe picture was rewarded or punished after there and not-there responses in about the same manner. Given the task's structure, only the presence or absence of the probe in the preceding list was relevant. As a result, the psychological signal allowing adaptive performance had to be something more abstract, like trace availability or trace activity. That is, the monkeys had to evaluate the trace strength of the memory location associated with the probe so that they could respond uncertain when probes were associated with memory traces that were indeterminately strong. These trial-specific, memory-based assessments are sophisticated, cognitively derived, and profoundly different from the signals usually available in operant situations, and they verge on demonstrating monkeys' capacity for memory monitoring. Indeed, Nelson, an important contributor to the metacognition literature for 2 decades, suggested that findings such as these are difficult to explain without metacognitive mechanisms (Nelson, 2003; see also Marino, 2003; Mazzoni, 2003; Son, Schwartz, & Kornell, 2003).

To feel this memory-monitoring strategy at work, consider which probe picture in Figure 6.4 contacts a stronger trace in memory and thus was more likely to have been a member of the list of pictures in Figure 6.2. Readers may choose the probe on the left because it was the last item in Figure 6.2's memory trial and received a recency boost of activation from being in that favored list position. Animals often responded there on this kind of probe. Readers may find that the probe on the right contacted a weaker trace in memory because it was presented in the second, weakest serial position. Animals often responded uncertain when they saw this kind of probe.

The memory result broadens the empirical picture to include positive uncertainty-monitoring results from both perceptual and memory tasks. Despite this broader empirical picture, some still prefer to adopt animal psychology's historical tendency to demote animals' performance (e.g., Shettleworth & Sutton, 2003; Wilkins, Cardaciotto, & Platek, 2003; Zentall, 2003; for replies to these commentaries, see Smith et al., 2003b, pp. 358–369). That is, they try to reassert that animals' uncertainty responses could be associatively learned or responsive to stimuli. To do so, they necessarily focus on the perceptual uncertainty-monitoring task (i.e., the dense–sparse task already discussed), emphasizing that this perceptual result could perhaps be explained using low-level associative mechanisms.

However, this narrow interpretative stance fails. The previous discussion made clear why the stimulus-based interpretation cannot explain the important result involving memory monitoring. These results indicate a higher level and more cognitive kind of monitoring. Given a task that requires a high-level cognitive interpretation and another that does not, it is not necessarily good science to force through an account that explains the tasks in qualitatively different ways. If we already know that animals know when they do not remember, why should they not also know when they cannot tell dense from sparse? Thus, one cannot interpret the animal's perceptual performance in a vacuum once one has strong evidence about the animal's memory-monitoring performance. If one does so, one grants the animal two indeterminacy-resolution systems, one of which is already high level. But then the parsimonious account would be to invoke one indeterminacy-resolution system but grant that it applies to memory traces and threshold perceptual impressions.

In fact, the narrow, associative species of parsimony is doubly unparsimonious. It focuses on a subset of the relevant performances (i.e., the perceptual performance) in pursuing the associative account, but then requires a second mechanism to explain the most important data (i.e., the memory-monitoring performance). It focuses on the animals' performance, but then requires a second mechanism to explain humans' metacognitive performance. It is not a truly comparative perspective because it is both task- and species-narrow. Yet the theoretical problem of metacognition and the existing empirical picture are truly comparative because they are task and species broad.

This discussion gives me a chance to credit Morgan (1906) for the sophistication of his views on animal behavior. He did not intend for us to cherry-pick results so that we could analyze them dismissively at a low level. He knew that the analytic situation changes when data from another task become available. King (2003) noted that it is unfortunate that Morgan's canon has been linked historically with the idea that the animal mind must be no more than a Cartesian automaton. Rather, King noted, Morgan meant to state the modest principle that animal behavior should be interpreted in terms of the simplest explanation consistent with the available evidence. (I also thank Robert Remez for his contribution to this discussion about Morgan in a personal communication.) That this was Morgan's intent is shown by the last sentence in his famous chapter:

> To this, however, it should be added, lest the range of the principle be misunderstood, that the canon by no means excludes the interpretation of a particular activity in terms of the higher processes, if we already have independent evidence of the occurrence of these higher processes in the animal under observation. (Morgan, 1906, p. 59)

This reminder applies well here. The memory-monitoring results provide independent evidence that the animals under observation are showing higher level cognitive processing—in fact, processing that is tantamount to memory monitoring. Accordingly, the parsimonious explanation is not to divide the tasks and multiply the explanatory mechanism. It is to unify the tasks and the mechanism and come to understand the psychological structure of the capacity that lets animals monitor their perceptual and memory states and respond to those states adaptively.

All in all, the lesson from cross-task analyses of cognition is the same as that from cross-species analyses. Given a single task like the perceptual uncertainty task, a low-level explanation may suffice because it does not conflict with anything else. In this case the historically common application of Morgan's (1906) canon may seem appropriate. However, it may not be appropriate as areas of comparative inquiry broaden their empirical bases to include tasks in different domains. Then the low-level explanation may contradict the interpretation of other tasks and multiply kinds of explanation wrongly. Then, too, we will be helped by a broader, more inclusive species of parsimony.

## Toward a More Inclusive Species of Parsimony

What will this broader explanatory framework be like? Lacking a full answer to this question, I have tried to find inclusive psychological ideas that bear useful explanatory weight across species and tasks. Here I offer three ideas that I have found helpful in understanding the performance of humans and animals in an integrative, not separative way.

First, it has helped me to have a common information-processing description of the tasks the species are performing. Signal detection theory (SDT) provides such a description (MacMillan & Creelman, 1991, pp. 209–230). Regarding the memory-monitoring task, for example, one can assume that the list items as they are presented create memory impressions that are more or less strong. These can be viewed as lying along the continuum of trace strength shown in Figure 6.5A. Then the probe picture queries the strength of one trace. Probes on not-there trials will generally point to weak traces, perhaps averaging 0.0 plus or minus the scatter of memory variability (distribution NT in the figure). Probes on there trials will point to stronger traces on average (though still with memory variability—the four T distributions in the figure). The overlap between the not-there and there distributions illustrates the SPR task's difficulty and uncertainty because it shows that probes on there and not-there trials will often feel equally well remembered to the participant. Facing this indeterminate memory situation, humans and animals must find a way to organize three response regions along the memory-trace continuum, hopefully being able to respond not there, uncertain, or there as the probe contacts a memory trace that is weak, indeterminate, or strong. The criterion lines delineating these response regions are drawn to illustrate SDT's description of Monkey Baker's decisional strategy. He completed the memory tests (i.e., he responded not there or there) that presented with weak or strong traces.

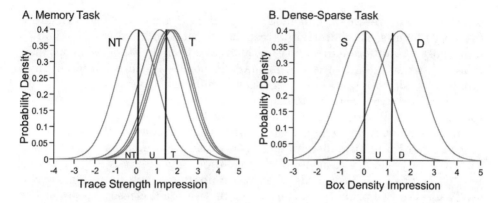

**Figure 6.5.** (A) A signal detection theory portrayal of Monkey Baker's decisional strategy in the serial probe recognition task of Smith et al. (1998). Unit-normal trace-impression distributions are centered at the locations along the trace-strength continuum corresponding to the animal's memory sensitivity on probes of the four serial positions in the memory lists (T) and at 0.0 for the not-there probes (NT). These normal curves are overlain by the decision criteria that define the monkey's three response regions (from left to right, NT, uncertain [U], and T). (B) A signal detection theory portrayal of a hypothetical decisional strategy in the dense–sparse task of Smith et al. (1997). Unit-normal density-impression distributions are centered at 0.0 for sparse (S) trials and at a positive density for dense (D) trials. These normal curves are overlain by decision criteria that define the three response regions (from left to right, S, U, and D).

He responded uncertain for indeterminate trace strengths that could have been caused by a there or a not-there probe.

The SDT description also describes with almost no change what humans and animals do in the dense–sparse task (Figure 6.5B). Now there are distributions of perceptual impressions created by sparse and dense trials, with the human or animal organizing sparse, uncertain, and dense response regions along the continuum of impressions running from clearly sparse through indeterminate to clearly dense.

Thus, one sees that the SDT processing description travels well across tasks and across species. It lets one think integratively about the decisional and criterion-setting problem that both species face in both tasks. It shows that there is no reason to multiply mechanisms across tasks—the same decisional mechanism could serve both tasks. It raises valuable questions, too, about how humans and monkeys set and use criteria, about the level those criterion-setting processes have in the cognitive system, and so forth.

Second, it has helped me to consider the grammar of my tasks as humans and animals may understand them. Humans' self-descriptions of performance reveal how they interpret the uncertainty-monitoring task. Humans know that these tasks have two primary input classes (dense–sparse, there–not there) and that one of these two events occurs on every trial. They know that every trial has a correct answer (dense–sparse, there–not there) if they could just discern it. Humans link the two primary responses in the task to the two input classes. They reserve the uncertainty response for when they cannot tell which input class was presented. This task construal has crucial psychological implications as follows. The uncertainty response is not associated with a stimulus input class. It is different in quality and character from the other two responses. It is in a sense meta to the two primary responses because it represents a judgment that neither of those responses is sufficiently warranted.

There are strong reasons to suppose that animals will have the same task construal that humans do, though more tacitly or nonverbally. We usually train animals with only the primary responses available, so the uncertainty response joins a mature discrimination in which two input class are already linked to two primary responses. We sometimes give animals a daily warm-up that begins with easy trials. This warm-up, during which animals exclusively use the two primary responses for the two input classes, reestablishes these linkages. We also sometimes give animals sessions in which they can only use the two primary responses, and this reestablishes the linkages as well. Finally, the uncertainty response, unlike the primary responses, never gets a reward or punishment and has the same neutral function in every context. It is likely that a response that is so functionally different will be psychologically different to animals, too.

If so, the theoretical interpretation is different, too. The associative interpretation of the dense–sparse task assumes that all three responses (sparse, uncertain, dense) are stimulus based, with the uncertain response dedicated to the middle stimulus region. That is, the three input classes (sparse, middle, dense) would be linked to the three responses. But this description is wrong if monkeys have a primary discrimination and a reserved response for when the discrimination fails. Then we are closer to cognitive monitoring. We are closer to what humans say. We reach a description of performance that works for perceptual uncertainty monitoring and for memory monitoring as performed by humans and animals. We gain parsimony and clarity, too. These come from thinking about the common psychological structure that a task has for humans and animals.

Third, it has helped me to think about the cognitive level of the decisional processes that humans and animals use in performing uncertainty-monitoring tasks. In the traditional operant situation (e.g., blue: left lever press; yellow: right lever press), the two stimulus input classes are qualitatively distinct and not confusable. Accordingly, the stimulus impression the organism registers is a consistent and reliable guide to behavior. Indeed, it would be adaptive for the animal to create strong associations between stimulus impressions and responses, perhaps welding them together so strongly that the response became automatic, attentionless, and less under voluntary control. Shiffrin and Schneider (1977) discussed this kind of task that has a consistent mapping between inputs and outputs and they explained why it is conducive to automatic processing.

In contrast, the uncertainty-monitoring tasks challenge the organism's capacity to resolve the input classes (i.e., dense–sparse; there–not there). Because of this, objective dense *and* sparse trials will often create in the organism the same subjective impression of density (Figure 6.5B). Likewise, objective there and not-there probes will often create in the organism the same impression of being remembered (Figure 6.5A). In these cases, the impression the organism registers will be an inconsistent and unreliable guide to behavior. No stable association between an impression and either primary response will be warranted. Instead, the inherent ambiguity will create the need for some additional kind of decisional processing that deals with the ambiguity and produces a decision about behavior. Shiffrin and Schneider (1977) also discussed this kind of task that has an inconsistent mapping between inputs and outputs and explained why it would be conducive to higher level controlled or attentional processing.

These considerations apply equally to humans and animals. The challenge to their perceptual and memory systems is the same. The inconsistent mapping of the tasks is, too. The impossibility of strong and stable associations between threshold impressions and primary responses is the same, as is the need for another process to decide behavior. Following Shiffrin and Schneider (1977), one can understand why this decisional process would be controlled and voluntary, occurring on a trial-by-trial basis. One can also understand why this decisional process could not be automatic or reflexive.

Suddenly, there is a lot of common psychological ground between humans and animals, and there is a description of performance (the controlled resolution of threshold indeterminacy) that applies to what both species do in both uncertainty-monitoring tasks. Once again we gain parsimony and clarity, too, and these come from considering humans' and animals' performance integratively.

Thus, I am recommending in this chapter a species of parsimony that is inclusive of tasks and species and that strives to find common, unifying psychological principles across these. In my domain, one element of this unifying psychological description is the decisional, criterion-setting nature of the uncertainty-monitoring task. Another element is the inconsistent-mapping problem posed by uncertainty-monitoring tasks that require a more controlled level of cognitive processing. Another element is the construal of the uncertainty-monitoring task that humans and animals may share in which the uncertainty response has a different quality and character from the primary discrimination responses.

## Strengths of an Integrative Comparative Perspective

I believe this integrative comparative perspective has many strengths. First, it has the potential to depolarize our discussions of human and animal behavior (e.g., language, tool use) that so often elevate humans and denigrate animals. Instead of this polarized perspective, and especially given a strong cross-species analogy like that shown in Figure 6.3, a carefully reasoned middle ground may be preferable to a collision between opposed explanatory frameworks at high and low psychological levels.

Second, the inclusive approach simplifies a broad empirical pattern that runs across tasks and species. Third, at the same time it grants parsimony to the common processing principles that are established.

Fourth, the inclusive approach constructively elevates the interpretation one makes of animals' performance in uncertainty-monitoring tasks, granting them some of the cognitive sophistication they deserve (i.e., controlled processing). Moreover, this elevation occurs in a principled way without doing harm to the goal of careful analysis in comparative research.

Fifth, however, the inclusive approach still lets one not grant animals the ultimate level of cognitive sophistication (i.e., consciousness, self-awareness, and full-blown explicit, declarative metacognition). That is, one might claim, or research might show, that animals are engaging in a kind of uncertainty monitoring that is less conscious and alive in immediate awareness than that of humans. (Sometimes even humans' uncertainty monitoring has this more implicit and less conscious character; Reder & Schunn, 1996.) King (2003) also noted that monkeys might have some of the full-blown metacognitive capacity but not all of it. Certainly, there might be common processing principles but also experiential differences across species. In fact, it would be a fascinating map of this area of comparative cognition if one could catch the human capacity building in stages across phylogenetic levels. Even the idea of this map emerges from the integrative perspective that begins with cross-species common ground.

Sixth, the integrative approach is productive of science. It raises new questions. Does it feel the same to a human and monkey to face indeterminacy in a perceptual or memory task? Do homologous brain areas regulate uncertainty monitoring and responding in both species? Would less cognitively sophisticated species than monkeys lack this capacity for resolving indeterminacy adaptively using controlled cognitive processes? Is uncertainty monitoring or uncertainty responding helped by consciousness? Is consciousness so necessary to uncertainty monitoring or responding that we are entitled to draw a profound inference about consciousness in monkeys? In contrast, I believe that the narrow, associative species of parsimony is less productive. What does one do after concluding that humans and animals perform these tasks in completely different ways—walk away when there is so much of interest to know? The separative approach presents, in a way, a disheartening dead end.

Seventh, the integrative approach makes productive theoretical connections. For example, it turns out that there is a close connection between my research and Rumbaugh's current interest in the emergent behaviors that animals choose in novel, untrained situations (for discussion, see Rumbaugh, Beran, & Pate, 2003; Smith et al., 2003b, pp. 366–367). In both the emergent and the uncertain situation, the animal faces an indeterminacy that the trained associative gradients—regarding approach, avoidance, dense, sparse, and so forth—do not resolve. In both situations, the animal is presumed to go to a higher and more cognitive level of information processing and response selection, in the one case to resolve the uncertain trial's indeterminacy, in the other case to construct a creative behavioral solution. Rumbaugh et al. even suspected that it could be the uncertainty that animals monitor in difficult situations that gives them the pause and the cause to reflect on the problem they face. They noted in particular that one subclass of emergent behavior by animals (i.e., insightful problem-solving behaviors as observed in chimpanzees, e.g., by Kohler, 1925) is often accompanied by a behavioral hesitation that is indicative of uncertainty. Their point is that the controlled processing seen in uncertain and in emergent situations could be closely and even causally related. These relationships are seen far more easily if one adopts the integrative perspective discussed here.

## Conclusion

It is my hope that the 2nd century of comparative cognition research will adopt in other domains this broader explanatory framework that is inclusive of tasks and species and that seeks common psychological principles. I believe that this approach represents one of the truest kinds of comparative analysis and one that would be productive of research and theory. Perhaps the integrative species of parsimony discussed in this chapter could even become a useful part of the "rational" behaviorism that Duane Rumbaugh has sought (e.g., Rumbaugh, 2002). If so, it would be an honor to contribute to Rumbaugh's theoretical program, just as he has contributed to so many within comparative psychology.

## References

Beran, M. J., Smith, J. D., Redford, J. S., & Washburn, D. A. (in press). Rhesus macaques (*Macaca mulatta*) monitor uncertainty during numerosity judgments. *Journal of Experimental Psychology: Animal Behavior Processes*.

Castro, C. A., & Larsen, T. (1992). Primacy and recency effects in nonhuman primates. *Journal of Experimental Psychology: Animal Behavior Processes, 18*, 335–340.

Deruelle, C., Mancini, J., Livet, M. O., Casse-Perrot, C., & de Schonen, S. (1999). Configural and local processing of faces in children with Williams syndrome. *Brain and Cognition, 41*, 276–298.

Donnai, D., & Karmiloff-Smith, A. (2000). Williams syndrome: From genotype through to the cognitive phenotype. *American Journal of Medical Genetics: Seminars in Medical Genetics, 97*, 164–171.

Flavell, J. H. (1979). Metacognition and cognitive monitoring: A new area of cognitive–developmental inquiry. *American Psychologist, 34*, 906–911.

Griffin, D. R. (2003). Significant uncertainty is common in nature. *Behavioral and Brain Sciences, 26*, 346.

Hampton, R. R. (2001). Rhesus monkeys know when they remember. *Proceedings of the National Academy of Sciences USA, 98*, 5359–5362.

Harlow, H. F. (1949). The formation of learning sets. *Psychological Review, 56*, 51–65.

Inman, A., & Shettleworth, S. J. (1999). Detecting metamemory in nonverbal subjects: A test with pigeons. *Journal of Experimental Psychology: Animal Behavior Processes, 25*, 389–395.

Karmiloff-Smith, A. (1998). Development itself is the key to understanding developmental disorders. *Trends in Cognitive Sciences, 2*, 389–398.

King, J. E. (2003). Parsimonious explanations and wider evolutionary consequences. *Behavioral and Brain Sciences, 26*, 347–348.

Köhler, W. (1925). *The mentality of apes*. New York: Liveright.

Koriat, A. (1993). How do we know that we know? The accessibility model of the feeling of knowing. *Psychological Review, 100*, 609–639.

Laing, E., Hulme, C., Grant, J., & Karmiloff-Smith, A. (2001). Learning to read in Williams syndrome: Looking beneath the surface of atypical reading development. *Journal of Child Psychology and Psychiatry, 42*, 729–739.

MacMillan, N. A., & Creelman, C. D. (1991). *Detection theory: A user's guide*. Cambridge, England: Cambridge University Press.

Marino, L. (2003). Can we be too uncertain about uncertainty responses? *Behavioral and Brain Sciences, 23,* 348–349.

Mazzoni, G. (2003). Animals show monitoring, but does monitoring imply awareness? *Behavioral and Brain Sciences, 26,* 349–350.

Morgan, C. L. (1906). *An introduction to comparative psychology.* London: Walter Scott.

Nelson, T. O. (Ed.). (1992). *Metacognition: Core readings.* Needham Heights, MA: Allyn & Bacon.

Nelson, T. O. (2003). Relevance of unjustified strong assumptions when utilizing signal detection theory. *Behavioral and Brain Sciences, 26,* 351.

Nelson. T. O., & Narens, L. (1990). Metamemory: A theoretical framework and new findings. *Psychology of Learning and Motivation, 26,* 125–141.

Reder, L. M., & Schunn, C. D. (1996). Metacognition does not imply awareness: Strategy choice is governed by implicit learning and memory. In L. M. Reder (Ed.), *Implicit memory and metacognition* (pp. 45–78). Hillsdale, NJ: Erlbaum.

Roberts, W. A., & Kraemer, P. J. (1981). Recognition memory for lists of visual stimuli in monkeys and humans. *Animal Learning and Behavior, 9,* 587–594.

Rumbaugh, D. M. (2002). Emergents and rational behaviorism. *Eye on Psi Chi, 6,* 8–14.

Rumbaugh, D. M., Beran, M. J., & Pate, J. L. (2003). Uncertainty monitoring may promote emergents. *Behavioral and Brain Sciences, 26,* 353.

Rumbaugh, D. M., & Pate, J. L. (1984). The evolution of cognition in primates: A comparative perspective. In H. L. Roitblat, T. G. Bever, & H. S. Terrace (Eds.), *Animal cognition* (pp. 569–587). Hillsdale, NJ: Erlbaum.

Rumbaugh, D. M., Richardson, W. K., Washburn, D. A., Savage-Rumbaugh, E. S., & Hopkins, W. D. (1989). Rhesus monkeys (*Macaca mulatta*), video tasks, and implications for stimulus–response spatial contiguity. *Journal of Comparative Psychology, 103,* 32–38.

Rumbaugh, D. M., Savage-Rumbaugh, E. S., & Washburn, D. A. (1996). Toward a new outlook on primate learning and behavior: Complex learning and emergent processes in comparative perspective. *Japanese Psychological Research, 38,* 113–125.

Rumbaugh, D. M., Washburn, D. A., & Hillix, W. A. (1996). Respondents, operants, and emergents: Toward an integrated perspective on behavior. In K. Pribram & J. King (Eds.), *Learning as a self-organizing process* (pp. 57–73). Hillsdale, NJ: Erlbaum.

Schwartz, B. L. (1994). Sources of information in metamemory: Judgments of learning and feelings of knowing. *Psychonomic Bulletin and Review, 1,* 357–375.

Shettleworth, S. J., & Sutton, J. (2003). Animal metacognition? It's all in the methods. *Behavioral and Brain Sciences, 26,* 353–354.

Shields, W. E., Smith, J. D., Guttmannova, K., & Washburn, D. A. (2005). Confidence judgments by humans and rhesus monkeys. *Journal of General Psychology, 132,* 165–186.

Shields, W. E., Smith, J. D., & Washburn, D. A. (1997). Uncertain responses by humans and rhesus monkeys (*Macaca mulatta*) in a psychophysical same–different task. *Journal of Experimental Psychology: General, 126,* 147–164.

Shiffrin, R. M., & Schneider, W. (1977). Controlled and automatic human information processing: II. Perceptual learning, automatic attending, and a general theory. *Psychological Review, 84,* 127–190.

Smith, J. D., Beran, M. J., Redford, J. S., & Washburn, D. A. (in press). Dissociating uncertainty states and reinforcement signals in the comparative study of metacognition. *Journal of Experimental Psychology: General.*

Smith, J. D., & Schull, J. (1989). [A failure of uncertainty monitoring in the rat]. Unpublished data.

Smith, J. D., Schull, J., Strote, J., McGee, K., Egnor, R., & Erb, L. (1995). The uncertain response in the bottlenosed dolphin (*Tursiops truncatus*). *Journal of Experimental Psychology: General, 124,* 391–408.

Smith, J. D., Shields, W. E., Allendoerfer, K. R., & Washburn, D. A. (1998). Memory monitoring by animals and humans. *Journal of Experimental Psychology: General, 127,* 227–250.

Smith, J. D., Shields, W. E., Schull, J., & Washburn, D. A. (1997). The uncertain response in humans and animals. *Cognition, 62,* 75–97.

Smith, J. D., Shields, W. E., & Washburn, D. A. (2003a). The comparative psychology of uncertainty monitoring and metacognition. *Behavioral and Brain Sciences, 26,* 317–339.

Smith, J. D., Shields, W. E., & Washburn, D. A. (2003b). The comparative psychology of uncertainty monitoring and metacognition: Reply to the commentaries. *Behavioral and Brain Sciences, 26,* 358–369.

Smith, J. D., & Washburn, D. A. (2005). Uncertainty monitoring and metacognition by animals. *Current Directions in Psychological Science, 14,* 19–24.

Son, L. K., Schwartz, B. L., & Kornell, N. (2003). Implicit metacognition, explicit uncertainty, and the monitoring/control distinction in animal metacognition. *Behavioral and Brain Sciences, 26,* 355–356.

Thorndike, E. L. (1911). *Animal intelligence: Experimental studies.* New York: Hafner.

Tolman, E. C. (1927). A behaviorist's definition of consciousness. *Psychological Review, 34,* 433–439.

Washburn, D. A., & Rumbaugh, D. M. (1992). Testing primates with joystick-based automated apparatus: Lessons from the Language Research Center's Computerized Test System. *Behavior Research Methods, Instruments, and Computers, 24,* 157–164.

Washburn, D. A., Smith, J. D., & Shields, W. E. (in press). Rhesus monkeys (*Macaca mulatta*) immediately generalize the uncertain response. *Journal of Experimental Psychology: Animal Behavior Processes.*

Wilkins, V. M., Cardaciotto, L., & Platek, S. M. (2003). Uncertain what uncertainty monitoring monitors. *Behavioral and Brain Sciences, 26,* 356–357.

Wright, A. A., Santiago, H. C., Sands, S. F., Kendrick, D. F., & Cook, R. G. (1985, July 19). Memory processing of serial lists by pigeons, monkeys, and people. *Science, 229,* 287–289.

Zentall, T. R. (2003). Evidence both for and against metacognition is insufficient. *Behavioral and Brain Sciences, 26,* 357–358.

# 7

# The Significance of the Concept of Emergence for Comparative Psychology

*Gary Greenberg, Ty Partridge, and Elizabeth Ablah*

Since the adoption of the scientific paradigm and methodology in the late 19th century, psychology has had a severe case of "physics envy." At first this manifested itself in the use of the atomistic metaphor of structuralism, the identification with positivism, and the reliance on Western science's materialism and clockwork universe. Indeed, the 20th century is understood by many to be the century of behaviorism: materialistic, positivistic, and reductionistic. Haggbloom et al. (2002) identified B. F. Skinner as the most eminent psychologist of the period. Just as this approach in physics and the other sciences has failed to live up to its initial promise and has since given way to a more holistic, field-oriented, and contextual paradigm (Davies & Gribbin, 1992; Goodwin, 1994; Kauffman, 2000; Sheldrake, 1995), so too psychology has begun to give up its adherence to an old-fashioned physics in favor of this newly emerging scientific paradigm (Chorover, 1990).

This new perspective, and its extremely broad application, can be summarized as follows:

> Since the 1960s, an increasing amount of experimental data ... imposes a new attitude concerning the description of nature. Such ordinary systems as a layer of fluid or a mixture of chemical products can generate, under appropriate conditions, a multitude of *self-organisation phenomena* on a macroscopic scale—a scale orders of magnitude larger than the range of fundamental interactions—in the form of spatial patterns or temporal rhythms. ... [Such states of matter] provide the natural archetypes for understanding a large body of phenomena in branches which traditionally were outside the realm of physics, such as turbulence, the circulation of the atmosphere and the oceans, plate tectonics, glaciations, and other forces that shape our natural environment; or, even, the emergence of self-replicating systems capable of storing and generating information, embryonic development, the electrical activity of the brain, or the behavior of populations in an ecosystem or in economic development. (Nicolis, 1989, p. 316)

As we show in this chapter, such ideas apply equally in psychology.

Of course, as with other developments in science, these ideas applied to psychology did not spring up de novo but rather have their own history. We see seeds of this line of thinking in Lloyd Morgan's (1923) emergent evolution, J. R. Kantor's interbehaviorism (1959; Pronko, 1980), and T. C. Schneirla's behavioral levels hypothesis (Aronson, Tobach, Rosenblatt, & Lehrman, 1972). Contemporary biologists such as Brian Goodwin (1994) and Stuart Kauffman (1993, 1995) have elucidated the linkages between developmental psychobiology and newly emerging concepts of complex adaptive systems and self-organization (Prigogine & Stengers, 1984).

These somewhat diverse positions are linked by three crucial ideas: the important organizing principle of integrative levels, the idea that there is a tendency toward increased complexity with evolutionary advance, and the contextual nature of behavioral events. These ideas lead to a developmental perspective in which behavior is seen to be the result of the fusion of biological and psychosocial factors, by probabilistic epigenetic events rather than by preprogrammed genetic or other biochemical ones (Gottlieb, 1992, 1997; Kuo, 1967). Nonlinear dynamic systems theory provides a theoretically consistent language with which to describe and analyze behavioral development (Michel & Moore, 1995). Nonlinear dynamics contains a lexicon of concepts pertaining to change processes over time that does not exist in any other known theoretical system. Dynamical models allow us to compare and contrast seemingly unrelated phenomena that often share common dynamical structures. Nonlinear dynamics and complex systems analysis are continuing to help revolutionize our understanding in many of the life sciences, though these ideas are just beginning to find their way into mainstream psychology (Boker, 2001; Damon & Lerner, 2001; Newell & Molenaar, 1998; Sulis & Trofimova, 2000). This situation was summarized by Stuart Kauffman (1993), a leading figure in the widespread application of these ideas, as follows: "Eighteenth-century science, following the Newtonian revolution, has been characterized as developing the sciences of organized simplicity, nineteenth-century science, via statistical mechanics, as focusing on disorganized complexity, and twentieth- and twenty-first-century as confronting organized complexity" (p. 173).

A crucial idea is the view that the universe is ordered as a family of hierarchies in which natural phenomena exist in levels of increasing organization and complexity. Indeed, the sciences themselves have been divided into areas of study based on these qualitative changes in complexity of organization, with physics and chemistry addressing the lower levels of complexity and biology, psychology, and sociology addressing the higher levels of complexity (see Feibleman, 1954). This is illustrated in Exhibit 7.1 and summarized by Aronson (1984), who said that the levels concept

> is a view of the universe as a family of hierarchies in which natural phenomena exist in levels of increasing organization and complexity. Associated with this concept is the important corollary that these successions of levels are the products of evolution. Herein lies the parallel with anagenesis. (p. 66)

**Exhibit 7.1.** Hierarchies

| Anthropology | Organism |
|---|---|
| Sociology | Organ |
| Psychology | Tissue |
| Biology | Cell |
| Chemistry | Organelle |
| Physics | Molecule |
| Mathematics | Atom |

Anagenesis recognizes the role played in psychology by the evolution of increasing complex biological forms, especially nervous systems (Greenberg, 1995; Greenberg, Partridge, Weiss, & Haraway, 1999).

For our purposes, it is important to emphasize that complexity is so pervasive a phenomenon that some have likened it to a second law of evolution after natural selection (Saunders & Ho, 1976, 1981, 1984). Many evolutionists have adopted this line of thinking, including Stebbins (1969), who suggested that we can recognize at least eight major levels of complexity in the evolution of life (and behavior), and Maynard Smith (Maynard Smith & Szathmáry, 1995), who at different times has identified five or eight levels of complexity, though he associated each with degrees of organization of genetic material. The important point is that there is a hierarchy of levels of increasing complexity and organization in the evolution of life, not five or eight such levels. This was recognized earlier by Pringle (1951), who noted, "The characteristic of living systems which distinguishes them most clearly from the non-living is their property of progressing by the process which is called evolution from less to more complex states of organization" (p. 175).

Schneirla's concept of behavioral levels, described systematically in a paper written with his colleague and student Ethel Tobach (Tobach & Schneirla, 1968), is derived from the concept of integrative levels. We have used this concept as an organizing theme to explain the full range of behavior across the animal kingdom (Greenberg & Haraway, 2002). The behavioral levels are separated into two groups, one at which biological factors dominate behavior and one at which psychological principles become important. The five levels originally proposed are as follows:

1. *Taxis*: At this level, behavior is under immediate stimulus control, such as in the case of a moth flying toward a light source.
2. *Biotaxis*: At this next, higher, level behavior is influenced not only by the immediate presence of a stimulus but also by the presence of biochemical sequelae from other organisms, that is, by the presence of stimulation that is a concomitant of the presence of other organisms. An example is the sexual attraction of male moths to pheromones secreted by females.
3. *Biosocial*: At this level, the social interaction of groups of animals plays an important role in organizing and regulating behavior. Among

Schneirla's research contributions was his analysis of the behavior of army ants, whose cyclic activity was seen to be a result of reciprocal social stimulation provided by the enormous number of individuals in an ant colony. One might study the behavior of individual ants fruitlessly to discern the source of their cyclical behavior pattern, which is displayed only when ants are together in large numbers (e.g., Gordon, 1988, 1997).

4. *Psychotaxis*: At this level, mediation by past experience enters into the behavioral equation, and behavior is no longer tied only to the immediate presence of a stimulus. Thus, an animal's current behavior may be affected by an earlier history of experiential effects. In their analysis of cat and kitten behavior, Rosenblatt and Schneirla (1962) showed that the relationship between infant and mother is founded in biotactic responses—the kitten orients to the mother by means of tactile and olfactory stimuli—but higher order phenomena such as learning and reinforcement play an important role in later stages of that relationship.

5. *Psychosocial*: Behavior organized at this level is represented by the complex social bonds and social behaviors that are characteristic of advanced vertebrates. For example, among primates, lasting social bonds result from complex biosocial and biotactic interactions between an infant and a mother such as those involved in rocking, providing contact comfort, and nursing.

This approach solves a serious problem in comparative psychology by providing a meaningful and objective framework within which to make behavioral comparisons between and among species. Comparisons may be made in a variety of ways within this framework. No matter what comparisons are chosen, they are informed by the relations, within the levels framework, of the species being compared. We agree with Tobach (1976) that the questions being asked should direct the choice of species selected for comparison. One might wish to ask how species at the same level, but subjected to different ecological pressures, may have evolved different solutions to particular ecological problems, such as how to orient themselves within their environments, how to procure food, or how to attract and maintain possession of a mate.

An example of the application of this behavioral taxonomy may be drawn from feeding behavior and its complexity across phyla and species. As animals became more complex, their nervous systems and feeding behavior became increasingly diverse and flexible, involving alterations in sensory use, locomotor activity, foraging style, food type, and so on. Exhibit 7.2 lists major animal groups and shows at which behavioral grades their feeding behavior is organized. This notion is directed at the classification of behavior. Note that for most groups, feeding behavior is organized at more than one level and that individual species can function behaviorally at more than one level. Each animal then should be classified at its highest level of behavioral complexity in respect to a behavior, with the idea that a higher classification subsumes the levels below it. In Protozoa, such as the amoeba, feeding is regulated solely by the presence of appropriate chemicals at appropriate intensities; they are thus

**Exhibit 7.2.** Levels of Behavioral Complexity Displayed by Major Animal Groups in Feeding Behavior

| Major group | Behavioral level |
| --- | --- |
| Protozoa | Taxis |
| Porifera | Taxis |
| Cnidaria | Biotaxis, biosocial |
| Echinodermata | Biotaxis, biosocial |
| Platyhelmenthes | Biotaxis, psychotaxis |
| Molusca | Biosocial, psychotaxis |
| Arthropoda | Biosocial, psychotaxis |
| Osteichthyes | Biosocial, psychotaxis |
| Amphibia | Biosocial, psychotaxis |
| Reptilia | Biosocial, psychotaxis |
| Aves | Psychotaxis, psychosocial |
| Mammalia | Psychotaxis, psychosocial |

*Note.* For a full explanation, see Greenberg and Haraway (2002).

organized taxically for this behavior. Among the Cnidaria, such as Hydra, feeding is mostly a taxic process, though the ability of these animals to distinguish among living or recently living foodstuffs for prey suggests some biotaxic organization. Feeding by molluscs shows even higher organizational processes at work, allowing learning to become an influence of their feeding behavior. Among the vertebrates, feeding complexity, the influence of conspecifics, and many components of learning (e.g., the remarkable caching ability of some birds) show their feeding behaviors to be organized at the highest levels. It recently has been shown how this organizational system can be applied to the full range of behaviors (Greenberg & Haraway, 2002).

## Emergence

The demarcations and transitions between the levels of organization listed in Exhibit 7.1 and the behavioral levels described by Tobach and Schneirla (1968) are nonlinear and probabilistically discontinuous. As the number of components or events at each level increases (i.e., as complexity increases), a critical ratio of component number and component interconnectivity results. At this critical ratio, the system displays molar-level stability and micro-level instability. In other words, the behavior of individual components within the system is volatile, but the global "structure" of the system as a whole is stable. This property of global stability and internal instability allows these systems, which Kauffman (1993, 2000) referred to as being "poised at the edge of chaos," to be quite adaptable to changing environmental pressures and contingencies, making them ideal for flourishing under principles of natural selection. There are thresholds of organizational complexity at which small quantitative increases result in qualitative discontinuities (i.e., levels), resulting in the appearance of new levels. In dynamic systems this is known as a *phase transition*.

A phase transition results in a new and more complex level of organization. No new inputs are required. Rather, the new level arises from a reorganization of the old elements and is characterized by a new whole, even though there are no new elements. The new whole demonstrates new properties not apparent before the transition. These new properties are said to emerge from the reorganization of the old elements. Emergence is often thought of as some sort of mystical concept. For example, the notion that consciousness emerges from complex neural functioning is understood by some to invoke a "vital" force that is somehow added to the mix. This misunderstanding stems from a confusion in ascribing properties to levels.

When asking someone favorable to the notion of emergence about the concept, they often reply that it means "a whole is different from the sum of its parts." Although this statement is true, it contributes to the misunderstanding. Language, for example, can be understood to be an emergent property of the dynamic interplay of a number of factors, including neocortical size relative to body size, social and cultural complexity, and abstract reasoning ability. Taken improperly, emergence in this context is taken to mean that language would be due to the additive effects of these factors plus the "emergent factor." The emergent factor here would be taken to be an independent property that is added to the mix as the necessary component to produce language and is often ascribed vitalistic properties. To the contrary, however, the emergent property is not a property at the level of individual components of a system. Rather, it is a property of the entire system. In the absence of that system, there is no emergent property. Put another way, concepts that are referred to as emergent (i.e., language, social behavior, symbolic thought, etc.) are not entities but rather are processes of collections of entities. Contemporary cognitive scientists understand mental phenomena from this perspective; thus, mind is conceptualized by some to be an emergent property of the organism's nervous and other critical systems (Bunge, 1980; Sperry, 1987).

The unique properties of water that arise when hydrogen and oxygen are combined two parts to one and catalyzed with a spark is one common example of emergence. These chemicals simply become water in the right context (i.e., the presence of electricity). The emergent properties of water are not separate from but are part of the system hydrogen–oxygen–electricity. The properties of water are not inherent in the properties of any component of that system.

This, then, is what we mean when we refer to "emergent properties" (e.g., "A significant aspect of nonequilibrium physics and self-organisation is the emergence of *new levels of description* brought about by the underlying dynamics"; Nicolis, 1989, p. 341). Another crucial aspect of what we mean when we talk about emergence is what could be referred to as the *emergent event*. The concepts of state-space, attractors, and phase transitions provide a useful language for understanding what we mean here. All systems are composed of organized components. Each component has variable attributes, but for simplicity's sake let us assume that each component has only one variable attribute and that that attribute varies between only two states (e.g., on and off). We can define a state-space for that system as the total number of combinations of component values. For example, a system with only two components, each of which could be either on or off, could only exist in the following states: on–on, on–off, off–on, off–off. This would be the total state-space for that system. However, most complex systems do not exist—and are not capable of existing—in every possible state in their state-space. For example, in our simple system, "on–on" may not be very likely.

The subset of states most frequently occupied by a system are referred to as *attractors*. To use the water example again, water, as an organized system of hydrogen, oxygen, and electricity, has three attractors: ice, liquid water, and steam. A phase transition takes place when the system jumps from one attractor to another, as when water goes from liquid form to gaseous form (steam). This transition can be referred to as "emergent" because there is no in-between stage, and the properties of steam and liquid water are qualitatively different. Thus, when we conceptualize language as emergent from increased brain complexity, symbolic reasoning ability, and social complexity, we are concluding that there is a qualitative leap from nonlanguage, or protolanguage, to "true" language.

As we stated above, although emergence is only now finding its way into the routine lexicon of science, the idea is not new. In 1941, the increasing influence of the concept of integrative levels resulted in a symposium at the University of Chicago, published in *Levels of Integration in Biological and Social Systems: Biological Symposia* (see Redfield, 1942). Papers presented there included Hyman's "The Transition From the Unicellular to the Multicellular Individual," Gerard's treatment of "Higher Levels of Integration," Jennings's "The Transition From the Individual to the Social Level," Carpenter's "Societies of Monkeys and Apes," and several others. More recently, Goldstein (1999) discussed the history of emergence as a scientific construct in a new journal, aptly titled *Emergence*. One author of the present chapter (Gary Greenberg), having worked from this perspective for the entirety of his career, is quite at home with these ideas. Others still have difficulty accepting the idea of emergence, suggesting that it has an almost "mystical" quality. Goldstein (1999) provided a workable definition: "Emergence . . . refers to the arising of novel and coherent structures, patterns, and properties during the process of self-organization in complex systems" (p. 49). Even the idea of self-organization troubles some contemporary scientists. However, these ideas are now well accepted in contemporary physics (Davies, 1989). It is the application of such concepts to new areas of science and to scientists unfamiliar with them that leads to skepticism. Even Einstein wrestled with the new ideas of quantum physics, unable to accept that some events are indeterminate and that there must be some as yet undiscovered underlying causes of all events. There is a parallel here with emergence and complex systems as alternatives to reductionistic analyses. Where do these new wholes come from, these newly emerging properties and structures, and how can events self-organize? It was Newton (1726/1972) who said he made no hypotheses about gravity—it just is. And so it is with self-organization. An implication of Big Bang cosmology is that given enough time, hydrogen and helium become thinking animals. This, however, does not imply that no causes are responsible for the changing events, as the following makes clear:

> It has to be stressed that the existence of chaotic outcomes of this kind does not involve an abandonment of causality *in principle*. If we could measure to the degree of accuracy we need then we could model the system, albeit in non-linear terms, and then we could predict what the outcome of changes would be. *In practice* we can't. It is precisely this practical limit—that word "limit"—which seems to set a boundary on science and science derived technology. (Byrne, 1998, p. 19)

We refer the reader to Goldstein's treatment for a thorough discussion of this aspect of the topic.

For our purposes, applied to psychology, we can identify emergent phenomena by several criteria. They display radical novelty (i.e., features not present in the underlying complex system); they display coherence or correlation (i.e., they have a unity over time); they exist at the global or macro level and not at all at the underlying micro level; they are dynamical, arising as a result of the dynamic interplay of underlying micro events; and they are ostensive (i.e., they really exist and are observable).

The contentious debate between reductionistic and holistic orientations regarding the fundamental nature of scientific explanations is one of the more longstanding debates in the philosophy of science. The concept of emergence, as it turns out, is perhaps the most important construct in the entire debate, but not in the manner in which we may have thought. As with most longstanding debates, the argument between determinism and holism has been characterized by straw man attacks and a difficulty in finding a clear articulation of the exact nature of the conflict. Frequently, the debate has been couched in terms of deterministic versus probabilistic models of causality. The reductionist approach is allied with a deterministic understanding of causality, and the holistic approach applies a more probabilistic interpretation of causality. However, this debate tends to be obfuscating. There are few adherents to strict determinism still operating in science. Indeed, even the most fundamental of all reductionistic scientific enterprises—the search for a universal theory in particle physics, which would unite our theoretical understanding of the physical universe from the very smallest of scales to the very largest of scales— is firmly entrenched in a probabilistic field calculus.

The other common debate relates to the flow of causal information. Reductionist approaches argue that the flow of causality runs from the parts to the whole. Considering brain–behavior relationships, for example, a reductionistic account could use a probabilistic causal model and even a complex "causal network" among brain regions, yet functional neural structure will always be considered primal in terms of causal relationships with behavioral functions. In contrast, a holistic approach would incorporate such neural structures into a broader multilevel causal network in which causal interdependencies would flow from both part to whole (i.e., neural organization to behavior) and from whole to part (i.e., behavior to neural organization)—that is, "downward causation" (Campbell, 1990). This latter type of argument comes closer to reflecting the fundamental differences between reductionistic and holistic approaches to science. However, the concept of emergence, which is the keystone to most holistic approaches, implies a much deeper distinction between these two philosophical approaches. This more central distinction has to do with the computability of causal relations from parts to wholes.

The principal thesis of emergentist theorists and philosophers has been that even with a full knowledge of all the lower order parts and their potential relationships, the laws of the higher order wholes cannot be deduced. For example, emergentists often argue that even armed with the full sequence of the human genome and a full understanding of the multiplicity of regulatory networks involved in protein synthesis, a full understanding of even morphological phenotypes let alone behavioral phenotypes could not be ascertained. Likewise, even if all of the neural circuits of the human brain could be sketched in a grand schematic and all of the probabilistic rules governing the synaptic flow of information could be catalogued, emergentists claim that neuroscience would be no better prepared to predict behavior. This has always been the defining claim of emergent holism, and for the better part of scientific history it has been the Achilles' heel of the position. The central and tacit assumption of the reductionist claim is that if we had a full knowledge of the parts of a system such as the genome or the brain, along with the rules governing their interrelationships, then an algorithm could be derived that could be used to predict the macrobehavior of the system—even if only in probabilistic terms. Serving the reductionist argument is a long history of empirical success at doing just this. Indeed, the entirety of the symbiotic relationship between mathematics and science has been based on this assumption. With this symbiotic relationship, reductionistic science has had the advantages of a formalized logical system in which to base its claims.

Emergentist approaches, on the other hand, have had to argue largely from intuition and empirical observation. However, the exponential growth in computational power across the last two decades has given rise to not only a new domain of science—the study of complex dynamic systems—but also an entirely new formalized logic system. Modern science has largely been built on a calculus composed of continuous functions. To use these functions, one needs to make enormous simplifying assumptions about the uniqueness of the constituent parts being studied. When studying a system made up of even a few discrete but interdependent units, this branch of mathematics becomes nearly intractable. Consider the difficulty inherent in the three-body problem, for example. Newton's laws of gravity work quite well for two bodies, but put a third body into the system and it becomes nearly impossible to solve.

Wolfram (2002) argued that computational algorithms provide an alternative symbolic system from which to analyze scientific problems. The primary advantage of the algorithmic approach is that it can model complex interactions among multiple discrete entities in a much more tractable manner than differential calculus. In investigating the broad utility of an algorithmic approach to science, Wolfram (2002) identified what is referred to as the *principle of computational irreducibility*. Part of the success of differential calculus in science is that it provides scientific laws in the form of a symbolic shorthand. In other words, a mathematical evaluation of the equations allows the scientist to know with a relative degree of certainty what the long-range behavior will be of the system the equation describes. For example, using Newton's laws of motion, we can calculate the state of our solar system 200 billion years from now rather than having to wait 200 billion years to find out. Unfortunately, the principle of computational irreducibility states that in complex dynamic systems, even if the rules governing local interactions are deterministic and simple, the long-run behavior of the system as a whole cannot be determined without running the system—we have to wait and see. Thus, the behavior of the system is just not predictable from a complete knowledge of the system components and their interactions. This is true even in computer simulations in which the programmer defines the rules!

We conclude this section with a comment by Goldstein (1999):

> In effect, there seems to be no end to the emergence of emergents. Therefore, the unpredictability of emergents will always be one step ahead of the ground won by prediction and, accordingly, emergence will always stay one step ahead of the provisionality argument. As a result, it seems that emergence is here to stay. Of course, this doesn't mean that there will be no great inroads into making the unpredictability of emergence more predictable. Rather, it goes along with the general reframing of the entire issue of predictability in scientific explanation that complexity theory has begun. Similar to the role of the uncertainty principle in quantum physics, the nonlinearity of the complex systems under investigation by complexity theory introduces a degree of unpredictability that even in principle will not completely yield to more and more probing. (p. 60)

## Rumbaugh's Contributions

Duane Rumbaugh, arguably one of comparative psychology's most important figures, has made contributions to the understanding of the development of complex behaviors that are more numerous and more significant than most of those from recent past decades. His contributions mirror and in some instances surpass those of previous comparative psychologists such as Harry Harlow, John Garcia, James McGaugh, Robert Rescorla, William Mason, and others. Rumbaugh is a comparative psychologist whose research and comprehensive analytic reviews over several decades on a range of different species have melded empirical work with scholarly theoretical formulations.

In the late 1960s, Rumbaugh was at the San Diego Zoo and San Diego State University conducting important primate learning research with a range of species. At the zoo he adapted Harlow's Wisconsin General Test Apparatus for use in the home quarters of the great apes and completed a series of studies on complex learning in the context of learning sets. This work at the zoo was augmented by research on monkeys in his laboratory at San Diego State University. To date, his publications from that time give the best and most complete picture of complex multiproblem learning in nonhuman primates. Much research in contemporary comparative psychology can be seen to be logical extensions of the work he began during that period.

## Rumbaugh and the Intractable Concept of Learning

Simple stimulus–response type learning has been demonstrated across many phyla and does not correlate well with evolutionary increases in brain size or organization. However, several forms of complex learning do correlate strongly with brain size and increased complexity of neural structures (Masterson & Berkley, 1974; Masterson & Skeen, 1972; Rumbaugh & Pate, 1984a, 1984b). Although at first glance the absence of a relationship between simple stimulus–response learning and brain complexity appears to be an anomaly, it is in fact just what the integrative levels concept predicts. Lower levels of organization and their functions are not "replaced" by higher levels but rather are subsumed at the higher levels; they become integral parts of the new, more highly organized, system. Lower functions thus are conserved at the higher levels. So it is expected that organisms at many levels of neural complexity would display simple stimulus–response learning capabilities. Organisms with simple neural organization are limited to stimulus–response learning, whereas for organisms with more complex neural organization it represents only a base level of learning capability. An early formulation of this idea identified 11 grades (or levels) of learning capacity and complexity, from nonassociative habituation to complex symbolic learning (Razran, 1971; see also Pringle, 1951). Jerison (1973, 1976, 1994) has written extensively on the relationships between brain evolution and intelligence.

Deacon (1990) also addressed this relationship, although from a perspective somewhat different from those of Razran (1971) and Jerison (1973, 1976, 1994). Deacon identified three levels of complex learning on the basis of nervous system complexity. The least complex form is characterized by an ability to reference objects in the environment to symbols that represent those objects only, in a one-to-one manner. One token or symbol can represent only one object. This level of learning permits simple and inefficient communication only. To be able to reference 20 objects, the organism needs 20 separate symbols. At the next level of learning complexity, objects and events can be represented by patterns of symbols, allowing a relatively large number of objects or events to be represented by a relatively small number of symbols arranged in different combinations. A good example is the Arabic numeral system in which only 10 symbols can be combined and recombined to represent an infinite number of events. However, it is only at the highest level of learning complexity that symbolic learning becomes possible. Symbols here are not restricted to object referencing, and meaning can be extracted from relationships between and among symbols. This is exemplified by algebraic concepts in contrast to a static numbering system. Deacon suggested that these learning levels represent levels of cognitive complexity, with language becoming possible only when the highest level is achieved, something we discuss below.

Of course, complex learning can be measured in several ways. It is possible to identify measures of learning that correlate learning complexity to measures of nervous system advance. The transfer index (TI), discussed extensively by Rumbaugh and Pate (1984a, 1984b; see also chap. 11, this volume), provides a suitable measure of learning complexity. In this paradigm, an animal learns a series of discrimination learning sets and is subsequently tested on its performance when the discriminanda are reversed. The TI "is determined by performance on the reversal trials . . . in relation to the prereversal criterion level" (Rumbaugh & Pate, 1984b, p. 226). The TI is widely understood to provide a measure of the animal's ability to abstract symbolic relationships from reference–index relationships. Most psychologists would reject the idea that any single quantitative measure could measure intelligence across species. However, a reading of the publications on the TI shows that Rumbaugh has come close to achieving this presumably impossible goal. His work on the TI stands as a unique quantitative formulation for cross-species cognitive capacities that is unparalleled in complementing the early Harlow learning set approach.

Eleven categories (levels) of learning proposed by Razran? Three proposed by Deacon? How many levels of learning are there? Do such formulations clarify our understanding of learning or lead to greater obfuscation? With his recent formulation of "emergents" as a higher order level of learned responses, Rumbaugh (2002; Rumbaugh & Washburn, 2003) has found a way of addressing and simplifying this issue, especially in terms of criticisms of Skinner's stimulus–response formulations. This draws on the evolutionary correlation of brain complexity and level of cognitive capacity in which animals functioning at Schneirla's highest behavioral level, the psychosocial, show the emergence of higher order cognitive skills. With this theoretical breakthrough, Rumbaugh has linked the study of cognition to the most contemporary formulations of the newly developing sciences of complexity and dynamic systems, providing more mathematical and empirical alternatives to traditional thinking about cognitive behavior. The point is that "emergent learning" proposed by Rumbaugh describes the long-range outcomes of a learning system that cannot be reduced to its respondent and operant elements; emergent learning thus constitutes a separate class of learning origins.

## Complex Behaviors as Emergents

Not only is this approach germane to psychological phenomena, but it provides a potential explanation for the appearance of complex behavioral phenomena, such as the development of language and culture in *Homo sapiens* (Greenberg et al., 1999; Savage-Rumbaugh et al., 1993; Savage-Rumbaugh & Sevcik, 1984) and childhood temperament (Partridge, 2002, 2003; Partridge & Lerner, in press). Rather than searching for single causes or taking an analytic approach, we can understand the development of much complex human behavior as an emergent property of the dynamic interplay of several sets of systems—biological, physiological, psychological, and sociocultural (i.e., ecological context)—the results involving not just one possible outcome. As noted by Byrne (1998),

> The issue is that in the social world, and in much of reality including biological reality, causation is complex. Outcomes are determined not by single causes but by multiple causes, and these causes may, and usually do, interact in a non-additive fashion. In other words, the combined effect

is not necessarily the sum of the separate effects. It may be greater or less, because factors can reinforce or cancel out each other in non-linear ways. ... In essence the complexity is locked away in the interaction term. (p. 20)

In a word, complexity is the result of emergence.

One major example of the emergence of complex behavior is that of language by human beings. Rumbaugh's analyses have permitted us at last to understand how this development may have come about in *Homo sapiens*, the result of the dynamic interplay of biological and cultural evolution as well as the contribution of ecologic factors (Greenberg et al., 1999). Although Skinner (1957) argued that human language development is the result of a long history of operant and respondent histories, it is, in fact, difficult to identify this historical sequence. Thus, although some, perhaps much, traditional learning may indeed be involved, the acquisition of language can be best understood as an emergent phenomenon resulting from processes similar to those involved in protolanguge acquisition by bonobos at the Language Research Center—the result of the dynamics of their unique social environment, in which they interacted almost constantly with their human caretakers. Parker and Russon (1996) identified common features shared by the great apes, including prolonged gestation, longer infancy and juvenile periods, a long developmental period, a long life span, and a large brain. These features ally the apes to humans and distinguish them from gibbons and Old World monkeys that show shorter gestation periods, shorter infancy and juvenile stages, shorter life spans, and smaller brains. Let us think of these elements as essential features, as components of a single dynamic system, much as hydrogen–oxygen–electricity are for the emergence of water. In our species, biological adaptations, socialization, and unique cultural experiences lead to a phase transition from protolanguage in chimpanzees and bonobos (and perhaps even in Neanderthals), to true language.

## Conclusion

There are several important implications of this point of view for understanding organismic behavior. A long-standing difficulty in psychology is understanding the relationship of behavioral phylogeny and ontongeny. Historical trends toward increased neuroarchitectural complexity and behavioral capacity need not be the result of "lucky genes" or a teleological process. Rather, as evolution led to more complex neurophysiologies, more neural integration was possible (e.g., Dean, 2000). The changes in behavioral capacity that seem to correspond with this increase in complexity and integration mirror the behaviors we find in dynamic systems. We see a similar process on an ontological time scale, the difference being one of magnitude. The behavioral diversity over a life span is much smaller than that over an evolutionary time scale.

Over the past 2 decades, there has been a burgeoning of theoretical developments across a diverse set of disciplines, including developmental psychology, sociology, developmental epidemiology, psychobiology, and embryology, that have a common conceptual foundation—and in many cases methodological approach—with the comparative psychology we have outlined here. Although these theoretical formulations differ in specifics, they share a core set of common assumptions such that Bronfenbrenner referred to these interdisciplinary advances as an "emergent convergence and isomorphism" (quoted in Cairns, Elder, & Costello, 1996, p. ix). Indeed, theoretical frameworks such as developmental contextualism (Lerner, 1998), ecological psychology (Bronfenbrenner, 1977), life-span psychology (Brim & Kagan, 1980), person-centered psychology (Magnusson, 1995), transactional psychology (Sameroff, 1983), and developmental psychopathology (Cicchetti & Cohen, 1995) have such a degree of commonality that Cairns has proposed incorporating them under the umbrella concept of developmental science (Cairns et al., 1996). It is in this context that Rumbaugh's proposals of emergents and of a new approach in psychology that he has referred to as rational behaviorism will play an important role in coming generations.

The following quotation from Stewart (2002) demonstrably links psychology and the other behavioral sciences with theoretical development in science in general:

> We know that our universe obeys simple low-level rules—laws of nature, including rules for subatomic particles and for space and time. We also know that life behaves in ways that do not seem to be built explicitly into those rules. Life is flexible; life is free; life seems to transcend the rigidity of its physical origins. This kind of transcendence is called "emergence." Emergence is not the absence of causality; rather it is a web of causality so intricate that the human mind cannot grasp it. We cannot understand how a frog works by listing the movement of every atom in it. In some sense, the atoms are the cause of the frog's behavior—but that's a totally useless way to approach frog biology. In order to understand the deeper significance of life we need an effective theory of emergent features. (pp. 7–8)

And finally, we offer these comments by Goldstein (1999):

> An appeal to emergence is thus a way to describe the need to go to the macro level and its unique dynamics, laws, and properties in order to explain what is going on. The construct of emergence is therefore only a foundation on which to build an explanation, not its terminus. . . . Finally, there is the fact that complexity science is only in its infancy. As it matures, better quantitative tools will be coming forth that offer richer ways of studying emergent phenomena. (pp. 58, 68)

## References

Aronson, L. R. (1984). Levels of integration and organization: A re-evaluation of the evolutionary scale. In G. Greenberg & E. Tobach (Eds.), *Evolution of behavior and integrative levels* (pp. 57–81). Hillsdale, NJ: Erlbaum.

Aronson, L. R., Tobach, E., Rosenblatt, J. S., & Lehrman, D. H. (Eds.). (1972). *Selected writing of T. C. Schneirla*. San Francisco: Freeman.

Boker, S. M. (2001). Differential structural equation modeling of intraindividual variability. In L. M. Collins & A. G. Sayer (Eds.), *New methods for the analysis of change* (pp. 5–27). Washington, DC: American Psychological Association.

Brim, O. G., & Kagan, J. (1980). *Constancy and change in human development.* Cambridge, MA: Harvard University Press.
Bronfenbrenner, U. (1977). Toward an experimental ecology of human development. *American Psychologist, 32,* 513–531.
Bunge, M. (1980). *The mind–body problem.* Oxford, England: Pergamon.
Byrne, D. (1998). *Complexity theory and the social sciences: An introduction.* London: Routledge.
Cairns, R. B., Elder, G. H., & Costello, J. (1996). *Developmental science.* New York: Cambridge University Press.
Campbell, D. T. (1990). Levels of organization, downward causation, and the selection-theory approach to evolutionary epistemology. In G. Greenberg & E. Tobach (Eds.), *Theories of the evolution of knowing* (pp. 1–17). Hillsdale, NJ: Erlbaum.
Chorover, S. L. (1990). Paradigms lost and regained: Changing beliefs, values, and practices in neuropsychology. In G. Greenberg & E. Tobach (Eds.), *Theories of the evolution of knowing* (pp. 87–106). Hillsdale, NJ: Erlbaum.
Cicchetti, D., & Cohen. D. J. (1995). *Developmental psychopathology.* New York: Wiley.
Damon, W., & Lerner, R. M. (2001). *Theoretical models of human development.* New York: Wiley.
Davies, P. (Ed.). (1989). *The new physics.* Cambridge, England: Cambridge University Press.
Davies, P., & Gribbin, J. (1992). *The matter myth.* New York: Simon & Schuster.
Deacon, T. W. (1990). Rethinking mammalian brain evolution. *American Zoologist, 30,* 629–705.
Dean, A. (2000). *Complex life: Nonmodernity and the emergence of cognition and culture.* Aldershot, England: Ashgate.
Feibleman, J. K. (1954). Theory of integrative levels. *British Journal for the Philosophy of Science, 5,* 59–66.
Goldstein, J. (1999). Emergence as a construct: History and issues. *Emergence, 1,* 49–72.
Goodwin, B. (1994). *How the leopard got its spots: The evolution of complexity.* New York: Scribner.
Gordon, D. (1988). The development of flexibility in the colony organization of harvester ants. In G. Greenberg & E. Tobach (Eds.), *Evolution of social behavior and integrative levels* (pp. 197–203). Hillsdale, NJ: Erlbaum.
Gordon, D. (1997). Task allocation and interaction in social insect colonies. In G. Greenberg & E. Tobach (Eds.), *Comparative psychology of invertebrates: The field and laboratory study of insect behavior* (pp. 125–134). New York: Garland.
Gottlieb, G. (1992). *Individual development and evolution: The genesis of novel behavior.* New York: Oxford University Press.
Gottlieb, G. (1997). *Synthesizing nature–nurture: Prenatal roots of instinctive behavior.* Mahwah, NJ: Erlbaum.
Greenberg, G. (1995). Anagenetic theory in comparative psychology. *International Journal of Comparative Psychology, 8,* 31–41.
Greenberg, G., & Haraway, M. M. (2002). *Principles of comparative psychology.* Boston: Allyn & Bacon.
Greenberg, G., Partridge, T., Weiss, E., & Haraway, M. M. (1999). Integrative levels, the brain, and the emergence of complex behavior. *Review of General Psychology, 3,* 168–187.
Haggbloom, S. J., Warnick, R., Warnick, J. E., Jones, V. K., Yarbrough, G. L., Russell, T. M., et al. (2002). The 100 most eminent psychologists of the 20th century. *Review of General Psychology, 6,* 139–152.
Jerison, H. J. (1973). *Evolution of the brain and intelligence.* New York: Academic Press.
Jerison, H. J. (1976). Principles of the evolution of the brain and behavior. In R. B. Masterton, W. Hodos, & H. Jerison (Eds.), *Evolution, brain and behavior: Persistent problems* (pp. 23–45). Hillsdale, NJ: Erlbaum.
Jerison, H. J. (1994). Evolution of the brain. In D. W. Zaidel (Ed.), *Neuropsychology* (pp. 53–82). San Diego, CA: Academic Press.
Kantor, J. R. (1959). *Interbehavioral psychology: A sample of scientific system construction.* Bloomington, IN: Principia Press.
Kauffman, S. A. (1993). *The origins of order: Self-organization and selection in evolution.* New York: Oxford University Press.
Kauffman, S. A. (1995). *At home in the universe: The search for the laws of self-organization and complexity.* New York: Oxford University Press.

Kauffman, S. (2000). *Investigations.* Oxford, England: Oxford University Press.
Kuo, Z. Y. (1967). *The dynamics of behavior development.* New York: Random House.
Lerner, R. M. (1998). Developmental contextualism. In G. Greenberg & M. M. Haraway (Eds.), *Comparative psychology: A handbook* (pp. 88–97). New York: Garland/Taylor & Francis.
Magnusson, D. (1995). Individual development: A holistic integrated model. In P. Moen, G. H. Elder, & K. Luscher (Eds.), *Examining lives in context: Perspectives on the ecology of human development* (pp. 19–60). Washington, DC: American Psychological Association.
Masterson, R. B., & Berkley, M. A. (1974). Brain function: Changing ideas in the role of sensory, motor and association cortex in behavior. *Annual Review of Psychology, 25,* 277–312.
Masterson, R. B., & Skeen, L. C. (1972). Origins of anthropoid intelligence: Prefrontal system and delayed alternation in hedgehog, tree shrew, and bush baby. *Journal of Comparative and Physiological Psychology, 81,* 423–433.
Maynard Smith, J., & Szathmáry, E. (1995). *The major transitions in evolution.* New York: Oxford University Press.
Michel, G. F., & Moore, C. L. (1995). *Developmental psychobiology: An interdisciplinary science.* Cambridge, MA: MIT Press.
Morgan, C. L. (1923). *Emergent evolution: The Gifford lectures.* London: Williams & Norgate.
Newell, K. M., & Molenaar, P. C. M. (1998). *Applications of nonlinear dynamics to developmental process modeling.* Mahwah, NJ: Erlbaum.
Newton, I. (1972). *Philosophiae naturalis principia mathematica* [Mathematical principles of natural philosophy] (3rd ed.). Cambridge, MA: Harvard University Press. (Original work published 1726)
Nicolis, G. (1989). Physics of far-from-equilibrium systems and self-organisation. In P. Davies (Ed.), *The new physics* (pp. 316–347). Cambridge, England: Cambridge University Press.
Parker, S. T., & Russon, A. E. (1996). On the wild side of culture and cognition in the great apes. In A. E. Russon, K. A. Bard, & S. T. Parker (Eds.), *Reaching into thought: The minds of the great apes* (pp. 430–450). Cambridge, England: Cambridge University Press.
Partridge, T. (2002). Biological and caregiver correlates of behavioral inhibition. *Infant and Child Development, 12,* 71–87.
Partridge, T. (2003). Temperament: Developmental and ecological dimensions. In J. R. Miller, R. M. Lerner, & L. B. Schiamberg (Eds.), *Human ecology: An encyclopedia of children, families, communities, and environments* (pp. 678–682). Santa Barbara, CA: ABC-Clio.
Partridge, T., & Lerner, J. (in press). A quadratic growth model of difficult temperament. *Infant and Child Development.*
Prigogine, I., & Stengers, I. (1984). *Order out of chaos: Man's new dialogue with nature.* New York: Bantam Books.
Pringle, J. W. S. (1951). On the parallel between learning and evolution. *Behaviour, 3,* 174–215.
Pronko, N. H. (1980). *Psychology from the standpoint of an interbehaviorist.* Belmont, CA: Wadsworth.
Razran, G. (1971). *Mind in evolution.* New York: Houghton Mifflin.
Redfield, R. (1942). *Levels of integration in biological and social systems.* Lancaster, PA: Cattell Press.
Rosenblatt, J. S., & Schneirla, T. C. (1962). The behavior of cats. In E. S. E. Hafez (Ed.), *The behaviour of domestic animals* (pp. 453–488). London: Ballière, Tindall & Cox.
Rumbaugh, D. (2002). Emergents and rational behaviorism. *Eye on Psi Chi, 6,* 8–14.
Rumbaugh, D. M., & Pate, J. L. (1984a). The evolution of cognition in primates: A comparative perspective. In H. L. Roitblatt, T. G. Bever, & H. S. Terrace (Eds.), *Animal cognition* (pp 569–587). Hillsdale, NJ: Erlbaum.
Rumbaugh, D. M., & Pate, J. L. (1984b). Primates' learning by levels. In G. Greenberg & E. Tobach (Eds.), *Behavioral evolution and integrative levels* (pp. 221–240). Hillsdale, NJ: Erlbaum.
Rumbaugh, D. M., & Washburn, D. (2003). *Intelligence of apes and other rational beings.* New Haven, CT: Yale University Press.
Sameroff, A. J. (1983). Developmental systems: Contexts and evolution. In P. H. Mussen & W. Kessen (Eds.), *Handbook of child psychology: Vol 1. History, theory, and methods* (4th ed., pp. 237–294). New York: Wiley.

Saunders, P. T., & Ho, M. -W. (1976). On the increase in complexity in evolution. *Journal of Theoretical Biology, 63,* 375–384.

Saunders, P. T., & Ho, M. -W. (1981). On the increase of complexity in evolution: II. The relativity of complexity and the principle of minimum increase. *Journal of Theoretical Biology, 90,* 515–530.

Saunders, P. T., & Ho, M. -W. (1984). The complexity of organisms. In J. W. Pollard (Ed.), *Evolutionary theory: Paths into the future* (pp. 121–139). New York: Wiley.

Savage-Rumbaugh, E. S., Murphy, J., Sevcik, R. A., Brakke, K. E., Williams, S. L., & Rumbaugh, D. M. (1993). Language comprehension in ape and child. *Monographs of the Society for Research in Child Development, 58* (Whole Nos. 3–4).

Savage-Rumbaugh, E. S., & Sevcik, R. A. (1984). Levels of communicative competency in the chimpanzee: Pre-representational and representational. In G. Greenberg & E. Tobach (Eds.), *Behavioral evolution and integrative levels* (pp. 197–219). Hillsdale, NJ: Erlbaum.

Sheldrake, R. (1995). *Seven experiments that could change the world.* New York: Riverhead Books.

Skinner, B. F. (1957). *Verbal behavior.* New York: Appleton-Century-Crofts.

Sperry, R. M. (1987). Structure and significance of the consciousness revolution. *Journal of Mind and Behavior, 8,* 37–65.

Stebbins, G. L. (1969). *The basis of progressive evolution.* Chapel Hill: University of North Carolina Press.

Stewart, I. (2002). *Life's other secret: The new mathematics of the living world.* New York: Wiley.

Sulis, W., & Trofimova, I. (Eds.). (2000). *Nonlinear dynamics in life and social sciences.* Amsterdam: IOS Press.

Tobach, E. (1976). Evolution of behavior and the comparative method. *International Journal of Psychology, 11,* 185–201.

Tobach, E., & Schneirla, T. C. (1968). The biopsychology of social behavior of animals. In R. E. Cook & S. Levin (Eds.), *The biological basis of pediatric practice* (pp. 68–82). New York: McGraw-Hill.

Wolfram, S. (2002). *A new kind of science.* Champaign, IL: Wolfram Media.

# 8

# The Emergence of Emergents: One Behaviorist's Perspective

## M. Jackson Marr

> Learning set is said to reflect the affirmation or rejection of hypotheses. Hypotheses are generated by the learner's brain, not its muscles. Thus, learning-set research served to advance the perspective that even nonhuman primates think and that their thinking reflects the active processing of information accrued from efforts to solve problems. Their learning processes are not simply the strengthening of some motor responses over others. Hence, learning-set research served to advance studies of animals as rational agents. This trend is serving to supplant the radical-behavioristic models, formulated earlier this century, with models predicted on rational processes for animals' complex learning and behavior.
> —Rumbaugh (1997, p. 197)

The present volume resonates astonishingly well with both the content of my discussions with other psychologists over many years as well as my own particular interests in behaviorism and behavior analysis, the place of reductionism in science, and the conditions for what is deemed emergence in a great variety of phenomena in nonbiological and biological sources. This has included human creativity, particularly in the sciences and mathematics (Marr, 2003). As for "rational" behaviorism, I am unsure just what that might mean, although I will comment on it later. I do know that some readers would consider the expression *rational behaviorism* an oxymoron. The adjective *rational* joins a long list of descriptors, including "Watsonian," "neo," "radical," "molar," "modern," and, most recently, John Staddon's "theoretical behaviorism" as outlined in his book *The New Behaviorism* (2001). For those outside the behaviorist community, the most common descriptor is "dead"! But such a report continues to be a gross and certainly uninformed exaggeration, to say the least, and I

---

I thank David Washburn for his kind invitation to contribute to this volume in celebration of Duane Rumbaugh's fertile career. Over the many years Rumbaugh and I have known each other, I have enjoyed sparring with him over conceptual and empirical issues in psychology. Though we have not always seen the science of behavior in the same way, I have learned much from him and his bold and imaginative work.

would argue, perhaps to his surprise, that Duane Rumbaugh's work is an active denial of the demise of a behaviorist perspective. To be sure, in common with words like *democratic* or *Christian*, the term *behaviorism* occasions many, varied, and not infrequently contradictory meanings to different interests, including, most of all, those calling themselves behaviorists, so empirical and conceptual arguments go on, further attesting to the liveliness of the enterprise.

I take as my point of departure the Rumbaugh, Washburn, and Hillix (1996) article "Respondents, Operants, and Emergents." The principal antecedents to this article were presented more than a decade ago by Rumbaugh at the annual meeting of the Association for Behavior Analysis. I do not think the term *emergent* was in common use at the time among psychologists interested in learning, but among behavior analysts the concept had been at least implicitly acknowledged in a number of domains. In fact, what I attempt to argue in this chapter is that emergence in various guises has always been a necessary part of a behavior-analytic approach, even if not explicitly acknowledged. Sidman and his colleagues (e.g., Sidman, 1986, 2000; Sidman & Tailby, 1982), exploring procedures known as *equivalence relations*, had by the early 1980s explicitly used the term to describe certain effects in complex stimulus control, which I return to shortly. There are other active areas of behavior-analytic research focusing on what I would call "emergent repertoires" that I also mention later. Nonetheless, Rumbaugh and his colleagues were using the term in a seemingly much larger and bolder sense to describe a host of phenomena commonly characterized as cognitive or, indeed, rational. Of special interest is their assertion that emergents should be treated as a unique class of behaviors perhaps founded on, but qualitatively distinct from, what were described as the early Skinnerian categories of respondents and operants.

## The Dynamics of Emergence

Before addressing this assertion, I first discuss the problematic term *emergent*. The term is a byword in biology, for example, to describe the behavior of a host of hierarchical organizations from molecules to ecological systems. Indeed, Ernst Mayr (1982, 1988, 1997) in more than one essay has pressed the thesis that biology should be considered as a wholly distinct science from physics, in part on the basis of the prevalence of emergent phenomena in biological as opposed to nonbiological systems. To quote Mayr (1982),

> In biology one deals with hierarchies. One is represented by *constitutive hierarchies*, like the series macromolecule, cellular organelle, cell, tissue, organ, and so forth. In such a hierarchy the members of a lower level . . . are combined into new units . . . that have unitary functions and emergent properties. The formation of constitutive hierarchies is one of the most characteristic properties of living organisms. (p. 65)

Mayr (1982) also wrote, "Emergence is a descriptive notion that, particularly in more complex systems, seems to resist analysis. *Simply to say, as has been done, that emergence is due to complexity is, of course, not an explanation* [italics added]" (p. 63).

In a chapter published some time ago (Marr, 1997) on what I called the "mechanics of complexity," I argued that Mayr's (1982) distinctions could, in fact, be seen along quantitative as opposed to qualitative dimensions and thus may show connections in the kinds of processes yielding complexity and emergence among virtually all significant phenomena of nature—living or not. Although there are noncontroversial cases, words such as *simple, complex, qualitative, quantitative,* and *emergent* are terms whose use in the language game of science depends on a host of interacting contingencies between the system of interest and the observer's history. I have argued that the field bringing some quantitative unity to the apparent disparate worlds of physics and biology is nonlinear dynamical systems theory, devoted, as the name implies, to the quantitative analysis of change in nonlinear systems—simple or complex. Treated examples have included such phenomena as animal coloration, arms control systems, bending beams, biochemical pathways, cardiac arrhythmias, computer programs, earthquakes, evolution, genetic algorithms, international conflict, neural networks, predator–prey interaction, plant growth, population dynamics, percolation, quasars, river flows, sand piles, stock market prices, testosterone cycles, tornadoes, turbulence, and even the course of true love (e.g., Bak, 1996; Covey & Highfield, 1995, Marr, 1992, 1997; Stewart, 1998). All of these examples, whether animate or inanimate, can display the emergent, irreversible, selective, stochastic—in a word, complex—behavior thought by Mayr (1982) to be characteristic only of living beings.

One common feature of nonlinear dynamical complexity is emergence. The *Oxford English Dictionary* has given one definition of *emergence* as "the process of coming forth, issuing from concealment, obscurity, or confinement" (*Oxford English Dictionary*, 1989). Generally, then, the term applies to phenomena whose properties are not predictable directly from known properties of constituents; in other words, these properties cannot be simply reduced to such known properties. Crystal formations of certain minerals or snowflakes are common examples. Emergent phenomena even occur in such putatively deterministic fields as Newtonian mechanics. A compelling example is the three-body problem, or the case of three gravitationally interacting bodies. This problem, first productively addressed by Poincaré (see Barrow-Green, 1996) in the early 20th century, served as a stimulus for the development of modern dynamical systems theory. Without going into the details (see Marr, 1997), even in the simplest cases the three bodies can exhibit extraordinarily complex behaviors—orderly and chaotic, stable and unstable—each with exquisite sensitivity to initial conditions. Of particular relevance, however, is that through certain variations in parameters, the behavior of the three-body system can demonstrate an intricate orderliness emerging from the interactions among the bodies. Such interactions render the system irreducible, a keynote of emergence. Moreover, the behavior, whether orderly or not, may be unpredictable in the sense that one must wait to see what will happen under some given set of initial conditions—it is full of surprises, another essential feature of emergence.

Stability and instability in complex systems typically depend on delicate interplays of positive and negative feedback. Whereas negative feedback contributes to stability, positive feedback leads to change. The proper combination of these two can lead to the emergence of intricate dynamic patterning. The behavior of neural systems, even a single neuron, illustrates the importance of both positive and negative feedback processes to achieve change as well as stability, essential characteristics in a physiological system embedded in and responding and adapting to an ever-changing environment. Moreover, neural systems manifest irreducibility in that the different "parts" act in coordinated relations to each other. This has been called the *emergence of long-range correlations*. A similar effect is seen when huge flocks of birds or schools of fish dart about as if these collections were single organisms.

Having mentioned the possibility of chaos in dynamical systems, I should emphasize at this point that even allowing for considerable and common stochastic influences, chaos, in any technical sense, is rarely if ever seen in behavior dynamics. Evolutionary processes selecting mechanisms for behavioral change would be most unlikely to have yielded truly chaotic activity in the behavior of organisms. In any case, emergence in behavioral systems of interest to most behavioral scientists is characterized by order, not chaos.

## Behavior Dynamics

A fundamental explanatory concept in the experimental analysis of behavior analysis is that of contingency. A history of relations between antecedent and consequent events dictates the orderly patterns of behavior of organisms, and these patterns are foundational to the very possibility of a science of behavior. In a chapter published in another book, I wrote,

> In the operant conditioning laboratory, we see the development and maintenance of complex patterns of responding under various contingencies of reinforcement. How do these patterns emerge? There is discernible structure at several levels of analysis from sequences of inter-response times to inter-reinforcement intervals to overall day-to-day patterns, and beyond. The effect of reinforcement is to induce change through selection. Reinforcement effects depend on the initial states of the system, for example, where in time, or what features of responding are occurring. As this continues, the system is changing, so reinforcement acts on a different pattern and so on. The patterns of behavior emerging and the pattern of reinforcement delivery are in a kind of dynamic dance, a flowing partnership between the effects of patterns of reinforcement on patterns of responding and the counter effects of patterns of responding on patterns of reinforcement. Together they typically produce some metastable distributions of behavior we might identify with a particular schedule, for example, fixed-interval, random ratio, etc. (Marr, 1997, p. 77)

These examples and numerous others in behavior dynamics show that complexity and attendant emergence may arise from simply described arrangements of constituents, governed by equally simple rules or principles. There is a seeming magic in this.

The fundamental question in the science of learning is the origin of novel behavior. Our task as behavior scientists is to explore and discover basic principles with the assumption that complexity may arise from simplicity, given the appropriate initial and boundary conditions. Another towering example that shares close analogies with behavior change is organic evolution. Though the details are considerable, all of the organic beings living now, or that have ever lived, can be understood through the application of a relatively few elementary principles of selection from variation. What is largely shared with behavior change (learning, if you like) is the operation of selective contingencies acting on variation. This is a basic formula for emergence, but one may be unimpressed if one is able to see the effects of certain contingencies through time, because then the surprise of emergence may be lost. Conversely, as in a magic show, focusing on the emerging rabbit from a hat may blind a person to the conditions needed to make the trick work (see, e.g., Marr, 2003).

*Response differentiation* (or shaping) is an essential example. Behavioral variation is the *Anlage* from which new classes of operant behaviors emerge. Thus, novelty is inextricably embedded in the defining property of operant behavior, namely that behavior is controlled or selected by its consequences (Neuringer, 2002). Skinner (1938) long ago viewed response differentiation as a dynamic interplay among reinforcement, punishment, extinction, and response generalization. Punishment and negative reinforcement can also play roles in the differentiation process so that behaviors that are difficult, awkward, or effortful, as well as those differentially resulting in extended intermittency or delay of consequences, will tend to be selected out. In sum, one may shape very fine-grained repertoires through appropriate and adroit applications of precise and extended contingencies.

To develop extreme response values, or novel and complex performances, one can increase variability in emitted behavior by withholding reinforcement either through controlled intermittences or extinction. These procedures can then engender new or more extreme values that can then be differentially selected. By artfully applying this procedure, completely new behaviors never seen in the "natural" repertoires of organisms can be produced. Indeed, it is difficult to imagine how such behaviors could be produced by any other means. Even if provided motorcycles in the forest, bears would not ride them, as they are seen to do—as if by magic—in a Russian circus. A most instructive example is the "creative" porpoise trained by Karen Pryor and her colleagues (Pryor, Haag, & O'Reilly, 1969). In this demonstration, reinforcement depended on the porpoise emitting behaviors not previously emitted; in other words, only novel behaviors were reinforced. The porpoise thereby came to emit extraordinarily varied and elaborate activities never seen before and probably impossible to train otherwise.

Behavioral variation has also been selected in both humans and rats through similar procedures. Allen Neuringer and his colleagues (e.g., Neuringer, 1986, 2002) have published a large number of studies showing that human subjects can come to emit random numbers through differential reinforcement via feedback from common statistical tests of randomness. Similarly, rats may be trained to respond randomly on a set of concurrently available levers. Again, it is difficult to imagine how this might be accomplished by any other means than through careful differential shaping. Certainly, with humans, simple verbal instruction would be ineffective. The participants themselves have no idea how they are doing the task. Thus, they could hardly be said to be behaving "rationally," except in the uncommon sense of behavioral adaptation to prevailing contingencies. By this sense, a body falling in a gravitational field could also be said to be behaving rationally.

Neuringer's (1986, 2002) studies provide compelling evidence for behavioral variation as a *response class*, a key concept in emergent behavior. From reading Rumbaugh et al.'s (1996) chapter on emergence (and from various discussions of his work), I detected more than a bit of stereotyping of the character of operant and respondent conditioning as well as reinforcement. If I properly interpret their views, all these phenomena could be encompassed by some stimulus–response or stimulus–stimulus formulae from a long-past age, or the term *behavior* only applies to particular occurrences of actions such as key pecks or lever presses (what Rumbaugh has called "motor responses"; Rumbaugh et al., 1996, p. 60), or functional consequences have only one property (that, surely, would be a reinforcement constraint!), or organisms can operate in a consequence vacuum, or the concept of response class is an ad hoc invention. Skinner, for example, never thought in these terms (see, e.g., Skinner [1953, 1974], wherein he explicitly explained why he was not a stimulus–response psychologist). Moreover, he recognized very early that contingencies acted on as well as engendered functional classes of behaviors, and through either natural or applied contingencies, new behavioral units could be synthesized of almost any complexity under control of equally complex conditions. Such units, by definition, must show certain functional properties to have the status of a response class. Just what classes are demonstrated, or might be established, is fundamentally an empirical question, not a theoretical one. With response differentiation, for example, where you end up is nothing like where you started; you have little idea how far you might take it, and the whole process is essentially irreversible—all in conformity with my earlier discussion of dynamical systems. Molar approaches to behavior analysis as characterized by quantitative models such as the matching law, hyperbolic discounting, behavioral economics and ecology, and what Rachlin (2000) called "the science of self control," must go beyond some assumed or nominal behavioral "bits" to extended patterns or units integrated over time through dynamical and differential behavior–consequence relations and contextual control.

Thus, emergents are not necessarily a separate class somehow mysteriously standing on the shoulders of basic conditioning mechanisms but are the natural, inherent outcome of these conditioning processes. I have just mentioned behavioral variation as an example, but there are many others, including the virtually limitless patterns of responding induced by scheduling of consequences, abstraction, conceptual categorization, generalized imitation, adduction, the synthesis of new response classes from already established ones, naming, and a host of relational classes as exemplified by, for example, Hayes, Barnes-Holmes, and Roche's *Relational Frame Theory* (2001). These include not only equivalence but the whole range of relational possibilities, including analogies, modeling, and logical operations. To appreciate these in any useful or creditable way, I would have to discuss in a great deal more detail than space allows a number of other equally important phenomena explored by behavior analysts, such as various forms of complex stimulus control, contingencies for achieving fluency or automaticity, and rule-governed versus contingently controlled behavior. These processes are fundamental in accounting for the most complex of human behaviors, including those esteemed as insightful or creative (Marr, 2003).

## Whence Complexity: Is The Brain a Rational Agent?

A key argument in Rumbaugh et al. (1996) was that emergents—for example, those shown in learning set, latent learning, relational control, counting, protolinguistic behavior, and the like—are to be understood in terms of differential brain function. Of course, there is a sense in which this must be true, but it seems a long way from understanding just what sense is meant. As Mayr (1982) emphasized, complexity per se is not an explanation for emergence. At the least, to say complex behavior only comes from complex brains invites one to seek some metrics of complexity in both domains and to relate the two functionally. This is most certainly not a trivial task, especially in light of the fact that evolutionary processes are fundamentally interactive; thus, cause–effect relations between, say, putative complexity of brains and putative complexity of behaviors are likely to remain largely mysterious. At any rate, schooling fish; flocking birds; lengthy migratory patterns; elaborate mating rituals; spider webs; foraging patterns; predator–prey interactions; termite, ant, and bee colonies; and many other such wonders should give one pause in considering the origin and nature of behavioral complexity as it relates to nervous systems. Sara Shettleworth in her massive book *Cognition, Evolution, and Behavior* (1998) and, more recently, Clive Wynne's book *Animal Cognition* (2001) have provided many surprising examples. These authors typically avoided the sort of anthropomorphic or metaphorical mediational accounts all too common today among comparative psychologists (and especially behavioral primatologists), and they were relatively cautious, if not skeptical, on the issue of the so-called "simple versus complex" brains. What they did emphasize was how particular contingencies interacting with special characteristics of species can account for many of the phenomena discussed.

When it comes to animals as "rational agents," as an occasionally rational behaviorist I do not think anthropomorphic explanations are at all useful in a behavioral science applied to humans, never mind to other animals. Moreover, to assign such explanations to brains is a major category error—a part-to-whole fallacy. If I make a mistake, does that mean my brain made a mistake? What could a brain do to make a mistake? Lumps of retecious meat, no matter how complicated, do not think, plan, relate, intend, talk, choose, serve as executives, generate hypotheses, solve problems, or make mistakes. To talk this way is to make the whole brain into a mysterious homunculus. One might just as well assert that legs walk, eyes see, and stomachs get hungry.

More to the point, most, if not all the examples given by Rumbaugh et al. (1996) of complex, emergent behaviors run into difficulties if seen strictly from a "complex brain" perspective. Take learning sets, for example. Rats using auditory or olfactory stimuli, as opposed to visual, are as quick to pick up learning sets as most primates. Even the lowly chicken, hardly a paragon of rational panache, does well in this task. As an undergraduate student long ago, I compared human performance with published data on rats in a common maze-learning task; in many cases, the rats acquired errorless maze performance more rapidly than the humans. Latent learning, that poster child for "cognitive maps," has been shown to be the outcome of rats' ability to come under fine control of extra-maze cues by moving about the maze; thus, a cognitive map is not needed. So-called self-concept studies are, at best, really a mess to interpret, as are observational learning and tool use. With regard to the latter, Alex Kacelnik and his colleagues (Chappell & Kacelnik, 2002; Weir, Chappell, & Kacelnik, 2002) have shown the construction of a tool by a New Caledonian crow, a task that I believe many humans could not have done without help, if at all. Given a grub at the bottom of a long vertical tube and a straight wire, the crow first attempted to use the wire to draw the grub from the tube. After some failed attempts, the crow, holding the wire in its beak and using its foot, bent the wire into a hook, then successfully retrieved the grub. Magic—emerging from a brain perhaps one tenth the size of most primate brains! (See also Savage, 1995.) Robert Cook's Web page (http://www.pigeon.psy.tufts.edu) contains dozens of studies using pigeons and other birds demonstrating extraordinarily elaborate behaviors emerging from properly arranged contingencies. And, of course, there is Pepperberg's (e.g., Pepperberg, 2002) grey parrot Alex, whose behaviors would seem to confound any attempts to identify "complex" (i.e., large) brains with complex behaviors.

Relational control is another problematical case. Again, an important concept needed to understand such performances is that of higher order response classes established through special histories; for example, relating itself may become such a class under contextual control, given appropriate contingencies. This idea has been emphasized by Hayes et al. (2001) in their book *Relational Frame Theory*, summarizing dozens of studies in its support.

With regard to stimulus equivalence, discussed in Rumbaugh et al. (1996), Sidman's (1986) first major summary of his equivalence work, titled "Functional Analysis of Emergent Verbal Classes," and his last major review, titled "Equivalence Relations and the Reinforcement Contingency," appearing in 2000 in the *Journal of the Experimental Analysis of Behavior*, despite the scores of studies in this field and related areas, share in common the empirical demonstration that, to use his words, "emergent performances seem to be explainable in no other way than by inclusion of defined responses in equivalence classes that three-term contingencies generate" (Sidman, 2000, p. 138). He called this theory "prosaic" but "productive" (pp. 143–144).

I should emphasize that the necessary and sufficient conditions for, and interpretations of, complex stimulus control are matters of considerable theoretical and experimental debate among behavior-analytic researchers as well as others. This field is perhaps the most active in behavior analysis, with many studies appearing each year involving human as well as animal subjects. See, for example, the special issue of the *Journal of the Experimental Analysis of Behavior* (Critchfield, Galizio, & Zentall, 2002) on categorization and concept learning. With respect to equivalence, it is now known, contrary to some earlier indications, that it can be unequivocally demonstrated in a number of species, including, of all creatures, harbor seals. As Sidman (2000) commented, "there has been no definitive demonstration that any species sensitive to reinforcement contingencies is incapable of equivalent relations" (p. 144). He also added that "the present theory is neutral with respect to the relevance of neural structure and function, genetic factors, or developmental processes" (p. 144).

The reader must have been thinking throughout much of this chapter that surely contingencies cannot be everything—indeed, Duane Rumbaugh and I have discussed this issue many times. The answer is: no, of course not. No matter what contingencies I arranged, I could not teach a pigeon, or a bonobo, for that matter, to sing an aria from *Tosca* or to solve a differential equation, not even for the Newtonian two-body problem. Not only are there many mysteries as to how contingencies actually work, but as I also mentioned earlier, contingency–species interaction is important (and I certainly should include, a host of other variables related to developmental, motivational, and individual difference conditions). As with many other interactions, especially nonlinear ones, one might expect surprises. Human verbal behavior is perhaps the most compelling example—the ultimate emergent, as it were, from selection processes ultimately bringing the human vocal apparatus under control of its consequences. Given such interactions, however, one must be exceedingly careful about definitive assertions regarding what a given creature is or is not capable of. Moreover, even if the creature does display some interesting and mysterious behavior, one should be equally careful to specify the conditions under which this may or may not occur rather than simply appealing to emergence through complexity or to some rational agency. With respect to contingency–species interaction, the devil is surely in the details.

In general, I liken the issues of complex brains and rational agents to the putative qualitative distinctions between biology and physics espoused by Mayr (1982, 1988) that I mentioned earlier. In fact, complexity may be found in many domains within and across the expanses of the quick and the dead. There is seeming magic and certainly mystery to be found everywhere in Nature, and I express my special appreciation to Duane Rumbaugh for showing me some of that magic and mystery.

## References

Bak, P. (1996). *How nature works*. New York: Springer-Verlag.
Barrow-Green, J. (1996). *History of mathematics: Vol. 11. Poincaré and the three body problem*. Providence, RI: American Mathematical Society.
Chappell, J., & Kacelnik, A. (2002). Tool selectivity in a non-primate, the New Caledonian crow (*Corvus moneduloides*). *Animal Cognition, 5*, 71–78.
Covey, P., & Highfield, R. (1995). *Frontiers of complexity*. New York: Fawcett.
Critchfield, T., Galizio, M., & Zentall, T. R. (Eds.). (2002). Categorization and concept learning [Special issue]. *Journal of the Experimental Analysis of Behavior, 78*(3).
Hayes, S., Barnes-Holmes, D., & Roche, B. (Eds.). (2001). *Relational frame theory*. Reno, NV: Context Press.
Marr, M. J. (1992). Behavioral dynamics: One perspective. *Journal of the Experimental Analysis of Behavior, 57*, 249–266.
Marr, M. J. (1997). The mechanics of complexity: Dynamical systems span the quick and the dead. In L. Hayes & P. Ghezzi (Eds.), *Investigations in behavioral epistemology* (pp. 65–80). Reno, NV: Context Press.
Marr, M. J. (2003). The stitching and the unstitching: What can behavior analysis say about creativity? *The Behavior Analyst, 26*, 15–27.

Mayr, E. (1982). *The growth of biological thought.* Cambridge, MA: Harvard University Press.
Mayr, E. (1988). *Toward a new philosophy of biology.* Cambridge, MA: Harvard University Press.
Mayr, E. (1997). *This is biology.* Cambridge, MA: Harvard University Press.
Neuringer, A. (1986). Can people behave "randomly?": The role of feedback. *Journal of Experimental Psychology: General, 115,* 62–75.
Neuringer, A. (2002). Operant variability: Evidence, functions, and theory. *Psychonomic Bulletin and Review, 9,* 672–705.
*Oxford English Dictionary.* (1989). Emergence. Retrieved April 20, 2006, from http://www.oed.com/
Pepperberg, I. M. (2002). *The Alex studies: Cognitive and communicative abilities of grey parrots.* Cambridge, MA: Harvard University Press.
Pryor, K., Haag, R., & O'Reilly, J. (1969). The creative porpoise: Training for novel behavior. *Journal of the Experimental Analysis of Behavior, 12,* 653–661.
Rachlin, H. (2000). *The science of self-control.* Cambridge, MA: Harvard University Press.
Rumbaugh, D. (1997). The psychology of Harry Harlow: A bridge from radical to rational behaviorism. *Philosophical Psychology, 10,* 197–210.
Rumbaugh, D., Washburn, D., & Hillix, W. (1996). Respondents, operants, and emergents: Toward an integrated perspective on behavior. In K. Pribram & J. King (Eds.), *Learning as a self-organizing process* (pp. 57–73). Hillsdale, NJ: Erlbaum.
Savage, C. (1995). *Bird brains.* San Francisco: Sierra Club.
Shettleworth, S. J. (1998). *Cognition, evolution, and behavior.* New York: Oxford University Press.
Sidman, M. (1986). Functional analysis of emergent verbal classes. In T. Thompson & M. Zeiler (Eds.), *Analysis and integration of behavioral units* (pp. 213–245). Hillsdale, NJ: Erlbaum.
Sidman, M. (2000). Equivalence relations and the reinforcement contingency. *Journal of the Experimental Analysis of Behavior, 74,* 127–146.
Sidman, M., & Tailby, W. (1982). Conditional discrimination vs. matching to sample: An expansion of the testing paradigm. *Journal of the Experimental Analysis of Behavior, 37,* 5–22.
Skinner, B. F. (1938). *The behavior of organisms.* New York: Appleton-Century-Crofts.
Skinner, B. F. (1953). *Science and human behavior.* New York: Macmillan.
Skinner, B. F. (1974). *About behaviorism.* New York: Knopf.
Staddon, J. E. R. (2001). *The new behaviorism.* Philadelphia: Taylor & Francis.
Stewart, I. (1998). *Life's other secret: The new mathematics of the living world.* New York: Wiley.
Weir, A., Chappell, J., & Kacelnik, A. (2002, August 9). Shaping of hooks in New Caledonian crows. *Science, 297,* 981.
Wynne, C. D. L. (2001). *Animal cognition.* New York: Palgrave.

# 9

# The Perception of Emergents

*David A. Washburn*

In 1971, Donald M. MacKay, the noted Scottish vision scientist, physicist, philosopher, and apologist, debated B. F. Skinner on issues ranging from determinism to theology on the television show *Firing Line*, hosted by William F. Buckley. Among the points made in that debate, MacKay accused Skinner of the fallacy of "nothing but-tery." That is, Skinner was accused of erroneously claiming that behavior is determined by "nothing but" circumstance and contingency—that determinism obviates the existence of free will. In other writings, MacKay (1988) explained this criticism:

> It looks, in fact, as if the old debate as to whether man is "nothing but" a mechanism or "something more" arose not because people were disagreed about facts but rather because both sides were agreed in accepting a wrong formulation of the problem. Our suggestion is that it needn't be a matter of "either . . . or" at all, but only one of deciding which description is appropriate in which circumstance. (p. 13)

MacKay believed that the choice between determinism and volition represents a false dichotomy and that in fact both may be accurate under some circumstances. One might take the liberty of extending the same argument to the behaviorism–cognitivism rift, although in both cases the logic was rejected by Skinner and others who claimed that determinism and behaviorism eliminate any need or scientific basis for free will and cognitive psychology, respectively. Although that determinism–free will debate is only tangentially related to the topic of the present chapter, here I propose yet another instance of Skinner's nothing but-tery and suggest that both sides of this present argument may have again accepted an incorrect formulation of the problem.

Skinner (1937) described two categories into which all learned behavior was presumed to fall: respondents and operants. Over 6 decades later, the accumulated data supporting the existence of operant and respondent conditioning are unassailable. There is no denying that behavior is influenced by

---

Preparation of this chapter was supported by National Institute of Child Health and Human Development Grant HD-38051.

associations, contiguities, contingencies, and consequences. The reasons or mechanisms by which behavior is altered may be debated, but the fact is evident in organisms ranging from the activities of the simplest animals to the most complex human behaviors. However, there is also a growing corpus of evidence that an additional class of behavioral changes also exists—instances of learning that are neither operant nor respondent. These data contravene the contention that learning is "nothing but" classical or instrumental conditioning.

Denying that all changes in behavior are classically or instrumentally conditioned creates the burden of indicating what other possibilities exist, however. This third category of acquired behaviors has been called *emergent* (Rumbaugh, 2002; Rumbaugh & Washburn, 2003; Rumbaugh, Washburn, & Hillix, 1996) in deference to the way these competencies emerge or appear without (necessarily, at least) clear and causal antecedents or contingencies. This is not to suggest that emergent competencies are unrelated to experience, any more than it is to suggest that the capacity for emergents is unrelated to biological constraints. In contrast, we do believe that the probability of an emergent behavior emerging is influenced by an animal's experience and computing capacity. However, it is the claim that this experience (the antecedents and consequences that surround emergents) provides information to be perceived, not the causal agent for changing behavior. Emergents are the instances of learning, as reflected by changes in behavior, that result from the processing of this information. They are more than mere alterations in the probability of a particular behavior in a particular context; rather, they may in fact reflect quite improbable behaviors—novel, innovative, and efficient solutions that are not predicted by prior stimulus–stimulus or stimulus–response associations. As Rumbaugh (2002) described,

> Emergents can be manifested as new behavior patterns that are noted for being synergistic, integrative, and clever. Emergents also can be manifested as new capabilities, such as speech comprehension, that are not to be accounted for as responses or behaviors altered by basic conditioning procedures.... Emergents reflect the natural operations of the brain as being comprised of keen pattern-detection and synthesizing systems. (p. 9)

The notion of emergents as a third class of behaviors (complementing respondents and operants) has been criticized as being underspecified, unconvincing (Whiten, 2003), antireductionist (Thomas, 2001), and insufficiently distinct from operants and respondents (Marr, chap. 8, this volume). It is certainly possible that the former two criticisms are accurate, and the present chapter may do little to ameliorate the criticisms. However, the latter two criticisms need not be valid.

## What Emergents Are Not

Thomas (2001) distinguished between two types of emergentism. One is consistent with material reductionism and seems to reflect primarily one's inability to explain the mechanisms that underlie a behavior at a particular time. That is, one may believe that behavior is emergent because no specific explanation can presently be offered, although there is an "in principle" acceptance that a material explanation will one day be produced. In this sense of the term, emergent behaviors are simply unexplained operants or respondents, occupying a role in science much like the homunculus occupies in Attneave's (1961) gradualist view—a placeholder for behaviors gradually to be explained by future research.

The other form of emergentism is anathema to scientific psychology, according to Thomas (2001), in that it supposes irreducible forces and agents that alter behavior—similar to the life forces that distinguish living from nonliving beings according to biological vitalism. Thomas noted that antireductionist, emergentist views hinder the progress of psychology.

The present use of the term *emergents* does not fit within either of these categories of emergence, although it is certainly closer to the former than the latter. The notion of emergents does not imply an absence of physiological and biochemical correspondence with cognitive phenomena. Neither is the notion of emergents as it is intended here antithetical to disciplined scientific investigation (Rumbaugh, 2002). Unlike vitalism in the biological domain, which Thomas (2001) examined insightfully for the purpose of gleaning lessons for psychology, terming a class of learned behaviors *emergents* need not imply an extra, immaterial substance or force that produces these behaviors. Wetness is an emergent property of water, not apparent in either the hydrogen or the oxygen that constitute it; however, wetness is not a mystical substance that is added to the elements to form water. That said, the characteristics of water are not simply the sum of the characteristics of the elements; it is the sum of the elements and their interactions that constitute the emergent substance. Thus, Thomas need not have worried about statements that emergent ideas are "more than just a composite of the simple ideas on which they were based" (Rumbaugh et al., 1996, p. 59). The whole may indeed be more than the sum of its parts, and noting this merely acknowledges the contribution that the interaction term adds to the behavior. However, everything within this equation is subject to scientific psychological inquiry and to corresponding physical implementation in biology, chemistry, and physics. Indeed, I fully expect the notion of emergents to be a catalyst for this scientific discovery and join others in the field as being challenged to identify the social, cognitive, and neural mechanisms of emergent behaviors.

## What Behaviors Are Not Emergents

Some changes in behavior are not emergent. They are the product of stimulus associations or are alterations in the probability of some behaviors by the consequences of these actions in an organism's personal history. However, many behaviors—those that typically interest cognitively oriented researchers—defy explanation by simple conditioning. Although the notion of emergents does not imply that the behaviors cannot be linked to neural or chemical correlates, the notion of emergents does imply that the behaviors cannot be reduced to operant or respondent mechanisms. Indeed, *reduction* is not even the appropriate term here, because the category of emergents occupies the same level of analysis as operants and respondents. Emergents, operants, and respondents are complementary categories of learned behaviors. As Skinnerian-type conditioning cannot be "reduced" to Pavlovian-type conditioning, so emergent behaviors cannot be explained in terms of these associative classes of behaviors. These behaviors are not "something more" than stimulus–stimulus and stimulus–response associations. They are something different.

That said, one is still obligated to investigate the mechanisms that underlie the putative class of emergents. Thomas (2001) was correct in warning that the new name risks reification, in much the same way, I would argue, that terms like *reinforcement* and *stimulus control* have come to be reified as explanatory vehicles for complex behaviors. It is easy to demonstrate that a so-called "reinforcer" alters the probability of some simple behavior in a simple context and then to make two "in principle" leaps in inference: first, that reinforcers "reinforce" or act to cause the change in behavior and, second, that chains of similarly simple associations can account for any behavior, whether simple or complex, whether the chain is observed or just assumed. In any case, cognitive accounts of behavior should resist circular reasoning and the nominal fallacy, even if behaviorist accounts have been similarly susceptible to those temptations. What are the characteristics and conditions that distinguish emergents from operants and respondents?

## How to Distinguish Emergents From Operants and Respondents

Rumbaugh and colleagues (e.g., Rumbaugh, 2002; Rumbaugh et al., 1996; Rumbaugh & Washburn, 2003) have attempted to determine the characteristics of emergents by listing behaviors that seem not to be explicable by conditioning mechanisms. That is, they have looked for competencies that appear without specific reinforcement or experience with particular stimuli. Relatively simple emergent phenomena (like learning set and stimulus equivalence) and more complex emergent competencies (e.g., self-recognition and language comprehension) do not appear in the absence of particular types of experience; however, learning to learn, the formation of generalized equivalence classes, and so forth are not themselves directly reinforced or conditioned in training. This is not to suggest that researchers in the experimental analysis of behavior are mute with respect to explanations for these phenomena. Indeed, their journals and books are replete with conditioning-based descriptions of the behaviors that we consider to be emergent.

Consider, for instance, the phenomenon of transposition. A variety of animals have been shown to respond correctly on tasks requiring rulelike selection between stimuli that vary on a single stimulus dimension (e.g., to pick the larger of two, or the smaller, or the brighter). For example, a pigeon might be presented with a circle that is 5 centimeters in diameter and a circle that is 10 centimeters in diameter. Pecks to the 10-centimeter circle result in access to a grain reward. Once the pigeon learns this discrimination, it might be presented with a novel trial of a 10-centimeter circle and a 15-centimeter circle. Although the bird's prior reinforcement history would suggest an associative response (i.e., selection of the previously rewarded 10-centimeter circle), it is likely to respond to this probe trial relationally (i.e., selection of the larger stimulus). This would seem to belie associationist explanation; however, Spence (1937) and many others have shown that transposition can indeed be accommodated readily within a conditioning perspective (see reviews in Lazareva, Wasserman, & Young, 2005; Riley, 1968).

However, Lazareva et al. (2005) recently challenged the long-held conclusion proffered by Spence (1937), providing compelling evidence that transpositional responding cannot be explained by associative learning and ancillary notions like stimulus generalization. Rather, the animals learn to respond according to the rules of the task (in this case, "select the larger stimulus" or "select the smaller stimulus"). I further argue that this is a lovely example of an emergent behavior. Relational learning was not required during training; the birds could simply have learned a list of "execute a particular response in the presence of specific stimuli" associations and thereby gained the rewards there were to garner. Nevertheless, performance on the novel test trials revealed that the birds learned much more than this: They learned to respond not to the specific stimuli but to the relation. Their behavior in this study certainly reflected operant and respondent conditioning, but it also reflected something else—something general, rulelike, cognitive, and emergent.

Rumbaugh, our colleagues, and I (e.g., Rumbaugh, 2002; Rumbaugh et al., 1996; Rumbaugh & Washburn, 2003) have attempted to distinguish such instances of learning from the long list of behavioral changes that are clearly operant or respondent in nature. In the latter instances, associations between stimuli or between stimuli and responses alter the probability of specific behaviors. In contrast, emergents are rulelike and generative competencies that appear from experience, often permitting an organism to defy the associative inclinations of operant and respondent conditioning and to respond in new, generalized, creative, and unpredictable ways.

## Meaningful Failures to Learn

Two studies illustrate this distinction. First, consider an experiment in which the animals could have learned relationally, but their responses revealed that they never escaped the bondage of stimulus–response associations and thus never picked up on the "rules of the game." Filion, Washburn, and Fragaszy (1995) tested rhesus monkeys on a task in which a line moved from the top-left corner of the screen linearly toward one of five squares spread evenly across the bottom of the screen. At some point before the line reached the squares, it stopped and the monkey had to indicate which square was being "pointed at" by the line (i.e., which box would have been hit had the line kept moving). When the monkeys reached criterion on this task, the experimenter moved either the point of origin (e.g., by having the line begin at the top-right corner) or the location of the squares (e.g., by having them spaced vertically across the left edge of the screen). In this way, the monkeys could be trained to criterion, tested with five novel probe trials; trained to criterion on these new trials, tested with five new probes; and so forth, until each possible location of the origin and the squares had been used.

The goal of this task was to determine whether the monkeys could extrapolate from the trajectory of the line to indicate the position that was being "pointed to." The monkeys rapidly learned each new screen configuration to a criterial level of performance. However, with each probe trial, the monkeys responded on a stimulus–response associative basis rather than on the basis of the relation that would have been correct. That is, the monkeys picked the square that had previously been associated with the end-position of the line, not the square that would have been hit had the line continued to move. On this task—in which the monkeys could have learned in the relational, emergent way that humans certainly would have manifested—the animals were, for some reason, constrained to respond associatively. I have termed instances in which organisms that are capable of relational learning nonetheless fail to use the rule *meaningful failures to learn* (Rumbaugh & Washburn, 2003). They are "failures to learn" only in the sense that the animals could have learned an efficient, generalized, relational basis for responding, but they learned instead a large number of psychologically more simple associations. (In doing so, learning is parsimonious from the perspective of Morgan's canon but unparsimonious in the sense of Occam's razor—a point to be discussed below.) Such instances are meaningful in that in studies like these, the experimenters do everything reasonable (and occasionally many other unreasonable things) to try to get the animals to respond relationally. I suggest that such occasions of *associative drift*—a term used to parallel Breland and Breland's (1961) notion of *instinctual drift*, in which animals abandon conditioned behaviors in favor of instinctual tendencies—help to highlight the distinction between conditioned responding (operants and respondents) and relational learning (emergents). It may not always be clear whether some behavior is emergent rather than operant or respondent in character; however, it is really noteworthy when animals fail to respond in some seemingly obvious relational way but show instead, by their performance on probe trials, that they are constrained by operant and respondent conditioning. Over the years, I have been accumulating more and more of these meaningful failures in the hopes that patterns will one day be evident to reveal why animals (including humans) who can learn relationally sometimes do not learn relationally (i.e., why emergents sometimes do not emerge).

By way of contrast, consider rhesus monkeys' competencies with numerical symbols (Washburn & Rumbaugh, 1991; see also Beran, Gulledge, & Washburn, chap. 13, this volume). Monkeys were presented with pairs of Arabic numerals and were rewarded proportionately irrespective of which numeral was selected (e.g., four pellets if the numeral 4 was selected, zero pellets if the numeral 0 was selected). The monkeys quickly learned to select the numeral corresponding to the greater value, even on novel probe trials with never-before paired numerals (e.g., 6 vs. 7, when the monkeys had seen each numeral paired with other stimuli but never with one another). Further, during training some monkeys had never seen the numeral 5 paired with a larger value; thus, selecting the 5 was the "correct" response on every training trial in which it appeared. Notwithstanding, the monkeys were significantly likely to select the numerals 6, 7, 8, or 9 when paired with a 5, even though they had no direct reinforcement history of 5 being incorrect. Of course, all of this is reasonable if the monkeys had learned the "rules of the game" (pick the numeral that represented the bigger value) and also the absolute—or at least the relative—quantity symbolized by each numeral. Whereas the former knowledge might have been obtained by conditioning, the latter requires more than stimulus–response associations. Indeed, the latter competency could be shown (as in the numeral-5 tests) to produce responses that are the opposite of what simple conditioning would suggest.

## The Case for Emergents

Many other examples of competencies have been demonstrated, and they are considered emergents because they are not predictable a priori on the basis of operant or respondent mechanisms. However, it must be acknowledged that thin ice undergirds this stance. That is, the case for emergents is made in this way because of the difficulty (and maybe the impossibility) of identifying the specific antecedents, the specific contingencies, or the specific experiences that may have made up an organism's conditioning history. Because the only behaviors with a $p = .00$ probability of occurrence are those behaviors that never occur, there is an "in principle" counterargument to the suggestion that emergents exist: the argument that what appears to be a new competency may well be a variant on a previously produced behavior that has been influenced by unspecified prior contingencies and ubiquitous stimulus condition. Marr (chap. 8, this volume) makes this counterargument for most of the emergents proffered by Rumbaugh and colleagues (e.g., Rumbaugh, 2002; Rumbaugh et al., 1996; Rumbaugh & Washburn, 2003). In other words, just because the antecedents and contingencies are not easily identified, this does not mean that they do not exist. (Of course, neither does it mean that they do!)

However, it is possible that the case for emergents has been constructed in the wrong location. Granting to the behaviorist explanation proprietary claim to terms like *stimulus*, *response*, and *reinforcement*, Rumbaugh, our colleagues, and I (Rumbaugh, 2002; Rumbaugh et al., 1996; Rumbaugh & Washburn, 2003) have attempted to demonstrate emergents by exclusion. That is, we have tended to label as an emergent only those behavioral competencies that cannot convincingly be explained by conditioning. Perhaps this is granting too much to the behaviorist perspective.

I agree that operants and respondents are the psychically simpler categories of behaviors, and thus Morgan's canon dictates that they be favored when all else is equal. However, there are many instances in which all else might not be equal and thus that the more cognitive and conative explanation inherent in the emergents category (relative to that of operants and respondents) is more parsimonious from the perspective of Occam's razor. That is, frequently the conditioning explanation of behavior requires one to accept myriad explanatory components (component behaviors, chained together by association and individually modified with respect to probability by experience with consequences and a network of potentially overlapping generalization gradients) when the "emergents" explanation involves a simpler explanation: Animals perceive invariance, regularity, and predictability in the world and respond according to this rulelike regularity.

Operants and respondents appear to have explanatory primacy because they are associated with more basic and preliminary forms of learning than are the cognitive abstractions captured by emergents. However, I contend that this supposed explanatory primacy is illusory and that it results from the fact that when psychologists think about learning, we begin at the wrong point. We begin with the organism's response to a stimulus. Whereas conditioning may account for how a rat comes to learn to press a lever, for example, conditioning is probably not how the animal comes to perceive the lever. Consider then the more basic question: How does an organism learn to regard sensory input as a stimulus?

## Perception

Cognitive research and theory on perception and pattern recognition continue to reflect disagreement about how organisms recognize stimuli. Accepted as a given are the descriptive principles of perception articulated by the Gestalt psychologists. Items that are similar, contiguous, or share a common fate are perceived as belonging together. Perception appears to strive for good form, closure, and continuity. Descriptive rules like these have been embraced in the influential theories by Marr (1982), Treisman and Gelade (1980), and others to indicate how features become bound together into recognizable objects. Less agreement has surrounded the nature and number of steps that intervene between the detection and recognition. That is, grouping of boundaries, depth cues, and other features, on the one hand, and the recognition of patterns as stimuli that are distinct from one another and that exist within networks of conceptual groupings, on the other, permit both within- and between-category relations.

The classic debate in the perception literature is the contrast between direct and indirect (or constructivist) theories. That is, is perception a stimulus-driven, bottom-up effect of the information that exists in the stimulus array as was championed by J. J. Gibson (1979)? Or is perception a top-down process of actively constructing reality from the interaction of expectation, experience, and exploration as was suggested by Neisser (1976) and many others? Although the constructivist view has been the more popular (in large part because it is more consistent with the data) in cognitive psychology, we may have lost the baby with the bathwater, as frequently happens in our discipline. This was never Neisser's goal, as he indicated in his own writing. Perception, pattern recognition, and categorization do indeed reflect active and constructive processing of information, but these constructs also benefit from—and indeed require—the wealth of information available in the stimulus array. Perception is occasionally inaccurate (as with visual illusions), suggestible (as with context effects), or selective (as is demonstrated in attention research), owing to the constructed nature of the percept. However, perception is generally veridical, providing a testimony to the importance of the information that is there in the sensory array to be perceived.

James J. Gibson (1979) noted that regularities in sensory information (he specifically studied visual perception, but presumably his theory extends to other sensory domains) are "picked up" or abstracted by the perceiver. He termed these regularities *invariants* and noted that the information available in sensory invariants does not require processing or construction, only detection. According to J. J. Gibson, the perceptual system resonates to or is attuned to the invariant structure of the environment, abstracting when necessary the regularity or similarity that exists between distinct objects or events. Of course, much of the information available "in the light" (or other sensory media) is not perceived. Organisms may not, and indeed do not, pick up all of the available information conveyed by the invariants. There are regular differences in the taste, appearance, and bouquet between merlot wines and pinot noir wines; however, these invariants have never been salient to me. I have never attended to these differences or learned to perceive the distinctions between these two varieties of libation. In contrast, a sommelier would perceive this difference easily, and additionally might perceive the invariants that allow for merlot wines to be recognized by year, by region, and so forth. Thus, the process of learning to perceive a stimulus, for example, to perceive the appearance of an apple or to perceive between the aromas of two different varieties of apples, is the process of becoming sensitized to, or perhaps of abstracting the regularity of, invariant information that is itself a part of the sensory array, because it is itself part of the objects and events that constitute the environment. As Sherlock Holmes chided Dr. Watson on numerous occasions, "You see but you do not observe" (Doyle, 1891). This suggests, as E. Gibson (1969) documented empirically, that the difference between skilled and unskilled perceivers lies not in what the former brings to the perceptual experience by way of information processing but rather in what the former takes from the perceptual event by way of sensory invariants.

For the contemporary behaviorist, such a formulation may afford little discomfort. Indeed, many accounts of conditioning are themselves descriptions of the information that is available in experience—information that according to the behaviorist does not require processing or manipulation. Consider the view of classical conditioning, in which the conditioned stimulus (CS) comes to signal to the organism that the unconditioned stimulus (UCS) is likely to appear, and thus produces the conditioned response (CR) not by a mechanism of stimulus substitution but rather by the information provided. The regularity of the CS–UCS is essential for the generation of the CR, according to this perspective; however, it is not the association strength that results from repeated temporal pairings of the CS with the UCS that seems to be important here. It is the predictive value of the CS with respect to the appearance of the UCS that results in respondent conditioning. Put another way, the behavior analyst might indicate that the organism's behavior is being controlled by the invariant (or at least the regular and predictive) relation that exists between CS and UCS.

J. J. Gibson (1979) directly addressed the difference between such a formulation and his direct perception theory:

> Note how radically different this [direct perception view] is from saying that if stimulus-event A is invariably followed by stimulus-event B we will come to expect B whenever we experience A. The latter is classical association

theory (or conditioning theory, or expectancy theory). It rests on the stimulus-sequence doctrine. . . . An event is only known by a conjunction of atomic sensations, a contingency. If this recurrent sequence is experienced again and again, the observer will begin to anticipate, or have faith, to learn by induction, but that is the best he can do. (p. 230)

J. J. Gibson (1979) contrasted such a framework with his concept of information pickup. He noted that learned organisms respond, not on the basis of associations or memory of how events have been regular in the past but rather on the basis of information that is available in present experience—information that can be perceived now by the educated organism, even if it was missed (though available) on previous occasions. "The process of pickup is postulated to be very susceptible to development and learning. The opportunities for educating attention, for exploring and adjusting, for extracting and abstracting are unlimited" (J. J. Gibson, 1979, p. 230). Of course, all of these processes, plus the very notion of invariance, require some comparison of present stimulation with past events.

## Pattern Recognition and Categorization

I join with Neisser (1976) and many others in believing that J. J. Gibson's (1979) theory provides an incomplete version of perception and that it specifically underplays the role of expectation, intention, and memory in perception. These omissions are particularly obvious when one ventures from the construct of perception (i.e., how does one come to perceive an apple?) to the domains of pattern recognition and categorization. How does one come to label this percept as an apple, to group it with other apples that may differ across a variety of dimensions, and to distinguish it from other concepts that are similar in appearance (e.g., pears), name (e.g., Adam's apples), and so forth?

The pattern recognition literature and the categorization literature are remarkably disconnected from one another—remarkable because the theories debated in each domain parallel one another (as indeed should be the case, given that the recognition of a pattern is functionally equivalent to the assigning of a stimulus to a category). Research in each literature is driven by the contrast between three theories. Pattern recognition, for example, is thought to be accomplished by feature detection, by template matching, or by prototype comparison. Categorization (or, more accurately, the acquisition of concepts, or mental representations of categories) is theorized to be accomplished by feature learning, by template (or exemplar) matching, or by prototype formation. One can clearly see why these literatures have remained separate!

Exemplar-based or template-based theories suggest that one recognizes a pattern on the basis of comparisons with memories of previously experienced stimuli. Each new apple one encounters is recognized by comparison with previously perceived fruit. If the new object is sufficiently similar to one of the exemplars stored in memory, the new object will be considered a member of that same category. This "recognition–categorization by specifics" theory has proved to be powerful for accounting for concept-learning data, particularly for small, highly familiar categories.

In contrast, prototype-based explanations involve the abstraction of an idealized representation and subsequent comparison of new objects with these prototypes, which themselves may never have been directly experienced or encountered. If a new stimulus resembles the prototypical apple, then it belongs in this category. Prototype theories of concept learning have also received extensive empirical support, particularly for relatively large and less familiar categories (for reviews, see Smith, 2002; Smith & Minda, 2000).

Note here, as in the discussion above, that both the exemplar and the prototype view depend on experience with stimuli in order for concepts to form and patterns to be recognized. (Feature-based theories may also require experience, but these models are less directly relevant to the present discussion, and in any case they almost certainly work in concert with exemplar and prototype mechanisms of recognition and categorization.) However, the consequences of this experience are very different for the two theories. In the template or exemplar case, each experience makes an individual contribution to subsequent performance, and how well a novel object is recognized will depend on the match between it and the stimuli in an organism's history. In prototype theories, family resemblance among the members of a category (i.e., the invariants or similarities that characterize the class) provides a rulelike abstraction on the basis of which subsequent judgments can be made. It seems reasonable to posit that there is a metaphorical, if not a mechanistic, parallel between the exemplar–prototype dichotomy in pattern recognition and category learning and the associative–relational distinction (or, for the present discussion, the operant-and-respondent/emergent distinction) in other learning domains.

The point is that humans and other animals are very good at recognizing patterns, at detecting regularities, and at grouping objects or events into classes. These abilities are evident across species and across the life span in our own species from birth onward. The process by which organisms learn—for instance, to form equivalence classes among previously unrelated stimuli, or to comprehend that a particular symbol represents some referent not present in time or space, and so forth—is fundamentally a pattern-recognition process. It is not a stamping-in of associations by the consequences of behavior, but rather the perception of regularities that are already stamped into the sensory array.

## Stimulus Control, Reinforcer Control, Brain Control

So what role do reinforcers play in this type of learning if they do not literally alter the strength of association between stimuli and responses? Given this pattern-recognition perspective, reinforcers serve two purposes: (a) They are part of the sensory array and consequently may be part of the invariant patterns that are available to be perceived, and (b) they are incentives that serve to motivate an organism to behave at all, thus providing both the reason why a human or other animal should care about the invariant and also the behavior that a researcher must use as a measure of learning. A rat may come to learn the spatial layout of a maze simply by navigating through it, that is, without specific reinforcement for traversing the maze correctly. However, there is no reason for the rat to manifest this recognition of the maze pattern in the absence of reinforcement—or more descriptively, reward—for moving through the maze efficiently. Without the bit of cheese in the goal box, why should the rat bother to attend to the spatial pattern of the maze? Even if the animal did abstract the pattern from its experience in the labyrinth, why in the absence of incentive would it respond on the basis of this pattern?

If reinforcement does not determine learning, then what does? The temptation is to note that the brain is a pattern detection and categorization machine, an organ that appears to be specialized for recognizing the familiar and for making sense of the novel. Of course, we have yielded to this temptation in previous writings, in part because it is absolutely (but trivially) true. That is, where else would learning be categorized except in the brain? The fact that we can learn in all of these varied contexts indicates that brains are engineered to do this very thing. We have suggested that perceptual learning is what brains do, in part, because it seems less inferential and less introspective than saying that perceptual learning is what minds do, although of course this too is exactly what we mean if indeed mental activity is understood to be the active and constructive interplay of expectations, intentions, and information available in the environment.

Most generically, what I want to suggest is that organisms detect invariants in the sensory array, whether this sensory array is the simultaneously available features of a visual stimulus exciting optic and cortical neurons or a sequence of events that itself forms a regular and perceivable pattern and that might contain information salient to the organism. That said, psychologists have tended to emphasize the importance of *brain* in this type of learning because instances of emergent learning are more easily generated for humans than for nonhuman animals, more easily generated for apes than for other nonhuman animals, and so forth. Although emergent forms of learning can be cited for a wide range of animals (with a correspondingly wide range of brain sizes and complexities, as was emphasized in Marr's contribution to this volume [chap. 8]), the number and cogency of examples of emergents seem roughly to correlate with brain size and complexity.

And what of stimulus control? First, I am confident that stimuli can elicit an involuntary shift of attention, can trigger the execution of automatic responses, and can alter one's expectations and the perception of future stimuli. However, I am equally confident that traditional behaviorist perspectives have ignored an important source of stimulation—self-directed inputs—that serves to make behavior purposive and flexible. Hershberger (1987) noted that "it is eminently conceivable that at least occasionally we too control our inputs so that at least some of our behaviors are neither emitted nor elicited responses, but self-controlled inputs" (p. 606). This notion of self-controlled inputs is a compelling one. Because self-directed inputs are certainly instated in material terms—these are not homunculi or mysterious forces but are as natural and neurologically based as are other forms of cognition like perception and memory—they provide a language for describing planning, monitoring, learning and response set, insight, symbol comprehension, and other emergent phenomena. The idea that self-generated stimuli may underlie some cognitive behavior also fits seductively with the theoretical assumption and preliminary findings that emergents are more common (but not, as Marr notes in chap. 8, this volume, the exclusive province of complex brains), particularly with respect to frontal-lobe development and linguistically competent animals.

Of course, these remain empirical issues, and continued research is required to understand the nature of self-controlled inputs and their relation to the perception of temporal or spatial patterns of regularity. The important thesis of this chapter, echoing the writings by Rumbaugh (e.g., 2002), is that this understanding can only be gained from an inclusive analysis of cognitive phenomena together with operants and respondents. To quote Hershberger again,

> Skinner's error, of course, is the empty organism, not the overt behavior. This error is not in attending too much to overt behavior, but rather in attending to too little overt behavior. He has overlooked the type of overt behavior that implies a nonempty organism (namely, purposive, self-controlled input). Therefore, cognitive science cannot expect to rectify Skinner's oversight by ignoring overt behavior. On the contrary, if cognitive science is to stand on Skinner's shoulders, as opposed to his face, it must consider fully all types of overt behavior, purposive included. In short, psychology must be a conative, as well as cognitive, science. (Hershberger, 1988, p. 823)

## Rational Behaviorism Versus Radical Behaviorism

The goal for this chapter, as with earlier discussions of emergents (Rumbaugh, 2002; Rumbaugh & Washburn, 2003; Rumbaugh et al., 1996), is to advance a framework that is more inclusive and integrative than those typical of either contemporary behaviorism or cognitive perspectives. Rumbaugh, our colleagues, and I have called this view *rational behaviorism*, although this term too has been criticized (e.g., Marr, chap. 8, this volume). We embraced the term *behaviorism* to emphasize the dependency on observable behavior as the only datum for psychological inquiry. It also serves to emphasize our acceptance of the basic conditioning phenomena of operants and respondents. However, we have called the framework *rational* to denote the inclusion of cognition. This certainly does not mean that animals always act logically and rationally. Indeed, in 2002 Daniel Kahneman became psychology's second-ever Nobel laureate for research on humans' lack of rationality in decision making (for a review, see Kahneman & Tversky, 1984). However, the term *rational behaviorism* provides a useful contrast with so-called "radical" behaviorism, the more extreme forms of associationism that were championed by Skinner (e.g., 1938) and many others. In truth, many contemporary behaviorists practice a less radical, more rational (as in "reasonable") behaviorism that allows for substantial cognitive inputs and mediators, although typically these are sanitized with terms more typical of the behaviorist tradition. There is tremendous merit in the trend for cognitive topics to receive the attention of the experimental analysis of behavior as well as for cognitive psychologists to acknowledge the findings of stimulus control. It is my hope that psychology will remain a behavioral discipline and that we will follow Rumbaugh's example in attempting to be more rational and integrative in our theories. Perhaps the acceptance of rational behaviorism and its triumvirate of operants, respondents, and emergents is itself an emergent—a pattern to be recognized (rather than an association to be reinforced) from the regularities that appear in the data on human and nonhuman animal learning.

## References

Attneave, F. (1961). In defense of homunculi. In W. A. Rosenblith (Ed.), *Sensory communication* (pp. 777–782). New York: MIT Press & Wiley.
Breland, K., & Breland, M. (1961). The misbehavior of organisms. *American Psychologist, 16*, 681–684.
Doyle, A. C. (1891). *Scandal in Bohemia*. Retrieved April 12, 2006, from http://www.4literature.net/Arthur_Conan_Doyle/Scandal_in_Bohemia/
Filion, C., Washburn, D. A., & Fragaszy, D. M. (1995, June). *Trajectory estimation by rhesus macaques*. Poster presented at the annual meeting of the American Psychological Society, New York.
Gibson, E. (1969). *Principles of perceptual learning and development*. East Norwalk, CT: Appleton-Century-Crofts.
Gibson, J. J. (1979). *The ecological approach to visual perception*. Boston: Houghton Mifflin.
Hershberger, W. A. (1987). Some overt behaviors are neither elicited nor emitted. *American Psychologist, 42*, 605–606.
Hershberger, W. A. (1988). Psychology as a cognitive science. *American Psychologist, 43*, 823–824.
Kahneman, D., & Tversky, A. (1984). Choices, values, and frames. *American Psychologist, 39*, 341–350.
Lazareva, O. F., Wasserman, E. A., & Young, M. E. (2005). Transposition in pigeons: Reassessing Spence (1937) with multiple discrimination training. *Learning and Behavior, 33*, 22–46.
MacKay, D. M. (1988). *The open mind and other essays: A scientist in God's world* (M. Tinker, Ed.). Leicester, England: Inter-Varsity Press.
Marr, D. (1982). *Vision*. San Francisco: Freeman.
Neisser, U. (1976). *Cognition and reality*. San Francisco: Freeman.
Riley, D. A. (1968). *Discrimination learning*. Boston: Allyn & Bacon.
Rumbaugh, D. M. (2002). Emergents and rational behaviorism. *Eye on Psi Chi, 7*, 8–14.
Rumbaugh, D. M., & Washburn, D. A. (2003). *Intelligence of apes and other rational beings*. New Haven, CT: Yale University Press.
Rumbaugh, D. M., Washburn, D. A., & Hillix, W. A. (1996). Respondents, operants, and emergents: Toward an integrated perspective on behavior. In K. Pribram & J. King (Eds.), *Learning as a self-organizing process* (pp. 57–73). Hillsdale, NJ: Erlbaum.
Skinner, B. F. (1937). Two types of conditioned reflex: A reply to Konorski and Miller. *Journal of General Psychology, 16*, 272–279.
Skinner, B. F. (1938). *The behavior of organisms: An experimental analysis*. New York: Appleton-Century.
Smith, J. D. (2002). Exemplar theory's predicted typicality gradient can be tested and disconfirmed. *Psychological Science, 13*, 437–442.
Smith, J. D., & Minda, J. P. (2000). Thirty categorization results in search of a model. *Journal of Experimental Psychology: Learning, Memory, and Cognition, 26*, 3–27.
Spence, K. W. (1937). The differential response in animals to stimuli varying within a single dimension. *Psychological Review, 44*, 430–444.
Thomas, R. K. (2001). *Hazards of "emergentism" in psychology*. Paper presented at the annual meeting of the Southern Society for Philosophy and Psychology, Louisville, KY. Retrieved April 9, 2006, from http://htpprints.yorku.ca/archive/00000011/00/HOE.htm

Treisman, A., & Gelade, G. (1980). A feature integration theory of attention. *Cognitive Psychology, 12*, 97–136.
Washburn, D. A., & Rumbaugh, D. M. (1991). Ordinal judgments of Arabic symbols by macaques (*Macaca mulatta*). *Psychological Science, 2*, 190–193.
Whiten, A. (2003, October 2). Thinking of apes. *Nature, 125*, 454.

# 10

# New Models of Ability Are Needed: New Methods of Assessment Will Be Required

## H. Carl Haywood

Psychologists are at this moment at a watershed in the history of intellective assessment and psychometrics. We do not simply need to change our approaches; rather, change is thrust upon us and will occur more or less regardless of what we do or do not do. We can, however, guide some of the inevitable change. I suggest that we have arrived at this moment for at least the following reasons.

1. The IQ as a predictor of school achievement has been enormously successful. Some psychologists believe that intelligence testing is what we do best as a profession. And why not? We have been conducting intelligence tests for more than a century, constantly refining our technology. Intelligence tests are at their best in predicting individual differences in school achievement, reaching 55% common variance with subsequent achievement tests (Anastasi, 1965). IQ is also a superb predictor of school nonachievement; indeed, from the very beginning of the intelligence testing movement, that has been one of its major goals (see, e.g., Binet & Simon, 1905; Haywood & Paour, 1992) and a source of great success. We need to ask how much further we can go in that direction and, even more important, how useful that accomplishment is in today's world.

2. Models of intelligence as an innate trait are useful for understanding the conceptual nature of intelligence, the latent variable, which is indeed what science is about, but such models have gone about as far as they can go in elucidating the nature of intelligence. In an applied sense, there are more important questions, such as how to program learning experiences to provide maximal benefits for all learners rather than to exclude those who have predictors that they cannot learn well from "normal" classrooms. In fact, I assert that intelligence as a major and preemptive construct has outlived its usefulness in both scientific and applied terms. Rather than restricting the task of psychometrics, that position broadens and expands our task greatly.

---

An earlier version of this chapter was presented in December 2002 at the Vanderbilt University/Kennedy Center Conference on Futures of Intellectual Assessment and Psychometrics, Nashville, Tennessee.

3. Although psychologists do it very well indeed, prediction of success in school, in specific learning situations, in vocational training, or in life in general, is no longer a useful goal. I could never quite understand psychology's fascination with prediction. Just what good does it do us or society? To be sure, it is useful to know which candidates will and will not succeed in expensive and sharply focused training programs, such as those for airplane pilots, perhaps air traffic controllers, and very specialized technical personnel. With respect to less sharply focused criterion variables, I suggest that psychometric prediction of individual differences in such criterion variables as school achievement may in fact do more harm than good. Beyond the obvious danger of constructing self-fulfilling prophecies, such a prediction more often predicts lack of success and poor performance and is not so good at predicting contributions of a notable nature from persons who score at the high end of psychometric scales. The essential problem, however, lies not in what intelligence test scores do but rather in what they do not do: They do not help much to specify the necessary conditions for better than predicted performance. From a social perspective, the cost of error in prediction is the measure of the success of intelligence tests. In the industrial sector, it is very costly to include in expensive training programs a large number of candidates who will not succeed. In the domain of education, it is far more costly to exclude or to track into inappropriate channels any number of learners who could have derived significant benefit from a more demanding tract. Avoiding false negatives is thus more important for education than for industry, and assessing abilities in such a way as to avoid such errors should constitute a major goal of the psychological assessment enterprise. I suggest in this chapter a model of ability and some psychometric methods that offer the hope of accomplishing that task.

4. Under the yoke of radical behaviorism, with its heavy debt to logical positivism, a simple stimulus–response paradigm seemed appropriate, and going beyond that association of antecedent and consequent variables was done at one's great peril. When I was in graduate school in the 1950s, and for some considerable time after that, the concept of cognition was discouraged on the ground that cognition was neither directly observable nor directly measurable. Even the great learning theorist Edward Tolman was held in some suspicion by psychological purists, because he seemed to be interested in unobservable variables. For the same reason, we were forbidden, as students, to use the term *potential*. Behaviorism has now made its modest contributions and psychology can move on, as it has done, to the study of variables whose very existence, and certainly whose characteristics, must be inferred from the behavior of related variables. In doing so, we have brought our philosophy of science up to the level of the physics of 1900. Given the freedom to study inferred variables, we are now capable of much richer concepts of human ability.

5. Assessment of human ability is not always guided by clear theories of the nature and development of intelligence. That is to say, in psychology's enthusiasm for empirical science, we have sometimes pursued technical excellence in assessment methods and the statistical treatment of test data in the absence of excellent contemporary theories or insights on the nature of what it is we are assessing so beautifully. The more successful we are at constructing

theories of the nature and development of intelligence, theories that can encompass the ever-emerging data from our empirical observations, the more possible it will become for us to construct and apply ever-better tools of assessment.

## A Transactional Perspective on Human Mental Ability

In this section I offer a brief description of a rather simple but more encompassing view of the nature of human mental ability that I refer to as a *transactional* perspective. This perspective, which has been evolving over the last 30 years or so, has the following major assumptions.

1. *Ability is multifaceted.* Although I would not want to go as far as Gardner (1999) has in this respect, it is necessary to observe that, in addition to the vital concept of "$g$," there are relatively specific manifestations of ability. No doubt, $g$ is what makes them correlated, but it may be useful to recognize that playing the cello, conceptualizing spatial relations, manipulating the language, and running the point-guard position on the basketball court may require and call on rather different kinds of ability, all of them correlated with general intelligence but occurring in different configurations in different individuals. Those who are good at several such activities do indeed tend to be good at other intellective functions as well, but they may also be rather bad at some. In my own case, as a bizarre example, I can play the piano (not the cello) and manipulate the romance languages reasonably well, but I find it difficult to conceptualize spatial problems, I manage to get lost when coming up out of the subway, and my only athletic injury has consisted of splinters from sitting on the bench. Studies of mathematical giftedness constitute a good example of the value of recognizing the multifaceted nature of ability and of studying its components (e.g., Benbow & Lubinski, 1996; Stanley, Keating, & Fox, 1974).

2. *Ability is multidetermined.* This assumption simply recognizes the nonexclusivity of the genetic determination of individual differences in intelligence while acknowledging the very heavy contribution of polygenic determination. It also asserts that the precise contribution of genetic determination is not the most important consideration even if individual differences in intelligence were 100% genetically determined (which might actually turn out to be the case!).

3. *Intelligence alone is not sufficient to account for individual differences in ability, learning effectiveness, or achievement.* It is not possible to have enough IQ points to escape the necessity for at least two other components of ability differences. These additional components are cognitive processes and motivation.

4. *It is essential to make a sharp distinction between intelligence and cognitive process.* It is not useful to consider them to be synonymous, and treating them as such only adds to the confusion that already exists about what constitutes human ability. I would be very happy to see psychologists stop using the terms *intellective* and *cognitive* as if they had the same meaning!

5. *Motivation plays a central role in determining ability differences.* This is especially true of what I and some others have called *task-intrinsic motivation*; that is, the motivation that inheres in information processing and action,

as Hunt (1963) referred to it, that enables one to derive satisfaction from cognitive activity for its own sake and as its own reward. Task-intrinsic motivation is the critical variable in producing lifelong independent learners as opposed to highly intelligent persons who learn when they have to and stop learning when formal education ends.

So, with apologies to Sternberg (1985), who has already co-opted all the "three" prefixes that I know, I offer a view of human ability that has three transactional components: intelligence, cognitive processes, and motivation.

## Intelligence and Cognitive Processes

I suggested earlier that it is useful to make a sharp distinction between concepts of intelligence and concepts of cognitive process. In my own conceptualizing, this distinction is similar to that between fluid and crystallized intelligence (see, e.g., Horn & Noll, 1997), especially with respect to the modifiability aspect, but in this case I am not trying to incorporate both into a single concept of intelligence. Table 10.1 shows a comparison of these two concepts with respect to six criteria of comparison: source, modifiability, character, assessment, composition, and developmental requirements.

To the extent that the concepts can be usefully differentiated on these six criteria, the distinction between them should prove to be beneficial. The first two are perhaps the most critical. Whereas the principal source of individual differences in intelligence is polygenic, individual differences in cognitive processes are acquired, often through learning, but always through experience. They differ as well with respect to their modifiability. The evidence so far suggests that intelligence is only modestly modifiable, that indeed the classical notion of the constancy of the IQ may be closer to the truth than we have recently suspected. Cognitive processes, on the other hand, having been acquired through successive encounters with the environment, are eminently

**Table 10.1.** Comparison of Intelligence and Cognitive Processes Along Six Dimensions

| Dimension of comparison | Intelligence | Cognition |
| --- | --- | --- |
| Source | Genetic (polygenic) | Acquired |
| Modifiability | Modest | Great |
| Character | Global, "g" | Generalizable, specific |
| Assessment | Achievement, past learning | Process, potential |
| Composition | Intellectual factors | Native ability, motives, habits, attitudes |
| Developmental requirements | Genes, nutrition, health, safety, fostering environment | Active, directed teaching; mediation of cognitive processes and cultural components |

*Note.* From "Multidimensional Treatment of Mental Retardation," by H. C. Haywood, 1989, *Psychology in Mental Retardation and Developmental Disabilities, 15,* p. 3. Copyright 1989 by the American Psychological Association.

modifiable, partly on the assumption that whatever is learned can be unlearned, relearned, and elaborated by way of further learning.

The character of the two constructs differs as well. Intelligence is typically thought of as consisting principally of a broad, general component ($g$), sometimes as having two major components (e.g., verbal and performance), and sometimes as having several or even many components (see Woodcock, 2002), all of them representing specific classes of mental operations. Cognition (read: logical thinking modes) consists both of specific mental operations and of broadly generalizable abilities that underlie performance in a vast array of content domains.

Most tests of intelligence are essentially achievement tests, based on the assumption that how much one has learned from previous opportunities to learn is a good indicator of one's learning ability and therefore of one's intelligence. The assessment of cognitive processes, on the other hand, is accomplished by way of process assessment, that is, systematic observation of how one learns in new learning situations, including those with teaching.

Intelligence is composed of essentially "intellective" variables, that is, components that have to do directly with ability itself. By contrast, cognitive processes represent a mix of native ability, motives, habits, and attitudes; in other words, of both intellective and nonintellective (emotional, social, motivational) variables.

In the development of intelligence, parents contribute genes, nutrition, health, safety, and a fostering environment. Cognitive processes, on the other hand, must be acquired by successive encounters with the physical and social environment, whether by the intergenerational transfer of culture, as Vygotsky (1934/1988) suggested, or by each individual's construction of his or her unique set of cognitive processes, as Piaget (1970) suggested.

To the extent that this distinction is useful, it should focus attention on the necessity to assess the state of development of fundamental cognitive processes as well as intelligence itself, and to combine the two sets of constructs into a comprehensive approach to assessment of human mental ability. Such fundamental cognitive processes, or logical thinking modes, consist of generalizable ways of perceiving, thinking, learning, and problem solving that underlie effective functioning in a possibly infinite variety of applications and content domains (Feuerstein, Rand, & Hoffman, 1979; Haywood, 1986).

Within this transactional perspective, my use of the term *cognitive processes* is intended to include metacognitive activity as well. The following is a brief discussion of these two concepts (see Brooks & Haywood, 2003).

The adjective *cognitive* refers broadly to thinking and knowing, and specifically to the processes of systematic, logical thought. A cognitive curriculum in education, for example, is a program of education whose focus is on development of systematic processes of logical thinking. The goal of cognitive curricula is to have the children acquire a set of logic modes that will enable them to think systematically, logically, and effectively by using the logic modes that they have acquired. Such logic modes might include the structures and operations identified by Piaget (1970); for example, those that constitute criteria for accession to the developmental stage of concrete operations, including comparison; classification; class inclusion; seriation; transitivity; and temporal, spatial, and quantitative relations. They might also include the psychomotor developmental accomplishments identified and assessed by Uzgiris and Hunt (1975). They include as well such habit structures as self-regulation, management of multiple sources of information, systematic search strategies in gathering essential information, recognition of the existence and nature of problems, and insistence on logical evidence (Brooks & Haywood, 2003; Feuerstein, Rand, Hoffman, & Miller, 1980; Haywood, 2003). The term *metacognitive* has two aspects. The first is to focus attention on one's own thinking processes and to become acutely aware of the processes that one uses to make order out of the multitudinous stimuli that impinge on one's senses at any moment. Piaget (1970), for example, observed that children often engage in little conversations with themselves in which they ask themselves such questions as "Have I seen anything like this before?"; "How did I do this the last time I saw such a problem?"; "How do other people do this?"; "Do I have all the information I need?"; and "How else could it be done?" These are metacognitive questions. In its second meaning, the term *metacognitive* refers to specific strategies of thinking—"tricks," if you will—that people use to organize their thoughts, to improve their memory, to focus their attention, or to think through a problem logically. Thus, most people recognize that they are more likely to remember a list of items, such as a shopping list, if they repeat the list over and over. Some people recognize that they are more likely to remember all the items on their grocery shopping list if they organize the list by categories, such as fruits and vegetables, dairy products, frozen food, canned food, and meat, just as these items are classified in the supermarket. Strings of numbers are recalled more accurately if they are grouped, as with telephone numbers and social security numbers. These are metacognitive strategies designed to improve memory. (For this discussion in the context of cognitive–metacognitive education, see Brooks & Haywood, 2003.)

## *Motivation*

Over the past 40 years, my students and I have repeatedly demonstrated the power of task-intrinsic motivation to influence the efficiency and effectiveness of learning, as well as such variables as task persistence, willingness to work, exploratory behavior, recall of learned information, preference for self-monitored behavior, school achievement, and the magnitude of benefits from cognitive education. We have also demonstrated positive effects of cognitive education on task-intrinsic motivation, suggesting a bidirectional relation of these variables. For details of these and other studies on the relation of individual differences in intrinsic motivation to a variety of performance criteria, see Haywood (1968a, 1968b, 1971), Haywood and Burke (1977), Haywood and Dobbs (1964), Haywood and Switzky (1985, 1986), Haywood and Wachs (1966), Haywood and Weaver (1967), Switzky and Haywood (1974, 1991, 1992), and Tzuriel and Haywood (1984). For an integrative summary of cognition and intrinsic motivation from a transactional perspective, see Haywood and Switzky (1992) and Haywood, Tzuriel, and Vaught (1992).

On the basis of these and many other studies, I have suggested previously (Haywood, 1992) that motivation and cognition are related in the manner of a biological symbiosis. When added to intelligence, these variables, exerting their simple and combined effects in a transactional manner, constitute a potentially powerful superconstruct that can encompass much of what we know about individual differences in human ability.

One might then ask what makes this perspective transactional. One might observe first that each component (i.e., intelligence, cognition, and motivation) is influenced by the other two, but such an observation only makes the model interactive. One can observe further that the effects of each component are influenced both independently and summatively by the effects of the other two, and that observation makes the model more complexly interactive. The transactional nature is found in the observation that each effect of one component on another changes the power and direction of effect that is possible for the "affected" (acted upon) component, in the manner of a developmental lattice. Although both intelligence and cognitive processes are required for one to be an effective learner (not even to consider motivation), the essential cognitive processes will be acquired more efficiently, with fewer repetitions of experience, the higher one's intelligence. Once the most basic cognitive processes have been acquired and elaborated, the expression of intelligence becomes more facile, so that acquisition of the next stage of cognitive processes becomes even easier than was the initial stage. When intrinsic motivation is added to the mix, the active seeking of potentially enriching experiences is promoted, and that leads, in its own turn, to more effective expression of intelligence and, ultimately, through vectors exerted by both the intelligence and the motivation components, to a faster growing, more accessible, and more readily applicable set of cognitive processes. Cognitive success leads in its turn to an enhanced task-intrinsic motivational orientation. Thus, there are at least three snowballs rolling at once, crossing each other's path, becoming changed by each crossing, and exerting different effects on the other snowballs each time they are changed by their encounters.

## New Methods of Assessment

It remains, then, to suggest what all of this says about how to assess individual differences in human ability. Here are some good beginning approaches, beginning with "fixing" the current methods by

1. adding process components;
2. taking seriously the important differences between intelligence and cognition;
3. abandoning exclusive emphasis on normative models, that is, those that stress comparing each individual's performance with the average performance of a normative group; and
4. adopting a more developmental approach by departing from the fixed trait notion and paying more attention to how ability develops than worrying about what it is or is not.

These steps would go a long way toward construction of a launching pad for the next period in the history of intellective assessment and psychometrics, but in my view they would not go far enough. I advocate a greater emphasis and much more research on what has come to be called *dynamic assessment*. A dynamic assessment approach would assess individual differences in ability, not merely in intelligence, by including assessment of specific cognitive and metacognitive processes as well as intrinsic motivation. The primary goal of testing would change from prediction to discovery of methods for defeating pessimistic predictions. The basic tactic would change from assessment of typical performance to assessment of maximal performance, as is already done, for example, in neuropsychological assessment. The referent for each person's performance would change from that of other, similar persons in a normative sample to that person's own performance, both before and after mediation (teaching) and across domains of knowledge and performance (i.e., from between persons to within persons). The basic datum would change from unassisted performance to the difference between unassisted and mediated performance (i.e., performance given teaching within the test). This means that we would be able to take advantage of currently available statistical methods by which the basic datum is the slope of a change curve. I suggest that the element of teaching within the test is the very essence of dynamic assessment. It is essential to establish a baseline, and that is precisely what one does with static, normative assessment. It is important to point out that in this paradigm, failure is not permitted; that is, examinees are given the help they need to succeed, and all instances of mediation are recorded. Tests given after teaching suggest the potential level of performance that would be possible following correction of some of the obstacles to each person's performance. Data from dynamic assessment would then be derived from these sources:

- Performance: baseline, same as static test scores.
- Mediation given: amount and kinds.
- Response to mediation: generalization and transfer.
- Obstacles to good performance, including ignorance, poor motivation, undeveloped cognitive processes, cultural difference: that is, many variables previously identified only as "test error."
- The zone of proximal development: that is, the difference between unassisted and assisted scores (performance and potential).

For descriptions and summaries of dynamic assessment, see Haywood (1997, 2001), Haywood and Lidz (in press), Haywood and Tzuriel (1992, 2002), Lidz (1987), Lidz and Elliott (2001), Sternberg and Grigorenko (2001, 2002), and Tzuriel (2001a, 2001b).

# References

Anastasi, A. (1965). *Individual differences*. New York: Wiley.
Benbow, C. P., & Lubinski, D. (Eds.). (1996). *Intellectual talent: Psychometric and social issues*. Baltimore: Johns Hopkins University Press.
Binet, A., & Simon, T. (1905). Le développement de l'intelligence chez enfants [The development of intelligence in children]. *Année Psychologique, 12,* 119–244.

Brooks, P. H., & Haywood, H. C. (2003). A preschool mediational context: The Bright Start curriculum. In A. S. S. Hoon, L. P. K. Hoon, & O. -S. Tan (Eds.), *Mediated learning experience with children: Applications across contexts* (pp. 98–132). Singapore: McGraw-Hill Education (Asia).

Feuerstein, R., Rand, Y., & Hoffman, M. B. (1979). *Dynamic assessment of learning potential.* Baltimore: University Park Press.

Feuerstein, R., Rand, Y., Hoffman, M. B., & Miller, R. (1980). *Instrumental enrichment.* Baltimore: University Park Press.

Gardner, H. (1999). *Intelligence reframed: Multiple intelligences for the 21st century.* New York: Basic Books.

Haywood, H. C. (1968a). Motivational orientation of overachieving and underachieving elementary school children. *American Journal of Mental Deficiency, 72,* 662–667.

Haywood, H. C. (1968b). Psychometric motivation and the efficiency of learning and performance in the mentally retarded. In B. W. Richards (Ed.), *Proceedings of the First Congress of the International Association for the Scientific Study of Mental Deficiency* (pp. 276–283). Reigate, Surrey, England: Michael Jackson.

Haywood, H. C. (1971). Individual differences in motivational orientation: A trait approach. In H. I. Day, D. E. Berlyne, & D. E. Hunt (Eds.), *Intrinsic motivation: A new direction in education* (pp. 113–127). Toronto, Ontario, Canada: Holt, Rinehart & Winston.

Haywood, H. C. (1986). On the nature of cognitive functions. *The Thinking Teacher, 3,* 1–3.

Haywood, H. C. (1989). Multidimensional treatment of mental retardation. *Psychology in Mental Retardation and Developmental Disabilities, 15*(1), 1–10.

Haywood, H. C. (1992). The strange and wonderful symbiosis of motivation and cognition. *International Journal of Cognitive Education and Mediated Learning, 2,* 186–197.

Haywood, H. C. (1997). Interactive assessment. In R. Taylor (Ed.), *Assessment in mental retardation* (pp. 103–129). San Diego, CA: Singular.

Haywood, H. C. (2001). What is dynamic "testing?" A response to Sternberg and Grigorenko. *Issues in Education, 7,* 201–210.

Haywood, H. C. (2003). Bright Start: Metakognitív, mediációs tanulási tanterv kisgyermekek számára [Bright Start: A metacognitive mediated learning curriculum for young children]. *Erdély Pszichológiai Szemle, 4,* 145–174.

Haywood, H. C., & Burke, W. P. (1977). Development of individual differences in intrinsic motivation. In I. C. Uzgiris & F. Weizman (Eds.), *The structuring of experience* (pp. 235–263). New York: Plenum Press.

Haywood, H. C., & Dobbs, V. (1964). Motivation and anxiety in high school boys. *Journal of Personality, 32,* 371–379.

Haywood, H. C., & Lidz, C. S. (in press). *Dynamic assessment in practice.* New York: Cambridge University Press.

Haywood, H. C., & Paour, J. -L. (1992). Alfred Binet (1857–1911): Multifaceted pioneer. *Psychology in Mental Retardation and Developmental Disabilities, 18,* 1–4.

Haywood, H. C., & Switzky, H. N. (1985). Work response of mildly mentally retarded adults to self versus external regulation as a function of motivational orientation. *American Journal of Mental Deficiency, 90,* 151–159.

Haywood, H. C., & Switzky, H. N. (1986). Intrinsic motivation and behavior effectiveness in retarded persons. In N. R. Ellis & N. W. Bray (Eds.), *International review of research in mental retardation* (Vol. 14, pp. 1–46). New York: Academic Press.

Haywood, H. C., & Tzuriel, D. (Eds.). (1992). *Interactive assessment.* New York: Springer-Verlag.

Haywood, H. C., & Tzuriel, D. (2002). Applications and challenges in dynamic assessment. *Peabody Journal of Education, 77,* 38–61.

Haywood, H. C., Tzuriel, D., & Vaught, S. (1992). Psychoeducational assessment from a transactional perspective. In H. C. Haywood & D. Tzuriel (Eds.), *Interactive assessment* (pp. 38–63). New York: Springer-Verlag.

Haywood, H. C., & Wachs, T. D. (1966). Size-discrimination learning as a function of motivation-hygiene orientation in adolescents. *Journal of Educational Psychology, 57,* 279–286.

Haywood, H. C., & Weaver, S. J. (1967). Differential effects of motivational orientation and incentive conditions on motor performance in institutionalized retardates. *American Journal of Mental Deficiency, 72,* 459–467.

Horn, J. L., & Noll, J. (1997). Human cognitive capabilities: Gf-Gc theory. In D. P. Flanagan & J. L. Genshaft (Eds.), *Contemporary intellectual assessment: Theories, tests, and issues* (pp. 53–91). New York: Guilford Press.
Hunt, J. McV. (1963). Motivation inherent in information processing and action. In O. J. Harvey (Ed.), *Motivation and social interaction* (pp. 35–94). New York: Ronald.
Lidz, C. S. (Ed.). (1987). *Dynamic assessment.* New York: Guilford Press.
Lidz, C. S., & Elliott, J. (Eds.). (2001). *Dynamic assessment: Prevailing models and applications.* New York: JAI/Elsevier.
Piaget, J. (1970). *Genetic epistemology.* New York: Columbia University Press.
Stanley, J. C., Keating, D. P., & Fox, L. H. (Eds.). (1974). *Mathematical talent: Discovery, description, and development.* Baltimore: Johns Hopkins University Press.
Sternberg, R. J. (1985). *Beyond IQ: A triarchic theory of human intelligence.* New York: Cambridge University Press.
Sternberg, R. J., & Grigorenko, E. L. (2001). All testing is dynamic testing. *Issues in Education, 7,* 137–170.
Sternberg, R. J., & Grigorenko, E. L. (2002). *Dynamic testing: The nature and measurement of learning potential.* New York: Cambridge University Press.
Switzky, H. N., & Haywood, H. C. (1974). Motivational orientation and the relative efficacy of self-monitored and externally imposed reinforcement systems in children. *Journal of Personality and Social Psychology, 30,* 360–366.
Switzky, H. N., & Haywood, H. C. (1991). Self-reinforcement schedules in persons with mild mental retardation: Effects of motivational orientation and instructional demands. *Journal of Mental Deficiency Research, 35,* 221–230.
Switzky, H. N., & Haywood, H. C. (1992). Self-reinforcement schedules in young children: Effects of motivational orientation and instructional demands. *Learning & Individual Differences, 4,* 59–71.
Tzuriel, D. (2001a). Dynamic assessment is not dynamic testing. *Issues in Education, 7,* 237–249.
Tzuriel, D. (2001b). *Dynamic assessment of young children.* New York: Kluwer Academic/Plenum.
Tzuriel, D., & Haywood, H. C. (1984). Exploratory behavior as a function of motivational orientation and task conditions. *Personality and Individual Differences, 5,* 67–76.
Uzgiris, I. C., & Hunt, J. McV. (1975). *Assessment in infancy: Ordinal scales of psychological development.* Urbana: University of Illinois Press.
Vygotsky, L. S. (1988). *Thought and language.* Cambridge, MA: MIT Press. (Original work published 1934)
Woodcock, R. W. (2002). New looks in the assessment of cognitive ability. *Peabody Journal of Education, 77*(2), 6–22.

*Part III*

# Learning and Cognition

# 11

# The Transfer Index as a Precursor of Nonhuman Language Research and Emergents

*James L. Pate*

In this chapter, I discuss several topics and, more important, the relations among them. In particular, I discuss classic learning theory and Rumbaugh's transfer index, nonhuman primate language research, particularly the research at the Language Research Center of Georgia State University (Atlanta), and emergents. I have collaborated with Duane Rumbaugh on publications involving the transfer index and language research, and we have discussed a variety of questions regarding emergents. First, I present a very brief history of classic learning theory as an historical context for some of the claims that I make.

## A Very Brief History of Learning Theory

In the golden age of learning theory, some of the major theorists (Clark Hull, Kenneth Spence, Edward Tolman, Edwin Guthrie, and B. F. Skinner—although Skinner might disagree with that classification) made some very strong, yet seldom-stated, assumptions. Implicitly, they assumed that any stimulus within an organism's sensory range could be used as the functional stimulus in a learning situation. They also assumed that any response could be used as the to-be-learned response, at least if the response were in the appropriate domain. Glandular and smooth muscle responses could be conditioned classically, and striated muscle responses could be conditioned instrumentally. In many of the theories, all learning was of one or the other of those two forms, although Miller (e.g., Miller & Dicara, 1967) spent much of his career attempting to demonstrate that even this elementary distinction was unnecessary. Furthermore, for most theorists, the animal to be studied was irrelevant; one could use mice, rats, chickens, pigeons, apes, or humans, depending on the availability of the creatures, because the same principles of learning applied to all of them.

Despite the popular but simple taxonomy of classical and instrumental conditioning, some theorists claimed that there were numerous forms of learning.

Tolman (1949), for example, claimed that there are six forms of learning, whereas Hull (1943) assumed that a single set of principles applied to all forms of learning. Even when more than one form of learning was postulated, the forms tended to be listed without any necessary or implied hierarchy. In one sense, the disagreements about what is learned are more important than the disagreements about the number of forms of learning, and I emphasize this distinction in the present discussion.

For many of the theorists, association was critically important. According to Hull (1943), an organism learns a connection or an association between a stimulus and a response; those are the elements of Hull's concept of habit. This approach to learning theory is similar to John Locke's association of ideas, except that Hull replaced ideas with stimuli and responses. Later British empiricist philosophers, such as James Mill, accepted the association of ideas and allowed for the combining of the ideas into larger units. However, nothing new was introduced by the combining of ideas into large units, which might be thought of as similar to a chemical mixture. Tolman (1967), in contrast, claimed that animals learn cognitive maps, *sign significate* (Tolman's term), and expectancies. It often was claimed that animals learn what leads to what, which clearly is very different from learning an association between a stimulus and a response. Researchers performed many experiments in which they attempted to demonstrate that animals learned stimulus–response connections or learned what leads to what, but neither group convinced the other of the correctness or incorrectness of their approach.

In traditional learning theory research, only a limited number of measures of a few responses were obtained. Typically, response frequency, response amplitude, and resistance to extinction were measured, but there was little interest in assessing the type of response that might be learned. In a demonstration in a learning class that I taught many years ago, a rat in a Skinner box developed a strange way of pressing the bar in an operant conditioning chamber: It more or less laid down under the bar and pulled the bar down. Given that I was attempting to demonstrate basic principles of instrumental conditioning in the operant chamber, I quickly instituted an extinction procedure and then retrained the rat to press the bar in the standard way.

All behavior was assumed to involve those simple associations or a composite of those simple associations. Complex behavior, according to this idea, was merely a composite of simple units that had been learned, an idea that is similar to James Mill's notion that nothing new is created in the combination of ideas.

Regardless of which theoretical conception was accepted, most of the researchers assumed that all animals in a given situation would learn the same thing, either stimulus–response connections or cognitive maps or some other simple entity. Other theorists formulated slightly different concepts with radically different names. In all of these theories, individual differences were ignored or were treated as a nuisance variable, but differences among organisms of the same species, and certainly differences among species, are critically important. I mention only a few instances of organism by situation interactions in which organisms exposed to the same situation and the same contingencies learn different connections, maps, or relations.

Early in my career, I investigated spontaneous alternation behavior in rats (Pate & DeLoache, 1970) and in children (Pate & Bell, 1971), but behaviors in the two species were not compared. Although no learning was involved, the importance of the situation by organism interaction was important. In a typical situation, a T-maze or a cross-maze was used, and the organism was given two trials. The question concerned the animal's choice in various configurations of the maze. For example, on Trial 1, the left arm of the maze might be white, and the right arm might be black. On Trial 2, the arms were reversed. On that second trial, the animal could repeat the response from the first trial and gain exposure to a new stimulus or could make a different response, which would lead to the same stimulus as on the first trial. Will animals alternate responses, or will they alternate stimuli? The proportion of response-alternating animals differed from the proportion of stimulus-alternating animals, but neither proportion was close to 100%. In support of my present thesis, I emphasize that animals that received the same exposure behaved in different ways. Some animals apparently were response alternators, and others were stimulus alternators. Or to use the terminology from that research domain, some animals sought titillation, but others avoided tedium.

In human memory research, Underwood, Ham, and Ekstrand (1962) performed an informative compound stimulus study. *Trigrams* (sets of three letters) were presented on colored cards that were paired with numbers, and the subjects, as they were called then, learned to say the number when a card was presented. After the subjects reached criterion, they responded to colored cards without trigrams or to trigrams printed on white cards. In general, subjects performed at a higher level when colored cards without trigrams were presented than when trigrams on neutral cards were presented, but not all of them responded in that way. The subjects could learn to respond to both the color and the trigram, to only the color, to only the trigram, or to neither. Underwood et al. wanted to determine which stimulus element was more effective or which stimulus was easier to learn. Although the number of correct responses was greater for the colors than for the trigrams, the subjects responded correctly to the color for some paired associates and responded correctly to the trigram for other paired associates. It is clear that the organisms, humans in this case, were selective; they did not learn to respond to all of the stimulus elements in a compound stimulus.

The findings that I have mentioned were surprising to learning researchers, but they should not have been. Most of the investigators were professors who should have known this principle. Students exposed to the same textbook and lecture material learn very different parts of the material even when study time is controlled. Some students learn material by rote and do not understand the basic ideas, but others learn and understand the fundamental ideas that are presented. Still others fail to learn the material in any way, although even the most unfortunate students usually learn some of the material. It is important to note that the learners in these examples do not differ simply in the amount that is learned; rather, some students learn only some ideas without understanding, and others understand the ideas and the relations among them.

## Transfer Index

Rumbaugh (1971) developed the transfer index, which was based, in part, on Harry Harlow's learning set idea, as a measure that could be used to compare behavior across species. In Harlow's (1949) research, a pair of items was presented to an organism for six trials. One of the items was positive, and one was negative. After six trials, a new pair of objects was presented. After several pairs had been presented, the organism began to respond correctly sooner than it had on the first few sets. Thus, the organism had learned that one of the two objects was positive; if the first one chosen was not the positive one, then the animal changed to the second object on the second trial. Obviously, the organisms were not learning stimulus–response connections; rather, they were learning relations among stimuli and behavior or were learning something akin to Tolman's (1967) expectancies.

Rumbaugh (1971) used a discrimination reversal task in which the valences of the two stimuli were changed after the organisms had reached one of two criteria. If the animals simply were learning stimulus–response pairings, then the reversal should have been more difficult for the organisms that had reached the 84% criterion than for those that had reached the 67% criterion. However, some species performed better on the reversal trials after 84% training than after 67% training, which means that these animals had learned something other than a stimulus–response association. They had learned the relation between the stimuli. Rumbaugh and I have summarized much of that research in publications in which we claimed that what the animal learns depends on the animal as well as on the training method and the situation (Rumbaugh & Pate, 1984).

The transfer index appears to be a quantification of how much is learned, the traditional measure in studies of learning, because it is based in part on the percentage of correct responses. However, it is, in fact, an assessment of what is learned rather than of how much is learned, at least for some species. When the amount learned is held constant, what has been learned differs from one species to another. Consequently, the transfer index studies are precursors of studies of language learning at the Language Research Center and of emergents. In language learning studies, the interesting attributes concern the types of things that are comprehended and that are produced; for example, we should be concerned with syntax, semantics, or pragmatics rather than with how many words or sentences or pages have been produced. Researchers should have known this even in the golden age of learning theory because important learning is not simply more responding or stronger responding or even responding that is more resistant to extinction; it is learning new and different responses or relations. For operant conditioning, the interest should have been in the initial production of the bar press rather than in the frequency with which bar presses occurred or in the pattern of the bar presses. To some extent, Skinner dealt with this concept in his studies of successive approximations and shaping (Skinner, 1938), but shaping did not receive the attention that schedules of reinforcement received, and in general, shaping was only a prelude to the study of interesting phenomena for the operant conditioning psychologists.

Although psychologists now call some theories of the 1930s through the 1950s learning theories, many of those theorists were attempting to construct theories of behavior. Unfortunately, they omitted many critical aspects of human and nonhuman behavior. Specifically, there was little discussion of problem solving, reasoning, thinking, talking, comprehending, knowing, or understanding. In part, those theorists and the investigators who based their research on their theories chose to investigate those attributes of behavior that could be measured easily. Their theories, despite the elegance of some of them, were not inclusive and concerned only the simplest behaviors. Most of that research was performed in a setting that allowed only a few options for responding.

To take this idea out of the laboratory, one should realize that adaptation does not involve doing something frequently; it often involves doing something that is different and novel. Problem solving and reasoning are forms of behavior that are adaptive, but generally speaking, making many responses or making responses rapidly is unlikely to be adaptive.

## Language

Language obviously is a complex form of activity, and attempts to account for language as mere composites of simple responses have not been successful. Early attempts to explain language were similar to John Locke's or James Mill's approaches to explaining the association of ideas. In those theories, the combinations of ideas produced nothing new, and in early studies of language, the sequences of units (e.g., phonemes, morphemes, and words) contained nothing other than the attributes of those units. More recent attempts to understand language are perhaps more like John Stuart Mill's approach than like the one favored by his father. Or, one might argue that Wilhelm Wundt's concept of creative synthesis should be used in accounting for complex behaviors. In the more modern form, the sequences of units contain more information and have different implications than do the individual units that were combined.

In language learning studies with chimpanzees or with other nonhuman organisms, one example is far more impressive and convincing than are tables of correct responses. Many examples of innovative uses of the lexigram keyboard to solve problems or to achieve some goals have been published. For example, the chimpanzee Lana's producing "Go Behind Room" when the feeding machine was malfunctioning is more important than was the production of many routine sentences. The use of lexigrams to request from another chimpanzee a tool that is needed to obtain food that has been placed in a container is more impressive than is the rapid production of many bar presses or even many correct choices in a T-maze. The use of symbols to indicate the number of stimuli in a display is another example of a type of behavior that is novel rather than simply more frequently occurring.

## Emergents

In one summary of transfer index research, Rumbaugh and I claimed that there was a shift from associative learning to mediational learning as one moves up the scale from prosimians to apes (Rumbaugh & Pate, 1984, p. 236). Some training paradigms produce very different results, depending on which animals are trained. Thus, one needs to know what animal is involved as well as what training paradigm is being used. For emergents, one must be concerned with the behavior of an organism rather than with learning per se.

Rumbaugh and I (Rumbaugh & Pate, 1984) perhaps were too conservative in claiming that the shift from basic learned responses to what Rumbaugh now calls emergents was a moderate one; Rumbaugh has made a more dramatic claim in his recent essay (Rumbaugh, 2002) about emergents than we made in 1984 or than Tolman made in 1967. Emergents are behaviors that cannot be explained on the basis of specific learning. They may involve, as does language, the combining of previously learned units, but they must be a compound rather than a mixture, to use again the distinction between the ideas of James Mill and those of John Stuart Mill. If the units merely are combined without any integration, the combination will not qualify as an emergent. Emergents clearly are related to earlier concepts such as the "aha" phenomenon in problem solving, generalization that involves relations rather than simple stimulus similarity, and some language attributes.

Obviously, I have allocated less space to the discussion of emergents than I have to the discussion of other concepts. This is not an implication about the importance of the concepts. Rather, the concept of emergents is relatively new, although Rumbaugh and I referred to the idea as early as 1984 (Rumbaugh & Pate, 1984, p. 237), and thus, I simply have less to write about it than about the other concepts.

In summary, Duane Rumbaugh's contributions to psychology are extensive and will be recognized even more as time goes by, because his contributions are emergents rather than mere rephrasing of old ideas.

## References

Harlow, H. F. (1949). The formation of learning sets. *Psychological Review, 56,* 333–342.
Hull, C. L. (1943). *Principles of behavior: An introduction to behavior theory.* New York: Appleton-Century-Crofts.
Miller, N. E., & Dicara, L. (1967). Instrumental learning of heart rate changes in curarized rats: Shaping and specificity to discriminative stimulus. *Journal of Comparative and Physiological Psychology, 63,* 12–19.
Pate, J. L., & Bell, G. L. (1971). Alternation behavior of children in a cross-maze. *Psychonomic Science, 23,* 431–432.
Pate, J. L., & DeLoache, J. S. (1970). Brightness and direction as cues for spontaneous alternation behavior. *Psychonomic Science, 18,* 27–28.
Rumbaugh, D. M. (1971). Evidence of qualitative differences in learning processes among primates. *Journal of Comparative and Physiological Psychology, 76,* 250–255.
Rumbaugh, D. M. (2002). Emergents and rational behaviorism. *Eye on Psi Chi, 6,* 8–14.
Rumbaugh, D. M., & Pate, J. L. (1984). Primate learning by levels. In G. Greenberg & E. Tobach (Eds.), *Behavioral evolution and integrative levels* (pp. 221–240). Hillsdale, NJ: Erlbaum.
Skinner, B. F. (1938). *The behavior of organisms.* New York: Appleton-Century.
Tolman, E. C. (1949). There is more than one kind of learning. *Psychological Review, 56,* 144–155.
Tolman, E. C. (1967). *Purposive behavior in animals and men.* New York: Appleton-Century-Crofts.
Underwood, B. J., Ham, M., & Ekstrand, B. (1962). Cue selection in paired-associate learning. *Journal of Experimental Psychology, 64,* 405–409.

# 12

# Monkeys Making a List: Checking It Twice?

## F. Robert Treichler

A seminal chapter by Rumbaugh, Washburn, and Hillix, published in Pribram and King's *Learning as Self-Organization* (1996), has stimulated renewed consideration and investigation of behaviors that reflect sophisticated information processing by infrahuman organisms. In that chapter, Rumbaugh et al. proffered the concept of the *emergent* as an adjunct to the traditional Skinnerian categories of respondent and operant behaviors. Inclusion of this category has served to enhance the range of integrative processes involved in behavioral modification (defined most inclusively). As conceived, emergents represent a class of operations too complex to be attributable to simple association and not typically involving specific reinforced responses. Although the behavioral category is applicable to all phyla, attempts to investigate performances that might fit the term *emergent* have most often been conducted with animals. Because uniqueness of human cognitive capability is widely presumed, testing for generality of the emergent has hinged on demonstrating that complex tasks, like concept formation, relational learning, or analogs of insight, are within the behavioral repertoire of animals. This context has also posed the adjunct issue of whether appropriate integrative support for such sophisticated learning is present in nervous systems less highly evolved than that of humans.

### Comparative Cognition

Of course, the history of comparative psychology is replete with controversy about parallels between human and animal performances, sometimes extensively grounded in anecdotal information (see Robert Boakes's *From Darwin to Behaviourism* [1984] for especially informative treatment of this issue). From a more contemporary perspective, such parallels have been treated in an array of thoughtful books and reviews that reveal the many-faceted issues included under the topic of comparative cognition. Among the books that provide some of the best overall treatments are Heyes and Huber's (2000) edited volume, *The Evolution of Cognition*, and proceedings of the Eighth Kent Psychological Forum, edited by Fountain, Bunsey, Danks, and McBeath (2002), titled *Animal*

*Cognition and Sequential Behavior*. The edited book *Animal Cognition in Nature* (Balda, Pepperberg, & Kamil, 1998) also contains excellent treatments of specific topics. One further overview deserves special comment: Wasserman's (1997) retrospective and prospective article titled "The Science of Animal Cognition: Past, Present, and Future."

An abiding issue treated in the above works is the extent of commonality among those integrative functions that support the behavioral capabilities of various species. Polarized views on this topic range from the radical behaviorist outlook that all phylogenetic differences are quantitative rather than qualitative to the kind of "cognitive ethology" that has postulated animal "minds" (Ristau, 1991). Although none of the contributors to the above-noted works espoused those extremes, the treatments do entail differential emphasis on conditioning vis-à-vis cognitive explanations of behavioral phenomena. Perhaps the most specific admonition about comparative study in these works was provided by Hulse's (2002) keynote to the Kent Forum. He stated that it seemed dangerous to attribute "mind" to animals (e.g., Cheney & Seyfarth, 1990; Griffin, 1976) because, he contended, such an approach represented regression to the anthropomorphism and analogical reasoning of the late 19th century. He suggested instead that a coalescence of experimental psychology, behavioral ecology, and modern evolutionary study could operate synergistically to enhance the understanding of comparative issues.

It is in answer to that same concern for better understanding of comparative contributions that the concept of the emergent may have its greatest utility. It provides a category of behavior that, although certainly acquired, may be generated covertly within the integrative network of an individual. Further, emergence occurs without a specific reward history for discrete responses yet typically relies on developing experience with a class of performances. The expression of an emergent may sometimes have a surprising aspect to the observer, but performances based on emergents do require specific and appropriate test environments for their manifestation. By virtue of each of these characteristics, the emergent offers a concept of especial value to those who wish to comprehend mechanisms related to memory organization.

## Organizing Factors in Monkey Memory

The issue of organization of memory has become a central theme of the research ventures in which my students and I are involved at the Primate Laboratory at Kent State University. This developed from earlier work testing monkeys on various acquisitions and retentions while examining the prospects for recovery of function after surgically induced cortical damage. However, memory assessment per se became the primary interest when we discovered that for both humans and animals (like monkeys) that serve as subjects for extended periods, realistic memory tasks require conjoint processing of both retained and contemporary information. Perhaps an appropriate analog of such real retention is provided by school curricula in which instruction on several different subject areas is given within a limited time frame, and the student is later asked to retain the different topics, integrate them with previous information on related issues, and recall or recognize them in a performance evaluation.

Translating this into the primate laboratory setting, we began to study characteristics of *concurrent discrimination* performance (as defined by Hayes, Thompson, & Hayes, 1953) using the Wisconsin General Testing Apparatus to train monkeys on several different two-choice object problems within the same session. Subsequently, retention was measured as a function of numbers of concurrently trained problems, the amount of time and information interposed between acquisition and retention, and the correspondence between numbers of objects paired in training and relationships between the trained and tested lists. This series of studies (Treichler, 1984; Treichler & Petros, 1983; Treichler, Petros, & Lesner, 1981; Treichler, Wetsel, & Lesner, 1977) revealed three overall properties that characterized monkeys' multiple problem memory. First, macaques were extremely proficient at retaining many (nearly 100) object problems over substantial time periods (90 or more days). Second, acquisition was much enhanced by contrasts in the frequency of appearance of correct and incorrect items. For example, a task with 4 correct and 16 incorrect items was much easier than one with 4 correct and 4 incorrect items, despite the larger number of items to be processed in the former. Apparently, differential frequency of appearance (i.e., often vs. rarely) served as a discriminative cue for long-term memory just as Parkinson and Medin (1983) had shown when using delayed matching to sample to test short-term memory. Third, retained information seemed organized into categorical lists, with the most salient feature of categorizing being designation of objects as members of a "correct list" (aided by the just-noted property of differential frequency of appearance). It did not matter how pairs were presented in training sessions, how long since trained, or what object another object had been paired with; if an object was correct in original training, it would most likely be selected again because it was on the "good" list. We even tried to build tasks in which it was advantageous to remember fewer incorrect items (with 4-to-1 ratios, like 32 vs. 8), but the pattern of training errors revealed that the monkeys always learned the items in the big, correct set. This is not to say they could not remember an incorrect list, because they would use such information to enhance performance on a transfer test in which a formerly incorrect list had its items paired with new and novel correct objects. In that case, monkeys showed less error than was required for acquisition of a control task using all new objects, and the associated error patterns were distinctively unlike those of learning a new task.

The summary message from the concurrent discrimination tests was that monkeys imposed organization on the information they retained to support long-term multiple problem memory. As a consequence of that finding, we considered a next appropriate step to be requiring memory organization that was more complex than the dichotomous (correct–incorrect) list assignment of concurrent discrimination but not radically different in procedure from the tests familiar to our subjects.

## Memory for Serial Information

Because of the training that our monkeys had undergone, the optimal candidate for our purposes would be a multiple-problem task that required organization yet could be administered in the simultaneous, two-choice format. Fortunately, such an instrument exists. It is the five-term (A–B–C–D–E) series task designed to assess the phenomenon of transitive inference in both developmental and, perhaps more generally, comparative settings. Under this procedure, each two adjacent list members are trained as two-choice problems, and subsequently, novel pairings of any two items may be presented in test. The procedure's origins are suggested to be in Bryant and Trabasso's (1971) tests of Piaget's (1928) contention that children younger than 7 years were unable to comprehend transitivity in a series. The task has been widely used with vertebrates in part because it incorporates some especially helpful features. First, the requisite information may be imparted by training on just four two-choice problems (termed the *premise pairs*, A–B, B–C, C–D, D–E), and, second, it provides a postacquisition test pair (B vs. D) that requires choice between items that have served as both rewarded and nonrewarded alternatives during original learning. That property helps to stem the criticism that test choices reflect mere preference or aversion of list-end items (as would be the case with a four-item list). Using the five-item task for our evaluations allowed progression from simple concurrent discrimination to the potentially more complex organization required to support concurrent, conditional performance. Here, *conditional* means that correct choice depends on the appearance of an item with some specific other item.

The earliest application of the five-item, concurrent, conditional test to primate performance was done by McGonigle and Chalmers (1977), who based their procedures on Bryant and Trabasso's (1971) developmental tests. Much to the credit of McGonigle and Chalmers, they have continued to be active investigators of serial integration issues with both animals (McGonigle & Chalmers, 1992) and humans (McGonigle & Chalmers, 1984) during the now more than quarter century since their pioneering work (for a recent report, see McGonigle & Chalmers, 2002). Throughout that period, the four premise-pair, five-item test has been applied to a wide variety of different species. These include (merely as representative examples) pigeons (von Fersen, Wynne, Delius, & Staddon, 1991), corvids (Bond, Kamil, & Balda, 2003), rats (Davis, 1992), squirrel monkeys (McGonigle & Chalmers, 1992), rhesus macaques (Rapp, Kansky, & Eichenbaum, 1996), and chimpanzees (Gillan, 1981).

A somewhat different procedure for testing serial list organization, the simultaneous chaining technique, has been used by D'Amato and Colombo (1988) with cebus monkeys and by several investigators associated with Terrace (e.g., Swartz, Chen, & Terrace, 1991) with macaques. Performance on simultaneous chaining requires specifically ordered serial choice of visual stimuli that all appear at trial initiation. The usual procedure requires touching first one and then the other ordered stimulus items with reward administered only on completion of correct choices on the entire list. The training usually starts with two stimuli and then adds further items until a list of some designated length is achieved. Evaluation of the properties of retained information (especially its organization) is achieved by presenting limited subsets of the ordered list items. If choice sequences on these tests conform to the trained order, especially when the subsets do not contain list-end items, it is presumed that the items are retained as an organized serial list. Incorrectly ordered test selections suggest that some other aspect of training (perhaps reward frequency or stimulus chaining) is responsible for retention.

## Organization Versus Association

As might be expected when using a variety of tests with different species under different procedures, there have been disagreements in interpreting outcomes. Seemingly, the core of controversy lies, once again, in the contrast between memory mechanisms supported by simple conditioning (associative) operations and those based on development of organized, symbolic representations of retained information. However, some interrelationships among and between the contrasting views may be noted. A comparative component has been involved in several interpretations. Proponents of organizational explanations have typically based their conclusions on data from primates; conversely, pigeon investigators have more frequently favored conditioning interpretations.

One intriguing explanation of the way pigeons treat the five-item (concurrent conditional) task has been offered by von Fersen et al. (1991). Their value transfer theory contended that a value accrues to any item (assume it to be the letter $D$ in a 5-item list) as a consequence of its rewarded choice, but this item's value also increases as a consequence of its paired appearance with another rewarded item, even though the original item (D) might now be nonrewarded. Thus, if the designated (D) item appears in a later test as a member of a novel pair, the likelihood of its selection is determined by the sum of both its direct and conditioned reinforcing (i.e., transferred) values. According to this explanation, value (and concomitant choice at test) is determined by an item's proximity to the always-rewarded "good" end of the serially ordered list. Choice on that critical test trial (the B vs. D one) is determined by whether the A or E end of the list had been rewarded consistently. If E was at the good end, then D would be chosen because it had appeared in training with E, and B would be rejected because it had appeared with the never-rewarded A.

Couvillon and Bitterman (1992) offered an alternative explanation grounded in stochastic learning and derived from the contention that von Fersen et al.'s (1991) training procedures had provided differential reward frequency for the B and D alternatives. In reply, Wynne, von Fersen, and Staddon (1992) agreed that the original outcome derived from differential reward of list items, but they maintained that this was an appropriate source of the values they postulated. So, the several investigators agreed that pigeons chose the alternative commensurate with list organization on the crucial B–D pair, but their choices were guided by conditioning properties established during acquisition and did not depend on remembering an organized list. Subsequently, Wynne (1998) reviewed a variety of other studies of serial organization using both animal and human subjects and contended that conditioning mechanisms are nearly universally sufficient for interpreting outcomes of tests with concurrent, conditional tasks.

However, when primate investigators have used the concurrent, conditional task to train lists, they have most often contended that retention was best characterized as an internal representation of the ordered series. McGonigle and Chalmers's (1977) early work with squirrel monkeys suggested that prospect, and the chimpanzee studies of Gillan (1981) and Boysen, Berntson, Shreyer, and Quigley (1993) were unanimous in their appeal to organizational interpretations. Rapp et al. (1996) reached a similar conclusion when testing rhesus monkeys after concurrent conditional training. Of course, proponents of conditioning views might claim that stimulus chaining, conditioned (secondary) reinforcement, or differential frequency of reward produced the observed outcomes. However, organizational interpretations garnered renewed support by results from Terrace's primate research (Swartz et al., 1991) and, most especially, his contrasts of primate and pigeon memory characteristics (Terrace, 1993; Terrace & McGonigle, 1994). The test instrument used in this research was the simultaneous chaining technique noted previously, and its virtue was that serial learning by this method was difficult to attribute to simple conditioning. Recall that in simultaneous chaining a list of $n$ items appears all at once in positions that vary between trials, and subjects must choose all items in correct order before being reinforced. So, no specific set of motor responses is required, and no informative reward occurs until the whole correct sequence is completed. That means that retention must incorporate the serial properties of the entire array. With the requirement that each item be picked at the appropriate time and place in the sequence, interitem association explanations encounter difficulty in accounting for such performances.

The contrast of monkey and pigeon test performances on these tasks has been especially revealing. Monkeys make accurate choices (i.e., ones conforming to the serial order) on subset test pairs composed of items from the middle of the list (B–C, B–D, or C–D), whereas pigeons are at chance on these and exceed chance only when an end-anchored item serves as a pair member. Further, differences in response latency patterns between pigeons and monkeys indicate distinctively different memorial encoding. Monkeys show latency differences related to original list position characteristics (latencies of first and second to-be-selected items vary as a function of item location in the trained series), whereas pigeons show no such differences. Presumably, they do not process five-item list position information. One further difference is that monkeys improve their acquisitions of simultaneously chained lists as a function of experience (Chen, Swartz, & Terrace, 2000), whereas no such improvement is seen in pigeons (Terrace, 1987). Accordingly, it has been proposed that there are qualitative differences in the way monkeys and pigeons represent serial information (Terrace, 1993) in simultaneous chaining tests.

The performances revealed when monkeys were tested by means of simultaneous chaining suggested that organized retention involved a covert, linear representation of the trained list. D'Amato and Colombo (1990) termed that kind of representation a *symbolic distance effect*, and if symbolic distances are generated when monkeys learn and retain the serial information imparted by the simultaneous chaining procedure, perhaps they do something analogous when a series is trained by means of the concurrent, conditional procedure. Unfortunately, results obtained from five-item concurrent conditional tests incur the criticism of being interpretable in terms of alternative conditioning explanations. However, the concurrent, conditional procedure does provide retention test trials in the same format as training trials (simultaneous two-choice problems), whereas the simultaneous chaining procedure requires subset tests with fewer items than were present in training. Although salience of that difference remains to be determined, correspondence of acquisition and retention methods is usually considered advantageous.

## Linking Lists

Our initial attempts to evaluate primate list organization were grounded in some presumptions about the symbolic distance effect and the possibility that the five-item, concurrent, conditional task might generate serial representation in a manner similar to that of simultaneous chaining. The training methodology used by Gillan (1981), sometimes called *successive inclusion*, seemed to appropriately meet our list requirements. Under that procedure, premise pairs are introduced in successive steps that provide new pairings only after competent performance is achieved on all previous problems (and concurrent combinations of problems). Both Gillan's (1981) and Boysen et al.'s (1993) chimps learned the required premise pairs by this procedure and revealed test choices indicative of list organization on the critical B versus D test, although in neither case was this accomplished by all subjects. However, Boysen et al.'s manipulations included some additional tests on novel pairs that supported their claim of list integration.

Our rationale was based on the premise that if concurrent conditional training yielded an ordered representation of otherwise neutral objects (i.e., ones without discernible relational differences), then it should be possible to join such lists by simply providing postacquisition training on a pair made up of an upper end object from one list and a lower end object from the other list. In this training, the formerly always correct, upper end item would be incorrect, and the formerly always-incorrect object would now be rewarded. So, if two 5-item lists had been learned, perhaps such training would link the independent lists to yield a serially ordered 10-item list. Of course, the test for this serial ordering would entail presenting novel two-choice combinations of items from both lists as well as novel pairs from within trained lists along with the original premise pairs.

Treichler and Van Tilburg's (1996) initial attempt to link 5-item lists trained under the concurrent conditional method seemed successful. Macaques appeared to retain and integrate the several phases of training as a 10-item serial list, and they did this in spontaneous fashion. Six days of postlinkage testing provided all possible combinations of object pairings (both from the different lists and from within originally trained lists), and equivalent performances were seen everyday. Right from the onset of testing, the overall level

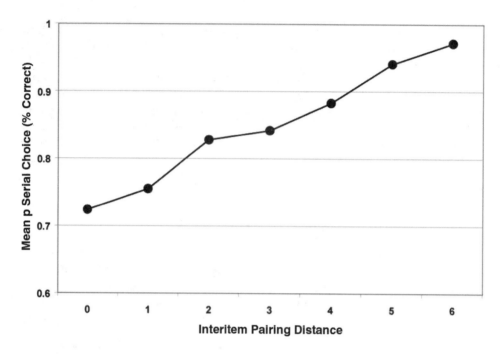

**Figure 12.1.** Mean proportion (p) serial choice (% correct) for test pairs representing the various possible serially separated locations in a 10-item list (interitem pairing distance).

of conformity to choices consistent with a 10-item list was quite high (80.9%). However, one aspect of the small amount of error detected in those tests was notable. The worst performances were obtained on those premise pairs that animals had seen throughout original learning and not on the never-before-experienced novel pairings (either within or between lists). Accordingly, it seemed clear that linking two lists posed little difficulty for the monkeys; rather, they found it harder to choose between two objects that were close to one another in the 10-item series. Projecting a systematic extension of that contrast led to generating the function included as Figure 12.1. This graph displays the likelihood of conformity to a 10-item (combined) serial list plotted as a function of the different numbers of items between objects paired in test. The resultant monotonic function reveals systematic reduction of error with increasing serial distance between elements paired in the test and has led to the conclusion that the retention by monkeys of conditional information is integrated into a serially ordered array. Additional support for representational arrays was provided by Treichler and Van Tilburg's (1996, Experiment 2) demonstration that monkeys would ignore infrequent violations of a retained serial order but could modify specific choices if the incidence of reward was sufficiently likely. Thus, in overview, generation of serial representation appeared to provide a common basis for retention under both concurrent conditional and simultaneous chaining training procedures.

## Knowledge of List Positions

Having concluded that representation was a common property in serial retention in monkeys, it was possible to ask what specific list characteristics were remembered. The usual answer from simultaneous chaining studies, both in reviews (see Conway & Christiansen, 2001) and from a variety of empirical tests (Chen, Swartz, & Terrace, 1997; Orlov, Yakovlev, Amit, Hochstein, & Zohary, 2002; Orlov, Yakovlev, Hochstein, & Zohary, 2000; Terrace, Son, & Brannon, 2003), has been that monkeys encode and retain knowledge of the specific ordinal position of each list item. It is as if they attach ordered labels or ratings to each item in a list and use these labels as selection guides in subsequent tests that require choices between items from different positions in the remembered array. Chen et al. (1997) demonstrated this characteristic when monkeys initially learned several different four-item lists and were retested after stimuli had been recombined into different lists. In test, the several original lists were reconstituted, but item order within lists was systematically manipulated so that stimuli either remained in their original serial positions (e.g., as first, second) or were shuffled into different serial positions (e.g., a second-position item became a fourth-position item). With maintained positions, there were few errors, but lists that mixed positions yielded as much error as learning a new list. Similarly, Terrace et al. (2003) trained macaques on four seven-item lists and then gave subset tests that provided novel pairings of items from both within or between the different lists. All pairings were of stimuli that had appeared at different positions in original learning, and on 94% of the test trials, subjects' choices conformed to the relative serial locations from initial training (selections on within- and between-list test pairs did not differ). Accordingly, the monkey's representation of a series seemed to depend on remembering the sequential positions of items in the list.

However, the list-linking result from Treichler and Van Tilburg (1996) was interpreted in a manner suggesting that list representations involved serial integration more complex than simply knowing the ordinal positions of list members. Treichler and Van Tilburg claimed that after two lists were acquired, isolated training that reversed the reward contingencies of just two appropriate end-items from 5-item lists would yield an inclusive serial ordering of all 10 stimuli. Of course, that interpretation did not deny that list positions were remembered, but, if correct, it meant that training on just one reconstituted pair informed subjects that list position alone was no longer the appropriate determinant for test choice. Further, this message about changing the ordering rules had to be imparted to the monkeys before retention tests were conducted. Did that really happen?

We became suspicious, especially after reexamining the test procedure of the 1996 study. Those daily retention tests had presented all of the possible object pairs from a 10-item list, arranged as $A < B < C < D < E+$ and $F < G < H < I < J+$ followed by link training, $E-F+$. The tests were composed of the possible 10 pairs from each of two 5-item lists and the 25 between-list pairings (although the pair that had been trained as a link was excluded). That meant 44 trials were administered each day. Of those 44, we noted that 30 would be correct if the item at the higher ordinal position were picked, whereas only 9 pairs required selection of an item from a lower ordinal position (5 pairs contained items from the same ordinal positions). Because the test provided a threefold bias in favor of alternatives at list positions indicating linear organization, it now seemed appropriate to question whether the original linkage interpretation might have been based on a procedural artifact.

## Linking Lists Revisited

At this juncture, a clear mandate in terms of research rationale was posed. The challenge was to assess the independent effects of two factors that might support serial list integration. These were (a) knowledge of list position and (b) list attachment training effects. Only after demonstrating the unique or interactive contributions of these influences could the factors underlying serial list performance on concurrent, conditional tasks be segregated. For this venture, the availability of four by now highly sophisticated performers on conditional discriminations in the Wisconsin General Testing Apparatus was most fortunate. In keeping with Chen et al.'s (2000) view of the role of experience in developing facility on serial tasks, our monkeys were sufficiently skilled that they could learn conditional lists without the drudgery of independent training-to-criterion on every premise pair and combination of pairs (as required by Gillan's [1981] successive inclusion procedure). With this greater procedural efficiency, we could train all four of the requisite premise pairs for a five-item list in just 15 daily 40-trial sessions. The tactical approach was to acquire three such lists (see Figure 12.2) and then, to ensure that all were remembered, give reminder training that presented each five-item list four times but with the orders of list appearance intermixed (for procedural details, see Treichler, Raghanti, & Van Tilburg, 2003).

On completion of the reminder phase, one of two different linkage-training conditions was imposed. Depending on assignment to condition, monkeys were initially either trained on the two pairs that linked between both the first and second (E–F+) and the second and third (J–K+) of the 5-item lists they had learned, or, alternatively, they got a control condition of no linkage training. In counterbalanced order, all monkeys experienced each linkage condition, immediately followed by its associated retention test. Influences of the linking manipulations were compared in 15-day retention tests that presented all of the possible pairings of objects from the three 5-item lists and rewarded choice of whichever of the 2 items was represented by the letter farther on in the alphabet. Choices were scored for their conformity to a 15-item list, serially ordered from A to O. Test pairings represented several different prospectively influential factors, including serial distance between items paired at test and contrasts of list positions from original acquisition. Each 3-day test block included a few appearances of the original premise pairs, some of the novel pairings from within original lists, and many pairings of items from the different 5-item lists. The pairings from different lists are depicted in Figure 12.2 to show the number and variety of pairs that represented the contrasts of list positions from original learning. Note that this does not include 18 instances per block that provided between-lists pairs

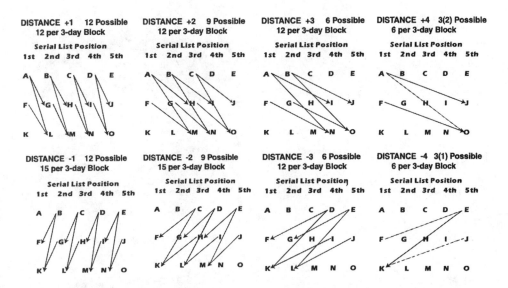

**Figure 12.2.** Various levels of the serial gap distance variable as represented by 60 possible interlist test pairings of objects from different original acquisition locations. Categories (+4 to –4) reflect the numbers of serial positions that intervene at test between positions from original learning as a 5-item list. Also noted are the actual and possible number of pairs for each test block at each serial gap. The rewarded object within any pair is indicated by an arrowhead. From "Linking of Serially Ordered Lists by Macaque Monkeys (*Macaca mulatta*): List Position Influences," by F. R. Treichler, M. A. Raghanti, and D. N. Van Tilburg, 2003, *Journal of Experimental Psychology: Animal Behavior Processes, 29*, p. 214. Copyright 2003 by the American Psychological Association.

from the same ordinal positions (termed E, for equivalent). The various position contrasts were considered to represent levels of an independent variable termed *serial gap distance* (that could vary from –4 to +4) and have used these levels to assess the influence of knowledge of list position.

When overall posttreatment performances on the retention tests were compared, the linking condition yielded greater conformity to the 15-item list on the first several 3-day test blocks, but those differences disappeared with continued testing or, perhaps more accurately, training. Further, the advantages associated with the linking procedure were seen only on the between-lists pairs. Pairs composed of objects from within lists, although relatively few in number, actually generated lesser error in test. The indication from this complex of results was that presence versus absence of linkage was not a simple effect and that other factors were influencing the outcomes of the retention–integration tests.

Of course, a major issue determining the rationale for this study was concern for the way knowledge of list positions from original acquisition influenced tests of list organization. To address that issue, the error proportions (relative to a 15-item list) associated with each of the possible serial gap distances were compared after both linked and nonlinked treatment conditions. Although these measures are exclusively derived from between-lists test pairs,

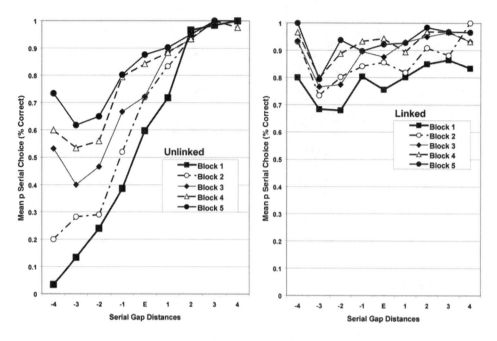

**Figure 12.3.** Mean proportion (p) serial choice (relative to a 15-item list) at each possible serial gap distance. Functions at each successive block are shown for both the linked (right panel) and unlinked (left panel) conditions. Serial gap distances range from –4 (with correct object four positions behind its training location) through 0 (objects at same list locations) to +4 (with correct object four positions ahead of training locations).

Figure 12.3 reveals a striking contrast of linkage effects at the different levels of the gap-distance variable. If pairs were maintained in their relative list positions from acquisition (the positive values), they were rarely missed, especially under the nonlinked condition. In corollary fashion, initial negative gap-distance values after nonlinkage also indicated that choices were determined by remembered list positions, but in that case, choices were wrong because they did not conform to the 15-item list. However, these graphs also indicate that there were changes in performance over the five 3-day test blocks, although the changes were almost exclusively associated with the nonlinked condition. That outcome is consistent with the monkeys coming to treat the objects as a 15-item ordinal list after experiencing reward for choice of items designated as being farther along in the list.

Accordingly, our serial gap-distance results provide unqualified support for Terrace and his colleagues' (Chen et al., 1997; Swartz et al., 1991; Terrace, 2001) emphasis on list position information as a choice determinant when items from independent lists are paired. However, just as emphatically, these results show that learning about a link between lists brings about a distinctive reorganization of serial memory. Figure 12.3 indicates that with linkage, choices reasonably closely conform to a 15-item list right from the onset of testing. The nature of list reorganization was to have link training tell animals that it was no longer appropriate to base their choices on old position information. Rather, they were informed that objects should now be treated as a single, 15-item serial list.

With that result in hand, questions could be raised as to whether monkeys really did encode a representational array that included all of the items from the three lists and, further, whether the nature of that representation could be generalized. To address that issue, measures analogous to ones used when assaying serial integration in the Treichler and Van Tilburg (1996) study were generated. It is possible to categorize every test pair by the number of items that intervene between the serial locations of its two objects in the 15-item list (e.g., a 2nd vs. 14th position represents an interitem distance of 11). Then, the extent of conformity to a 15-item list associated with each of these interitem pairing distances was plotted graphically (except for the farthest separated items that occurred only once or twice). Figure 12.4A displays results for the first 3-day test block under both linked and nonlinked conditions, and these are strikingly different. With linkage, the function shows a systematic increase in choices conforming to the list as the distance between items paired in test becomes greater. Without linkage, there is slightly less error on the "0" distance (training or premise pairs), but performances at all intermediate distances are clearly not in conformity to a combined list. Only with the great disparities provided by pairings of objects at opposite ends of lists do anomalous conforming choices appear. Accordingly, performance after linkage accords well with the view that monkeys retain the list as a paralogical array wherein items occupy ordered list positions, and difficulty of choice at test is related to the magnitude of difference in location of any two paired objects. Error at test is thus a consequence of inability to resolve location differences in the remembered array. That property seems sufficiently universal to allow comparison among results from investigations that have looked at memory for lists of various lengths. Figure 12.5 provides a comparison of interitem pairing distance functions derived from several of the studies conducted at Kent State. Correspondence among slopes of these functions is so striking as to suggest that a common characteristic, relative distance between the items paired at test, is a powerful determinant of ordinal choice. Certainly, the finding that the poorest performances are seen on the original, highly trained premise pairs offers a stern point of refutation to most associative interpretations of serial list learning.

In addition to the above-noted linkage effects, it is evident that even without links, monkeys do merge the three initially learned lists into a 15-item series if provided a sufficient term of exposure to appropriate reward contingencies. Although some investigators favor nondifferential reward during retention tests, results from our laboratory have indicated that this procedure may elicit confounding of performances in sophisticated primates (for discussion of the issue, see Treichler & Van Tilburg, 1999). In the Treichler et al. (2003) investigation, selective reward allowed detection of the unique and gradual course of error reduction on the negative gap distances seen in Figure 12.3 and also revealed that for both linked and unlinked conditions, interitem pairing functions (like Figure 12.4B) became nearly identical by the fifth test block. Those characteristics imply that both linking pair training and exposure to relatively long-term reward contingencies yield representations of serial order that are not different from one another.

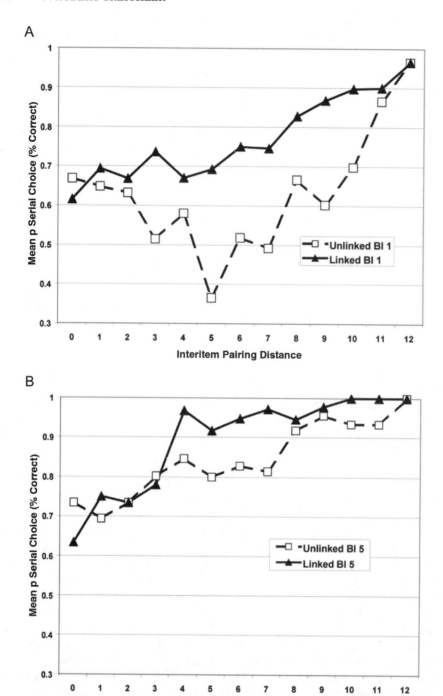

**Figure 12.4.** A. Mean proportion (p) serial choice at each possible interitem pairing distance (relative to a 15-item list) during the first 3 test days under linked and unlinked conditions. B. Mean proportion (p) serial choice at each possible interitem pairing distance (relative to a 15-item list) during the last 3 test days (13–15) under linked and unlinked conditions. Bl = block.

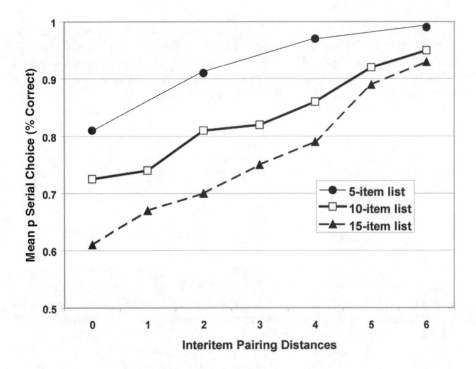

**Figure 12.5.** Mean proportion (p) serial choice as a function of interitem pairing distance for lists of 5, 10, or 15 items. The training pairs of each list are indicated by 0. Actual distances are shown for the 10-item list. Numbers for the 5-item list represent half its distances, and each point for the 15-item list represents the average of two adjacent distances.

Returning to the three-list linkage study's avowed goal of specifying the independent effects of knowledge of list positions and linkage training, that distinction now seems clarified. There was reaffirmation of Terrace's (2001) view that monkeys remember list positions from several independent lists and apply this information when tests pose novel object combinations. That property appears common to both simultaneous chaining and concurrent conditional procedures.

However, beyond that confirmation, this study provided clear, new evidence of a distinct and independent process derived from directional reward on just two, isolated, cross-list pairs. List-end linkage allowed three independently acquired lists to be merged almost immediately into an integrated 15-item series. Accordingly, the answer to the question posed (somewhat in jest) by the title of this chapter seems to be, "No, there's no need to check twice. Linking provides enough information to make a new list right away." Because revision of the array was so readily achieved, it seems unlikely that the earlier Treichler and Van Tilburg (1996) linkage result was an artifact of list position. Rather, both studies confirm that a unique organizational process allows attachments between ordered series. That demonstration, perhaps fortuitously, enhances the utility of concurrent, conditional discriminations as assessors of serial memory organization. Although the use of binary (two-choice) problems has been criticized as ineffective in differentiating serial organizational capabilities of monkeys and pigeons (DeLillo, 1996), linkage tests have neither been applied in that context nor attempted with pigeons. Similarly, although there is no a priori reason why linkage might not be attempted with simultaneous chaining, that too remains to be investigated. Serial linkage using concurrent, conditional, two-choice discriminations provides an especially useful instrument, in part because original training, linkage, and retention–integration tests all appear in corresponding formats. Fortunately, analogous organizational properties seem contributory to memory under both simultaneous chaining and concurrent conditional procedures.

## Organized Monkey Memory and Emergents

The phenomena treated in both our own and other reports of organization in multiple problem memory accord well with Rumbaugh's (2002) proposed emergent. When monkeys form learning sets, use frequency of appearance as a cue, or retain list positions as incidental learning, a class of behavior that is both covert and grounded in synthesis of experience seems involved. Further, if serial lists can be joined by simply informing monkeys about the relative values of end items, then the novel or surprising solutions characteristic of an emergent appear to contribute to such performances. The emergent also provides an integrative property that allows organizational characteristics as divergent as the gradual acquisition of list position information vis-à-vis the immediate appearance of list linking consequences to be viewed as related phenomena.

Of course, the ultimate merit of a descriptive concept, like the emergent, lies in its capacity to enhance comprehension and provide heuristic value. Rumbaugh's (2002) emergent shows promise in fulfilling these criteria. Certainly, the concerns it raises are abiding ones in the history of the discipline and seem likely to remain as such. Contemporarily, emergent phenomena are represented in comparative investigations that treat serial integration in species ranging from rodents (Roberts & Phelps, 1994) to adult humans (Colombo & Frost, 2001). Our own attempts to understand list linkage, especially in regard to its necessary and sufficient conditions (Treichler & Van Tilburg, 1999, 2002), reflect a continuing search for behaviors that emerge as a consequence of the complex interaction of experience and neuronal integration. Perhaps, appeal to the emergent may aid in alleviating Hulse's (2002) concern that comparative cognition might degenerate into a new mentalism. Rather, emergents can help to focus empirical efforts in support of Rumbaugh's plea for rational behaviorism.

## References

Balda, R. P., Pepperberg, I. M., & Kamil, A. C. (1998). *Animal cognition in nature: The convergence of psychology and biology in laboratory and field.* San Diego, CA: Academic Press.

Boakes, R. (1984). *From Darwin to behaviourism*. Cambridge, England: Cambridge University Press.

Bond, A. B., Kamil, A. C., & Balda, R. P. (2003). Social complexity and transitive inference in corvids. *Animal Behaviour, 65,* 479–487.

Boysen, S. T., Berntson, G. G., Shreyer, T. A., & Quigley, K. S. (1993). Processing of ordinality and transitivity by chimpanzees (*Pan troglodytes*). *Journal of Comparative Psychology, 107,* 208–215.

Bryant, P. E., & Trabasso, T. (1971, August 13). Transitive inferences and memory in young children. *Nature, 232,* 456–458.

Chen, S., Swartz, K. B., & Terrace, H. S. (1997). Knowledge of the ordinal position of list items in rhesus monkeys. *Psychological Science, 8,* 80–86.

Chen, S., Swartz, K. B., & Terrace, H. S. (2000). Serial learning by rhesus monkeys: II. Learning four-item lists by trial and error. *Journal of Experimental Psychology: Animal Behavior Processes, 26,* 274–285.

Cheney, D. L., & Seyfarth, R. M. (1990). *How monkeys see the world: Inside the mind of another species*. Chicago: University of Chicago Press.

Colombo, M., & Frost, N. (2001). Representation of serial order in humans: A comparison to the findings with monkeys (*Cebus apella*). *Psychonomic Bulletin and Review, 8,* 262–269.

Conway, C. M., & Christiansen, M. H. (2001). Sequential learning in non-human primates. *Trends in Cognitive Sciences, 5,* 539–546.

Couvillon, P. A., & Bitterman, M. E. (1992). A conventional conditioning analysis of "transitive inference" in pigeons. *Journal of Experimental Psychology: Animal Behavior Processes, 18,* 308–310.

D'Amato, M. R., & Colombo, M. (1988). Representation of serial order in monkeys (*Cebus apella*). *Journal of Experimental Psychology: Animal Behavior Processes, 14,* 131–139.

D'Amato, M. R., & Colombo, M. (1990). The symbolic distance effect in monkeys (*Cebus apella*). *Animal Learning and Behavior, 18,* 133–140.

Davis, H. (1992). Transitive inference in rats (*Rattus norvegicus*). *Journal of Comparative Psychology, 106,* 342–349.

DeLillo, C. (1996). The serial organisation of behaviour by non-human primates: An evaluation of experimental paradigms. *Behavioural Brain Research, 81,* 1–17.

Fountain, S. B., Bunsey, M. D., Danks, J. H., & McBeath, M. K. (2002). *Animal cognition and sequential behavior*. Boston: Kluwer Academic.

Gillan, D. J. (1981). Reasoning in the chimpanzee: II. Transitive inference. *Journal of Experimental Psychology: Animal Behavior Processes, 7,* 150–164.

Griffin, D. R. (1976). *The question of animal awareness: Evolutionary continuity of mental experience*. New York: Rockefeller University Press.

Hayes, K. J., Thompson, R., & Hayes, C. (1953). Concurrent discrimination learning in chimpanzees. *Journal of Comparative and Physiological Psychology, 46,* 105–107.

Heyes, C., & Huber, L. (Eds.). (2000). *The evolution of cognition*. Cambridge, MA: MIT Press.

Hulse, S. H. (2002). Perspectives on comparative cognition. In S. B. Fountain, M. D. Bunsey, J. H. Danks, & M. K. McBeath (Eds.), *Animal cognition and sequential behavior* (pp. 3–19). Boston: Kluwer Academic.

McGonigle, B. O., & Chalmers, M. (1977, June 23). Are monkeys logical? *Nature, 267,* 694–696.

McGonigle, B. O., & Chalmers, M. (1984). The selective impact of question form and input mode on the symbolic distance effect in children. *Journal of Experimental Child Psychology, 373,* 525–554.

McGonigle, B. O., & Chalmers, M. (1992). Monkeys are rational! *Quarterly Journal of Experimental Psychology, 45B,* 189–228.

McGonigle, B. O., & Chalmers, M. (2002). The growth of cognitive structure in monkeys and men. In S. B. Fountain, M. D. Bunsey, J. H. Danks, & M. K. McBeath (Eds.), *Animal cognition and sequential behavior* (pp. 269–314). Boston: Kluwer Academic.

Orlov, T., Yakovlev, V., Amit, D., Hochstein, S., & Zohary, E. (2002). Serial memory strategies in macaque monkeys: Behavioral and theoretical aspects. *Cerebral Cortex, 12,* 306–317.

Orlov, T., Yakovlev, V., Hochstein, S., & Zohary, E. (2000, March 2). Macaque monkeys categorize images by their ordinal number. *Nature, 404,* 77–80.

Parkinson, J. K., & Medin, D. L. (1983). Emerging attributes in monkey short-term memory. *Journal of Experimental Psychology: Animal Behavior Processes, 9,* 31–40.
Piaget, J. (1928). *Judgment and reasoning in the child.* London: Routledge & Kegan-Paul.
Rapp, P. R., Kansky, M. T., & Eichenbaum, H. (1996). Learning and memory for hierarchical relationships in the monkey: Effects of aging. *Behavioral Neuroscience, 110,* 887–897.
Ristau, C. A. (Ed.). (1991). *Cognitive ethology: The minds of other animals.* Hillsdale, NJ: Erlbaum.
Roberts, W. A., & Phelps, M. T. (1994). Transitive inference in rats: A test of the spatial coding hypothesis. *Psychological Science, 5,* 368–374.
Rumbaugh, D. M. (2002). Emergents and rational behaviorism. *Eye on Psi Chi, 6,* 8–14.
Rumbaugh, D. M., Washburn, D. A., & Hillix, W. A. (1996). Respondents, operants and emergents: Toward an integrated perspective on behavior. In K. Pribram & J. King (Eds.), *Learning as self-organization* (pp. 57–73). Hillsdale, NJ: Erlbaum.
Swartz, K. B., Chen, S., & Terrace, H. S. (1991). Serial learning by rhesus monkeys: I. Acquisition and retention of multiple four-item lists. *Journal of Experimental Psychology: Animal Behavior Processes, 17,* 396–410.
Terrace, H. S. (1987, January 8). Chunking by a pigeon in a serial task. *Nature, 325,* 149–151.
Terrace, H. S. (1993). The phylogeny and ontogeny of serial memory: List learning by pigeons and monkeys. *Psychological Science, 4,* 162–169.
Terrace, H. S. (2001). Chunking and serially organized behavior in pigeons, monkeys and humans. In R. G. Cook (Ed.), *Avian visual cognition.* Retrieved April, 12, 2002, from http://www.pigeon.psy.tufts.edu/avc/terrace
Terrace, H. S., & McGonigle, B. O. (1994). Memory and representation of serial order by children, monkeys and pigeons. *Current Directions in Psychological Science, 3,* 180–189.
Terrace, H. S., Son, L. K., & Brannon, E. M. (2003). Serial expertise of rhesus macaques. *Psychological Science, 14,* 66–73.
Treichler, F. R. (1984). Long-term retention of concurrent discriminations. *Physiological Psychology, 12,* 92–96.
Treichler, F. R., & Petros, T. V. (1983). Interference characteristics in concurrent discrimination performance by monkeys. *Bulletin of the Psychonomic Society, 21,* 206–208.
Treichler, F. R., Petros, T. V., & Lesner, S. A. (1981). Preference effects on acquisition and retention of concurrent discriminations by rhesus monkeys. *Animal Learning and Behavior, 9,* 401–405.
Treichler, F. R., Raghanti, M. A., & Van Tilburg, D. N. (2003). Linking of serially ordered lists by macaque monkeys (*Macaca mulatta*): List position influences. *Journal of Experimental Psychology: Animal Behavior Processes, 29,* 211–221.
Treichler, F. R., & Van Tilburg, D. (1996). Concurrent conditional discrimination tests of transitive inference by macaque monkeys: List linking. *Journal of Experimental Psychology: Animal Behavior Processes, 22,* 105–117.
Treichler, F. R., & Van Tilburg, D. (1999). Training requirements and retention characteristics of serial list organization by macaque monkeys. *Animal Cognition, 2,* 235–234.
Treichler, F. R., & Van Tilburg, D. (2002). Premise-pair training for valid tests of serial list organization in macaques. *Animal Cognition, 5,* 97–105.
Treichler, F. R., Wetsel, W., & Lesner, S. A. (1977). Some characteristics of concurrent discrimination and its retention by monkeys. *Learning and Motivation, 8,* 213–228.
von Fersen, L., Wynne, C. D. L., Delius, J. D., & Staddon, J. E. R. (1991). Transitive inference formation in pigeons. *Journal of Experimental Psychology: Animal Behavior Processes, 17,* 334–341.
Wasserman, E. A. (1997). The science of animal cognition: Past, present, and future. *Journal of Experimental Psychology: Animal Behavior Processes, 23,* 123–135.
Wynne, C. D. L. (1998). A minimal model of transitive inference. In C. D. L. Wynne & J. E. R. Staddon (Eds.), *Models for action* (pp. 269–307). Hillsdale: Erlbaum.
Wynne, C. D. L., von Fersen, L., & Staddon, J. E. R. (1992). Pigeon inferences are transitive and the outcome of elementary conditioning principles: A response. *Journal of Experimental Psychology: Animal Behavior Processes, 18,* 313–315.

# 13

# Animals Count: What's Next? Contributions From the Language Research Center to Nonhuman Animal Numerical Cognition Research

*Michael J. Beran, Jonathan P. Gulledge, and David A. Washburn*

Research into animal numerical cognition has as long a history as almost any other topic in comparative psychology. Counting behavior in nonhuman animals was investigated through a series of stages (Rilling, 1993). First, anecdotes were collected by naturalists. Next came demonstrations on pets and animals in zoos. Finally, experiments were carried out by scientists who controlled the lives and histories of their animals. During each of these stages, differing views emerged as to the extent to which nonhuman animals were sensitive to numerosity. For example, George Romanes and C. Lloyd Morgan, two of the principal researchers in the new field of comparative psychology, took opposite views of animal counting. Romanes stated that animals could count, but Morgan did not agree. Morgan reduced the problem of counting to a problem of timing (Rilling, 1993).

A number of experimental investigations offered evidence that perhaps Romanes was correct. Arndt (cited in Rilling, 1993) demonstrated the successive counting procedure with pigeons in which the birds were trained to eat only five peas from a cup, and Koehler (1951) used this procedure with other species of birds. Mechner (1958) used a technique called the *fixed consecutive number* procedure, which is a modification of the fixed ratio schedule. Rats were reinforced for a response on Lever B after a varying number of presses on Lever A. Rats performed correctly on this task with some regularity for 4, 8, 12, and 16 responses. However, there was greater variability in performance as the

---

Research and development of this chapter was supported by National Institute of Child Health and Human Development Grant HD-38051 and by the College of Arts and Sciences of Georgia State University, Atlanta.

required number of presses increased. In addition, there was no control for response latency as a possible nonnumerical cue. Levinson and Findley (1967) reported that baboons learned to produce a number of tones that varied from one to five depending on a presented visual stimulus. However, there was no mention of controls for temporal cues, and there was no clear description of the procedure. Thus, early research described as "animal counting" often was controversial. Later, valuable strides were made to control for nonnumerical cues that might lead to the appearance of numerical competence. For example, Ferster (1964) trained two chimpanzees to select one of two binary "numbers" to match a presented form with a quantity of shapes on it. The chimpanzees were observed in an apparatus that was entirely mechanized so as to eliminate problems of cuing by experimenters. However, it took the chimpanzees hundreds of thousands of trials to reach a high level of performance.

More recent experimental paradigms have continued to provide data that demonstrate nonhuman animals' sensitivity to numerosity while controlling the potential confounding variables and cues that sometimes occurred during earlier research (for reviews, see Brannon & Roitman, 2003; Davis & Perusse, 1988). It has been reported that pigeons (Emmerton, 1998; Emmerton, Lohmann, & Niemann, 1997), squirrel monkeys (Anderson, Awazu, & Fujita, 2000; Terrell & Thomas, 1990; Thomas, Fowlkes, & Vickery, 1980), tamarins (Hauser, Tsao, Garcia, & Spelke, 2003), rhesus monkeys (Brannon & Terrace, 1998; Hauser, Carey, & Hauser, 2000; Washburn & Rumbaugh, 1991), orangutans (Call, 2000; Shumaker, Palkovich, Beck, Guagnano, & Morowitz, 2001), and chimpanzees (Beran, 2001, 2004a; Beran & Beran, 2004; Boysen & Berntson, 1995; Boysen, Berntson, Hannan, & Cacioppo, 1996; Boysen, Berntson, & Mukobi, 2001; Boysen, Mukobi, & Berntson, 1999; Dooley & Gill, 1977; Rumbaugh, Savage-Rumbaugh, & Hegel, 1987) made accurate relative numerousness judgments in which the animals selected between two or more sets of items on the basis of their relative quantitative difference. Rats (Burns, Goettl, & Burt, 1995; Capaldi & Miller, 1988), raccoons (Davis, 1984), rhesus monkeys (Hicks, 1956), a parrot (Pepperberg, 1987, 1994), and chimpanzees (Biro & Matsuzawa, 2001; Boysen & Berntson, 1989; Matsuzawa, 1985; Tomonaga & Matsuzawa, 2002) evaluated stimulus sets as to their exact numerosity. Additionally, nonhuman animals have attended to and responded to arithmetic manipulations on sets (Beran, 2001, 2004a, 2004b; Boysen & Berntson, 1989; Call, 2000; Hauser & Carey, 2003; Rumbaugh et al., 1987; Sulkowski & Hauser, 2001; Terrell & Thomas, 1990). In some cases, claims of enumerative processes analogous to human counting have been made for nonhuman animals (Beran & Rumbaugh, 2001; Beran, Rumbaugh, & Savage-Rumbaugh, 1998; Biro & Matsuzawa, 2001; Boysen & Berntson, 1989; Capaldi & Miller, 1988; Davis & Bradford, 1986; Matsuzawa, 1985; Pepperberg, 1987, 1994; Rumbaugh, Hopkins, Washburn, & Savage-Rumbaugh, 1989; Xia, Siemann, & Delius, 2000). Thus, little doubt remains that numerosity is a salient aspect of the environment for nonhuman animals. The debate now centers on the mechanisms used to represent numerosity as well as the capacities of different species. Such a discussion would be lengthy, and given the focus of the present volume, our emphasis in this chapter is on the numerical cognition research conducted in Rumbaugh's laboratory and its significance for our growing understanding of nonhuman animal numerical competence. We do this not to ignore the important contributions from other laboratories but to emphasize the questions that have guided our research and to emphasize the data that have emerged from our investigations.

## Animal Counting

Rumbaugh and his colleagues initially were interested in demonstrations that nonhuman animals could count. This area of research had a particularly contentious history, beginning with the claims made for the now-famous horse, Clever Hans. Clever Hans was considered a mathematical genius because he could stamp his hoof to indicate numerical answers to all manner of questions (Davis, 1993). The scientific community accepted the feats as real and conceded advanced numerical cognition to the animal. However, more rigorous testing indicated that Clever Hans was nothing more than an astute observer of human behavior. With this embarrassing realization, psychologists took a critical and often cynical theoretical position regarding later numerical research with animals. Despite this, later well-controlled experiments confirmed that many nonhuman animal species were capable of countinglike feats. Capaldi and Miller (1988) demonstrated that rats counted the number of reinforced runs they had made through a maze. This ability was robust across a number of experimental manipulations, and independent research programs have confirmed that rats may be capable of countinglike skills (Burns et al., 1995; Davis & Bradford, 1986). Pepperberg (1987, 1994) reported countinglike skills in an African Grey parrot. Boysen and Berntson (1989) reported that a chimpanzee, Sheba, selected Arabic numerals that equaled the number of candies that were presented on a given trial. The chimpanzee also viewed Arabic numerals and indicated the placard with that number of dots on it. Using a computerized apparatus, Matsuzawa (1985) reported that a chimpanzee, Ai, mastered numerical naming of one to six items when she was presented with a stimulus set and had to select the Arabic numeral that corresponded to the number of items in that set (see also Biro & Matsuzawa, 2001; Tomonaga & Matsuzawa, 2002).

It was within this environment that Rumbaugh and his colleagues began to examine nonhuman primate numerical cognition. In the first investigation, Rumbaugh et al. (1989) reported the computerized training of a chimpanzee, Lana, to count. Lana interacted with the computerized apparatus by moving a cursor on a computer monitor so as to make contact with various stimuli. In the training phases, Lana had to learn the relation between a target numeral and the subsequent production of a number of responses related to this numeral. In the final test phases, trials provided no visual feedback (although there was auditory feedback), and Lana had to rely on her memory of her previous selections to terminate a trial correctly. Rumbaugh et al. (1989) concluded that Lana's skill entailed a reasonable approximation of an enumeration act that separated the counted from uncounted boxes; that determined when the set to be counted, as declared by the target's value on a given trial, had been counted; and consequently, that determined when counting should be stopped. Her task was not one of matching a number label for each of several arrays.

Rather, it was one of producing selection-box removal responses in accordance with the numerical value of a given target number (Rumbaugh, Hopkins, Washburn, & Savage-Rumbaugh, 1993; Rumbaugh & Washburn, 1993).

Beran et al. (1998) also reported counting by a chimpanzee in a computerized test paradigm. This chimpanzee, Austin, first learned to select in order the Arabic numerals from 1 to 9 by again moving a cursor on a computer monitor. Dots rather than Arabic numerals then were presented to Austin, and he had to select those dots, one at a time, to match a presented Arabic target numeral from 1 to 4. Initially, the dots always were presented in the same positions. However, in the final test phase, dots were positioned randomly, eliminating position and motor movements as cues. Overall, Austin was correct on 77% of the trials with the target numbers 1 through 4 (95% correct on Target 1, 76% correct on Target 2, 73% correct on Target 3, and 60% correct on Target 4). Austin's performance for the first 100 trials with each target number also was significantly better than chance. When Austin was incorrect on the first presentation of a trial, the same trial configuration was presented again. Austin changed his selection pattern frequently on these correction trials (40% of the time on Targets 3 and 4) while still completing those trials correctly.

The type of enumeration task used with Lana and Austin has been designated as *constructive enumeration* (Xia et al., 2000) because the subjects must construct, by selecting one at a time, a specific quantity of items from a larger set of items. This type of task differs from the numerical task of *responsive enumeration*, in which the entire stimulus set makes up the relevant numerical property for a given trial as in the research of Boysen and Berntson (1989); Matsuzawa (1985; Biro & Matsuzawa, 2001); Pepperberg (1987, 1994); and Xia, Emmerton, Siemann, and Delius (2001). In constructive enumeration, the stimulus set differs from the relevant number of items needed to perform a trial correctly, and this relevant number of items is represented by an Arabic numeral or other symbol. In these computerized constructive enumeration tasks, a chimpanzee's decision to end a trial was likely the result of its continuous comparison of the number of items it had selected with the referent target Arabic numeral.

A remaining question for chimpanzee "counting" within this experimental paradigm concerned the potential mechanism underlying this countinglike performance. A longstanding suggestion was that numerical representation by nonhuman animals was not an exact, symbol-based system like that used by children older than 5 years of age but was based on an approximate mechanism for representing numerosity (e.g., Gallistel & Gelman, 2000). One such mechanism has been called the *accumulator*. This accumulator produces less exact representations of numerosity for increasingly larger set sizes and thus represents quantities as magnitudes with scalar variability, or variability in the number of items selected that increases as a function of the numeral's value. There is evidence that young human children and adult humans may use such a mechanism in their numerousness judgments (Barth, Kanwisher, & Spelke, 2003; Huntley-Fenner, 2001; Huntley-Fenner & Cannon, 2000; Whalen, Gallistel, & Gelman, 1999).

Counting, however, is an enumerative process in which individual tags with a stable application order are applied, in one-to-one correspondence with items in an array, until all items have been tagged. The resulting absolute numerosity (or cardinal number) represents the sum total of items that were enumerated (Gelman & Gallistel, 1978). Older human children typically do not show scalar variability in their counting. However, claims of counting by nonhuman animals have been controversial because scalar variability sometimes is evident and also because the demonstrated range of counting is usually rather small (fewer than 10 items). In many cases, nonhuman animal counting falls within the range of numerosities for which a nonenumerative process, called *subitization* (Kaufman, Lord, Reese, & Volkman, 1949; Mandler & Shebo, 1982; Trick & Pylyshyn, 1994), is invoked that consists of the labeling of the numerosity of an array without regard to the enumeration of individual items within the array.

To better determine the potential mechanism that chimpanzees used in the constructive computerized counting task described above, an additional study was conducted (Beran & Rumbaugh, 2001). Two chimpanzees, Lana and Mercury, used a joystick to collect dots one at a time on a computer monitor and then ended a trial when the number of dots collected was equal to the Arabic numeral presented for the trial. The chimpanzees performed well in collecting a quantity of dots equal to the numerals 1 to 7. An examination of the errors that were made indicated that the number of dots selected became more variable as the target numerals increased. At present, it is our belief (e.g., Beran & Rumbaugh, 2001) that chimpanzee counting in this type of experimental task appears to be based on a continuous representation of magnitude rather than a discrete representation of number in that the chimpanzees' memory for magnitudes associated with each numeral is increasingly imperfect for successively larger numerals. One thing that is true is that the chimpanzees' memory for the values of these numerals is long lasting. When tested after a period of 3 years during which the computerized task was not available, both Lana and Mercury performed at levels significantly better than chance in creating sets of dots to match Arabic numerals from the very first session (Beran, 2004b).

## Chimpanzee Numerousness Judgments of Visible Sets

Of additional interest in Rumbaugh's laboratory were studies of nonhuman primates' ability to make judgments of *relative numerousness* (differences between sets). Dooley and Gill (1977) showed that Lana could select the larger of two quantities of up to nine cereal pieces. These data coincided with findings emerging from other laboratories that nonhuman animals were sensitive to differences in quantity and could select the larger or smaller of two sets (e.g., Boysen & Berntson, 1995; Boysen et al., 1996; Terrell & Thomas, 1990; Thomas et al., 1980). Rumbaugh also worked with the chimpanzees Sherman and Austin to study what he and his colleagues called *summation*. Initially, from one to four candies were placed in each of two wells in a modified version of the Wisconsin General Test Apparatus. The performance for both chimpanzees was greater than 95% correct in selecting the larger set. Later, pairs of food wells were presented, and the chimpanzees selected one of the two pairs of wells. This was done to examine the chimpanzees' competence for summing the quantities in both wells. Correct performance increased across time for both chimpanzees. However, performance was highly dependent on the closeness in number of items in the pairs of wells. When the pairs differed by only one item (especially when the total number of items was large, as in 9 vs. 10), neither chimpanzee performed very well. However, the chimpanzees succeeded even when a common quantity was assigned to each well, when the items were randomly arranged as well as arranged in geometric patterns, and when one well within the pair contained no items at all (Perusse & Rumbaugh, 1990; Rumbaugh et al., 1987; Rumbaugh, Savage-Rumbaugh, & Pate, 1988). Rumbaugh and his associates concluded from these studies that chimpanzees performed basic summation operations. Whatever the nature of the operation, it was something that the chimpanzees brought to the task. The chimpanzees did not have to be trained to respond, and there was no correction procedure. In addition, the chimpanzees were rewarded on every trial with the candies they had selected.

## Rhesus Monkey Numerousness Judgments of Visible Sets and Arabic Numerals

Investigations of numerosity judgments have not been limited to chimpanzees at the Language Research Center. During the time that the chimpanzees were being presented with sets of food items, the rhesus monkeys at the Language Research Center also were being taught the meaning of Arabic numerals. This research program was designed to determine the rhesus monkeys' understanding of the ordinal (or relational) meaning of symbols for numerosity as well as the exact numerical values of those numerals. Washburn and Rumbaugh (1991) first tested two rhesus monkeys on relative numerousness judgments. White Arabic numerals were displayed on a computer monitor, and the monkeys responded by contacting those numerals with a cursor via manipulation of a joystick. The number of food pellets corresponding to the selected Arabic numeral were dispensed to the animals when a numeral was contacted. Each trial, therefore, was reinforced regardless of the numeral selected. The numerals 0 to 9 were used. However, certain pairs were withheld for later probe trials. One monkey, Abel, responded at levels significantly greater than chance to all trials including probe trials. That Abel responded to pairs of numerals never before presented at levels comparable to familiar pairings indicated that he was responding on the basis of the relative value of the numerals in the pairs. Further tests were conducted to determine whether these monkeys would exhibit the capacity for ordinal and transitive judgments (i.e., the recognition that 8 > 7, 7 > 6, then 8 > 6). Novel arrays of up to five numerals were presented. Both monkeys chose the largest number available from these arrays at levels significantly better than chance, although performance varied somewhat as a function of both the number of available numerals and the quantitative difference between the numerals.

In a later study, Washburn (1994) also reported that rhesus monkeys selected the larger of two arrays of items even when the items were Arabic numerals that were incongruous (i.e., the larger array was composed of the smaller numeral such as seven 1s vs. six 2s). However, Stroop-like interference (Stroop, 1935) did occur, and this again suggested that the Arabic numerals had accrued quantitative meaning for the monkeys.

In a more recent investigation (Gulledge, 1999), the monkeys at the Language Research Center again were required to make numerousness judgments. However, rather than simply judging the greater of a pair of Arabic numerals as in Washburn and Rumbaugh's (1991) study, the monkeys also were required to judge the greater of two random dot patterns as well as to complete probe trials that paired Arabic numerals with the random dot patterns. Three stimulus conditions were used in this experiment. On numeral–numeral trials, two different Arabic numerals (1 to 9) were presented on each side of the joystick-controlled cursor. On dot–dot trials, two different randomly determined arrays of 1 to 9 randomly positioned white dots were displayed, one array on each side of the cursor. During the probe trials, numeral–dot trials were presented in which a randomly selected Arabic numeral was presented on the screen with a randomly selected (but different) quantity of dots. Overall accuracy averaged 84% for numeral–numeral comparisons and 82% for dot–dot comparisons. Symbolic distance effects, in which comparisons with greater numeric differences were performed at higher levels, were present for both trial types. For the probe trials (numeral–dot), performance was significantly better than chance and comparable across all monkeys. With these data, one can examine what happens when a target value is repeated on consecutive trials of the same stimulus type (e.g., when a 5:2 trial is followed, by chance, by a 5:4 trial, or when dot arrays of the same pairings are presented on consecutive trials). Moreover, one can examine what happens when the target value is repeated on consecutive trials of different stimulus types (e.g., when a target numeral 7 on Trial $N-1$ is followed by a trial in which seven dots is the correct answer, or vice versa). Responses were faster and more accurate (relative to the baseline condition in which the values did not repeat) when a target value was repeated in a single stimulus type. It is interesting that responding also was facilitated when a numeral–numeral trial was followed by a dot–dot trial with the same target value. In contrast, responding was slower and less accurate when dot–dot trials were followed by numeral–numeral trials with the same target value. Thus, dots primed dots, numerals primed numerals, and numerals primed dots; however, dots did not prime numerals. This pattern of results suggests that the monkeys, like the chimpanzees, may use an analog representational mechanism to solve these mixed-stimulus-type probe trials (i.e., when they see the numeral 3, for example, they think something like "•••").

Several conclusions follow from these studies. First, the monkeys understand that Arabic numerals represent specific quantities of items, because the monkeys accurately interdigitated Arabic numerals and random arrays of dots, even on the first exposure to these trials. Knowledge that the number 6 is bigger than the number 5 and bigger than five dots, which in turn is bigger than the number 4 and four dots, is not predicted by the view that what the monkeys have learned about numerals is a complex matrix of paired associations (e.g., choose 5 in the presence of 4 but not when paired with 6). Rather, these data support the conclusion that the monkeys have acquired knowledge about the absolute number of things represented by each Arabic numeral and could, even on probe trials, accurately compare these represented quantities with visible arrays of dots. Second, the priming data indicate that the monkeys perform this task, including probe trials, by translating the numerals into analog mental representations to be compared with the visible arrays and not by counting visible arrays and comparing mental symbols. The priming data indicate that the monkeys were not counting in that the comparison of numerals (like those that would be produced by formal counting) was not the common representational medium on which performance was based. These data indicate that the monkeys, like the chimpanzees, know something about the numerical symbols with which they have experience. The number 4 means more than "greater than 3 and less than 5." It represents a quantity that can be compared directly and accurately with visible arrays of stimuli. This competence would seem to satisfy the cross-modal or cross-class requirement for concept of number (Davis & Perusse, 1988; Seibt, as cited by Davis & Perusse, 1988) in that it entails the accurate comparisons of visible quantities and numerical symbols.

## Numerousness Judgments of Nonvisible Sets

We have continued to examine summation and other forms of arithmetic processes in the numerousness judgments of the Language Research Center chimpanzees. However, in these studies, the sets to be compared are never visible in their entirety. Rather, the presentation of each set is sequential in that each item is placed into an opaque container, one at a time. Here, the animals must represent individual items as part of an increasingly larger accumulation within a given nonvisible set. In one study (Beran, 2001), two chimpanzees, Lana and Sherman, were given a choice of receiving one of a pair of presented quantities of candy pieces. The pairs were constructed through sequential presentation of individual candy pieces that were placed into one of two opaque containers. After all pieces were placed into the two containers, the chimpanzees then selected one container, and they received the contents of that container. In Experiment 1, all candy pieces to go into the first container were placed into it before any pieces were placed into the second container. In this experiment, the number of candy pieces placed into a given container ranged from one to eight, with the differences between the two sums ranging from one to four pieces. Both chimpanzees performed at very high levels. In Experiment 2, a quantity of candy pieces was placed into the first container, and a second quantity was placed into the second container. Then, additional candy pieces were added, one at a time, to the first container. Finally, additional candy pieces were added, one at a time, to the second container, and the chimpanzees then made their selections (e.g., between the comparisons 3 + 3 vs. 1 + 4). Experiment 3 involved the additional step of adding a third set to each container before the chimpanzees made a selection (e.g., the comparison 1 + 3 + 3 vs. 2 + 4 + 2). The chimpanzees again were successful in selecting the larger sets. Beran (2001) concluded that the chimpanzees could sum multiple, sequentially presented sets in such a way that they could remember which of the two sets contained the larger quantity of items.

Another question that emerged from this experimental paradigm was whether the chimpanzees remembered only the relative difference between the two sets (i.e., which set was bigger) or also the relative size of the sets (i.e., how many items were in each set). An experiment was conducted in which two sequentially presented sets again were placed into opaque containers. A third set in its entirety then was revealed and remained visible during the chimpanzees' selections. Thus, the chimpanzees had to choose among two different nonvisible, sequentially presented sets and a third, fully visible, alternative. Both chimpanzees reliably selected the largest of the three sets presented in this manner (Beran, 2004a). This indicated that the chimpanzees were not assessing the sequential presentations solely on the basis of the relative difference of the two sets. The chimpanzees also were assessing the relative magnitude of, at least, the larger of those two sets. If the chimpanzees did not know approximately how many items were in the nonvisible sets, they could not determine when one of those sets was the better selection than the visible set (or when the visible set was the best option).

Decisions about which container holds more items also are made across longer time periods by these chimpanzees. In a study by Beran and Beran (2004), the one-by-one accumulation of items in two hidden sets required 20 minutes for the full presentation. During each trial, the first experimenter entered the test area several times with a single banana, placed that banana into one of the two opaque containers outside of a chimpanzee's cage area, and then left the test area. The chimpanzee never saw more than one banana at a time, and once a banana was placed into a container, it remained out of sight for the rest of the trial. At the end of the 20 minutes, a second experimenter, who did not know the total number of bananas in the containers, entered the test area. The second experimenter pushed the two containers forward so that the chimpanzee could make a selection by touching one of the two containers, and the chimpanzee received the bananas that were in the selected container.

It is important to note that within a trial the placement of bananas into the containers alternated between the left and the right container. Neither set was presented in its entirety at one time or in a single sequence (as in many of the previous studies with these chimpanzees described above). Rather, the final number of bananas in a given container was not established until the final banana had been presented. The relation of the containers to each other (i.e., the difference in the number of bananas) changed throughout a trial. As such, the chimpanzees had to update their memory for the set sizes in each container numerous times throughout a trial.

Eight comparisons were presented: 1 versus 2, 1 versus 3, 2 versus 3, 2 versus 4, 3 versus 4, 5 versus 8, 5 versus 10, and 6 versus 10. The interval between the placement of each banana within a trial ranged from 1 minute 20 seconds to 10 minutes (intervals between placements were equal within a trial). All 4 chimpanzees selected the larger quantity at levels significantly better than chance for almost all comparisons (one animal was not better than chance for trials with a 3 vs. 4 comparison). An important subset of trials was that in which the last banana was placed into the container with the smaller total number of bananas. If the chimpanzees simply were remembering the final location at which a banana was placed, performance should have been at or below chance levels on these comparisons. However, all 4 chimpanzees were significantly better than chance on these comparisons. These data indicate that over extended periods of time, chimpanzees are sensitive to arithmetic operations (specifically, additive operations).

## Summary

Researchers in laboratories around the world have contributed to the understanding of the role of numerosity in the lives and the behavior of nonhuman animals. In addition, programs of research have illuminated the extent to which nonhuman animals can represent numerosity either through some internal mechanism or through the use of an arbitrary symbol system such as that used by humans. The Language Research Center has played a significant role in this area, and Duane Rumbaugh's belief that an understanding of numerosity, like language, might be within the realm of nonhuman primate capability has proved to be true.

## References

Anderson, J. R., Awazu, S., & Fujita, K. (2000). Can squirrel monkeys (*Saimiri sciureus*) learn self-control? A study using food array selection tests and reverse-reward contingency. *Journal of Experimental Psychology: Animal Behavior Processes, 26,* 87–97.

Barth, H., Kanwisher, N., & Spelke, E. (2003). The construction of large number representations in adults. *Cognition, 86,* 201–221.

Beran, M. J. (2001). Summation and numerousness judgments of sequentially presented sets of items by chimpanzees (*Pan troglodytes*). *Journal of Comparative Psychology, 115,* 181–191.

Beran, M. J. (2004a). Chimpanzees (*Pan troglodytes*) respond to nonvisible sets after one-by-one addition and removal of items. *Journal of Comparative Psychology, 118,* 25–36.

Beran, M. J. (2004b). Long-term retention of the differential values of Arabic numerals by chimpanzees (*Pan troglodytes*). *Animal Cognition, 7,* 86–92.

Beran, M. J., & Beran, M. M. (2004). Chimpanzees remember the results of one-by-one addition of food items to sets. *Psychological Science, 15,* 94–99.

Beran, M. J., & Rumbaugh, D. M. (2001). "Constructive" enumeration by chimpanzees (*Pan troglodytes*) on a computerized task. *Animal Cognition, 4,* 81–89.

Beran, M. J., Rumbaugh, D. M., & Savage-Rumbaugh, E. S. (1998). Chimpanzee (*Pan troglodytes*) counting in a computerized testing paradigm. *Psychological Record, 48,* 3–19.

Biro, D., & Matsuzawa, T. (2001). Use of numerical symbols by the chimpanzee (*Pan troglodytes*): Cardinals, ordinals, and the introduction of zero. *Animal Cognition, 4,* 193–199.

Boysen, S. T., & Berntson, G. G. (1989). Numerical competence in a chimpanzee (*Pan troglodytes*). *Journal of Comparative Psychology, 103,* 23–31.

Boysen, S. T., & Berntson, G. G. (1995). Responses to quantity: Perceptual versus cognitive mechanisms in chimpanzees (*Pan troglodytes*). *Journal of Experimental Psychology: Animal Behavior Processes, 21,* 82–86.

Boysen, S. T., Berntson, G. G., Hannan, M. B., & Cacioppo, J. T. (1996). Quantity-based interference and symbolic representations in chimpanzees (*Pan troglodytes*). *Journal of Experimental Psychology: Animal Behavior Processes, 22,* 76–86.

Boysen, S. T., Berntson, G. G., & Mukobi, K. L. (2001). Size matters: Impact of item size and quantity on array choice by chimpanzees (Pan troglodytes). *Journal of Comparative Psychology, 115*, 106–110.

Boysen, S. T., Mukobi, K. L., & Berntson, G. G. (1999). Overcoming response bias using symbolic representations of number by chimpanzees (Pan troglodytes). *Animal Learning and Behavior, 27*, 229–235.

Brannon, E. M., & Roitman, J. (2003). Nonverbal representations of time and number in animals and human infants. In W. Meck (Ed.), *Functional and neural mechanisms of interval timing* (pp. 143–182). New York: CRC Press.

Brannon, E. M., & Terrace, H. S. (1998, October 23). Ordering of the numerosities 1 to 9 by monkeys. *Science, 282*, 746–749.

Burns, R. A., Goettl, M. E., & Burt, S. T. (1995). Numerical discriminations with arrhythmic serial presentations. *Psychological Record, 45*, 95–104.

Call, J. (2000). Estimating and operating on discrete quantities in orangutans (Pongo pygmaeus). *Journal of Comparative Psychology, 114*, 136–147.

Capaldi, E. J., & Miller, D. J. (1988). Counting in rats: Its functional significance and the independent cognitive processes that constitute it. *Journal of Experimental Psychology: Animal Behavior Processes, 14*, 3–17.

Davis, H. (1984). Discrimination of the number three by a raccoon (Procyon lotor). *Animal Learning and Behavior, 12*, 409–413.

Davis, H. (1993). Numerical competence in animals: Life beyond Clever Hans. In S. T. Boysen & E. J. Capaldi (Eds.), *The development of numerical competence: Animal and human models—Comparative cognition and neuroscience* (pp. 109–126). Hillsdale, NJ: Erlbaum.

Davis, H., & Bradford, S. A. (1986). Counting behavior by rats in a simulated natural environment. *Ethology, 73*, 265–280.

Davis, H., & Perusse, R. (1988). Numerical competence in animals: Definitional issues, current evidence, and a new research agenda. *Behavioral and Brain Science, 11*, 561–615.

Dooley, G. B., & Gill, T. (1977). Acquisition and use of mathematical skills by a linguistic chimpanzee. In D. M. Rumbaugh (Ed.), *Language learning by a chimpanzee: The LANA project* (pp. 247–260). New York: Academic Press.

Emmerton, J. (1998). Numerosity differences and effects of stimulus density on pigeons' discrimination performance. *Animal Learning and Behavior, 26*, 243–256.

Emmerton, J., Lohmann, A., & Niemann, J. (1997). Pigeons' serial ordering of numerosity with visual arrays. *Animal Learning and Behavior, 25*, 234–244.

Ferster, C. B. (1964). Arithmetic behavior in chimpanzees. *Scientific American, 210*, 98–106.

Gallistel, C. R., & Gelman, R. (2000). Non-verbal numerical cognition: From reals to integers. *Trends in Cognitive Sciences, 4*, 59–65.

Gelman, R., & Gallistel, C. R. (1978). *The child's understanding of number*. Cambridge, MA: Harvard University Press.

Gulledge, J. P. (1999). *Judgments of dot arrays and Arabic numerals by rhesus monkeys: Evidence of concept of number*. Unpublished master's thesis, Georgia State University.

Hauser, M. D., & Carey, S. (2003). Spontaneous representations of small numbers of objects by rhesus macaques: Examinations of content and format. *Cognitive Psychology, 47*, 367–401.

Hauser, M. D., Carey, S., & Hauser, L. B. (2000). Spontaneous number representation in semi-free-ranging rhesus monkeys. *Proceedings of the Royal Society of London B, 267*, 829–833.

Hauser, M. D., Tsao, F., Garcia, P., & Spelke, E. S. (2003). Evolutionary foundations of number: Spontaneous representation of numerical magnitudes by cotton-top tamarins. *Proceedings of the Royal Society of London B, 270*, 1441–1446.

Hicks, L. H. (1956). An analysis of number-concept formation in the rhesus monkey. *Journal of Comparative and Physiological Psychology, 49*, 212–218.

Huntley-Fenner, G. (2001). Children's understanding of number is similar to adults' and rats': Numerical estimation by 5–7-year-olds. *Cognition, 78*, B27–B40.

Huntley-Fenner, G., & Cannon, E. (2000). Preschoolers' magnitude comparisons are mediated by a preverbal analog mechanism. *Psychological Science, 11*, 147–152.

Kaufman, E. L., Lord, M. W., Reese, T. W., & Volkman, J. (1949). The discrimination of visual number. *American Journal of Psychology, 62*, 496–525.

Koehler, O. (1951). The ability of birds to count. *Bulletin of Animal Behavior, 9,* 41–45.

Levinson, P. K., & Findley, J. D. (1967). Counting behavior in baboons: An error-contingency reinforcement schedule. *Psychological Reports, 20,* 393–394.

Mandler, G., & Shebo, B. J. (1982). Subitizing: An analysis of its component processes. *Journal of Experimental Psychology: General, 111,* 1–22.

Matsuzawa, T. (1985, May 2). Use of numbers by a chimpanzee. *Nature, 315,* 57–59.

Mechner, F. (1958). Probability relations within response sequences under ratio reinforcement. *Journal of the Experimental Analysis of Behavior, 1,* 109–122.

Pepperberg, I. M. (1987). Evidence for conceptual quantitative abilities in the African Grey parrot: Labeling of cardinal sets. *Ethology, 75,* 37–61.

Pepperberg, I. M. (1994). Numerical competence in an African Grey parrot (*Psittacus erithacus*). *Journal of Comparative Psychology, 108,* 36–44.

Perusse, R., & Rumbaugh, D. M. (1990). Summation in chimpanzees (*Pan troglodytes*): Effects of amounts, number of wells, and finer ratios. *International Journal of Primatology, 11,* 425–437.

Rilling, M. (1993). Invisible counting animals: A history of contributions from comparative psychology, ethology, and learning theory. In S. T. Boysen & E. J. Capaldi (Eds.), *The development of numerical competence: Animal and human models—Comparative cognition and neuroscience* (pp. 3–38). Hillsdale, NJ: Erlbaum.

Rumbaugh, D. M., Hopkins, W. D., Washburn, D. A., & Savage-Rumbaugh, E. S. (1989). Lana chimpanzee learns to count by "NUMATH": A summary of a videotaped experimental report. *Psychological Record, 39,* 459–470.

Rumbaugh, D. M., Hopkins, W. D., Washburn, D. A., & Savage-Rumbaugh, E. S. (1993). Chimpanzee competence for counting in a video-formatted task situation. In H. L. Roitblat, L. M. Herman, & P. E. Nachtigall (Eds.), *Language and communication: Comparative perspectives* (pp. 329–346). Hillsdale, NJ: Erlbaum.

Rumbaugh, D. M., Savage-Rumbaugh, E. S., & Hegel, M. T. (1987). Summation in the chimpanzee (*Pan troglodytes*). *Journal of Experimental Psychology: Animal Behavior Processes, 13,* 107–115.

Rumbaugh, D. M., Savage-Rumbaugh, E. S., & Pate, J. L. (1988). Addendum to "Summation in the chimpanzee (*Pan troglodytes*)." *Journal of Experimental Psychology: Animal Behavior Processes,* 118–120.

Rumbaugh, D. M., & Washburn, D. A. (1993). Counting by chimpanzees and ordinality by macaques in video-formatted tasks. In S. T. Boysen & E. J. Capaldi (Eds.), *The development of numerical competence: Animal and human models* (pp. 87–106). Hillsdale, NJ: Erlbaum.

Shumaker, R. W., Palkovich, A. M., Beck, B. B., Guagnano, G. A., & Morowitz, H. (2001). Spontaneous use of magnitude discrimination and ordination by the orangutan (*Pongo pygmaeus*). *Journal of Comparative Psychology, 115,* 385–391.

Stroop, J. R. (1935). Studies of interference in serial verbal reactions. *Journal of Experimental Psychology, 18,* 643–662.

Sulkowski, G. M., & Hauser, M. D. (2001). Can rhesus monkeys spontaneously subtract? *Cognition, 79,* 239–262.

Terrell, D. F., & Thomas, R. K. (1990). Number-related discrimination and summation by squirrel monkeys (*Saimiri sciureus sciureus* and *S. boliviensus boliviensus*) on the basis of the number of sides of polygons. *Journal of Comparative Psychology, 104,* 238–247.

Thomas, R. K., Fowlkes, D., & Vickery, J. D. (1980). Conceptual numerousness judgments by squirrel monkeys. *American Journal of Psychology, 93,* 247–257.

Tomonaga, M., & Matsuzawa, T. (2002). Enumeration of briefly presented items by the chimpanzee (*Pan troglodytes*) and humans (*Homo sapiens*). *Animal Learning and Behavior, 30,* 143–157.

Trick, L. M., & Pylyshyn, Z. W. (1994). Why are small and large numbers enumerated differently? A limited-capacity preattentive stage in vision. *Psychological Review, 101,* 80–102.

Washburn, D. A. (1994). Stroop-like effects for monkeys and humans: Processing speed or strength of association. *Psychological Science, 5,* 375–379.

Washburn, D. A., & Rumbaugh, D. M. (1991). Ordinal judgments of numerical symbols by macaques (*Macaca mulatta*). *Psychological Science, 2,* 190–193.

Whalen, J., Gallistel, C. R., & Gelman, R. (1999). Nonverbal counting in humans: The psychophysics of number representation. *Psychological Science, 10,* 130–137.

Xia, L., Emmerton, J., Siemann, M., & Delius, J. D. (2001). Pigeons (*Columba livia*) learn to link numerosities with symbols. *Journal of Comparative Psychology, 115,* 83–91.

Xia, L., Siemann, M., & Delius, J. D. (2000). Matching of numerical symbols with number of responses by pigeons. *Animal Cognition, 3,* 35–43.

# 14

# Do Primates Plan Routes? Simple Detour Problems Reconsidered

## Emil W. Menzel Jr. and Charles R. Menzel

If one were to search the World Wide Web today using the key term "shortest path AND barrier," one would encounter more references than anyone could read in the next year. The topic is of interest and importance to many students of computer science, robotics, physics, neuroscience, geography, and a few psychologists and primatologists (for reviews by psychologists, see Collett, 2002; Gallistel, 1990). Nor is it coincidental, we believe, that a popularized book titled *Emergence* opens its introduction by describing how and why "Toshiyuki Nakagaki . . . trained an amoebalike organism called slime mold to find the shortest route through a maze" (Johnson, 2001, p. 11). The author is wisely albeit perhaps unwittingly following a good and very old psychological precedent.

In primate psychology, the classic reference on animals' relative abilities to cope with simple barriers is Köhler's *The Mentality of Apes* (1925; see especially chap. 1). Köhler's term for his tasks was *umwege*, which, his translator Ella Winter said (p. 11n), is also sometimes called *roundabout methods*; *detours*; *roundabout ways, paths, or routes*; and *indirect ways*. One of Köhler's major theoretical problems was whether such seemingly intelligent (*einsichtig*) behaviors are emergents, biologically or psychologically speaking, or are instead completely reducible to simpler processes. Other important earlier references here include Harlow (1949), Hebb (1949), Hediger (1964), Hull (1952), Lewin (1935), Lorenz (1971), and Tolman (1932), as well as Fraenkel and Gunn's (1961) review of still earlier work on animal orientation and forced movements that explicitly avoided the use of barriers and had little interest in learning or cognition and, so it would seem today, suffered the consequences.

---

Research was supported by National Institutes of Health Grants HD-38051, MH-58855, and NS-42867 and by National Science Foundation Grant SBR-9729485. We thank Duane Rumbaugh for support and discussions over a period of many years. We also thank Mary Beran, Stephanie Berger, John Kelley, and Jamie Russell for assistance with data collection. We dedicate the chapter to Harriet Anne Menzel in honor of her 50th wedding anniversary.

If by chance the problem of emergents has not yet been settled for all practical purposes, we can only hope that the present volume will administer the *coup de grace*. As William James (1890) said,

> Let us look at a few facts.
>
> If some iron filings be sprinkled on a table and a magnet brought near them, they will fly through the air and stick to its surface. A savage seeing the phenomenon explains it as the result of an attraction or love between the magnet and the filings. But let a card cover the poles of a magnet, and the filings will forever press against its surface without its ever occurring to them to pass around its sides and thus come into more direct contact with the object of their love.....
>
> If we now pass from such actions as these to those of living things, we notice a striking difference. Romeo wants Juliet as the filings want the magnet; and if no obstacles intervene he moves toward her by as straight a line as they. But Romeo and Juliet, if a wall be built between them, do not remain idiotically pressing their faces against its opposite sides like the magnet and the filings. Romeo soon finds a circuitous way, by scaling the wall or otherwise, of touching Juliet's lips directly. With the filings the path is fixed; whether it reaches the end depends on accidents. With the lover it is the end which is fixed, the path may be modified indefinitely. . . . The pursuance of future ends and the choice of means for their attainment are thus the mark and criterion of the presence of mentality in a phenomenon. (pp. 6–8)

We do not propose here to define precise boundary lines between intelligent and nonintelligent animals or psychological processes, let alone to offer a new version of the medieval *Scala Natura*, or continuous scale of nature, on which all things may be precisely ranked, from high to low (Lovejoy, 1936). In our opinion, all boundary lines between categories and concepts are fuzzy, and this is particularly true in the domain of psychology. However, as Figure 14.1 shows, the travel paths of normally reared chimpanzees and chickens in situations involving barriers, as portrayed by Köhler (1925), and the travel paths of inanimate objects, as described by James (1890), do indeed seem to form a straightforward continuum of sorts (see also the many fascinating paths described in Fraenkel & Gunn, 1961). Our purposes in the present chapter are threefold: (a) to define that continuum in more detail, objectively and quantitatively (Maxwell [1877/1991, p. 11] said, "I would advise those who study any system of metaphysics to examine carefully that part of it which deals with physical ideas"), and here our major idea is *space*; (b) to examine the performances of a few selected primates; and (c) to do so using a systematically designed set of simple barrier tasks, which happen fortuitously to include all three of the barrier patterns shown in Figure 14.1.

The tasks that we use enable dissection of various components of distance to the goal in a more systematic fashion than would be possible ordinarily with complex mazes (e.g., Fragaszy, Johnson-Pynn, Hirsh, & Brakke, 2003; Johnson-Pynn et al., 2001; C. R. Menzel, Savage-Rumbaugh, & Menzel, 1999). Furthermore, given that the tasks are video simulations in which both the barrier patterns and the joystick cursor's travel paths are digitized, the paths literally

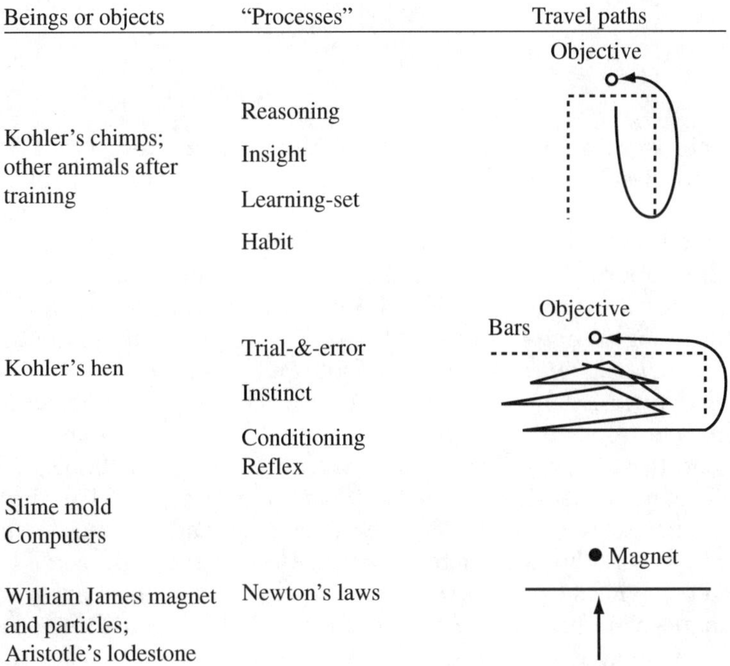

| Beings or objects | "Processes" | Travel paths |
|---|---|---|
| | | Objective |
| Kohler's chimps; other animals after training | Reasoning<br>Insight<br>Learning-set<br>Habit | |
| Kohler's hen | Trial-&-error<br>Instinct<br>Conditioning<br>Reflex | Bars    Objective |
| Slime mold<br>Computers | | |
| William James magnet and particles;<br>Aristotle's lodestone | Newton's laws | ● Magnet |

**Figure 14.1.** *Scala Natura* and travel paths.

may be assessed jump by jump, like the moves of a checker on a checkerboard (Rumbaugh, Washburn, Savage-Rumbaugh, & Hopkins, 1991; Washburn, 1992). We try to describe the tasks and the animals' behavior in sufficient detail that students of robotics (see, e.g., Braitenberg, 1984; Deutsch, 1960; Krieger, Billeter, & Keller, 2000; Walker & Miglino, 1999) who wish to simulate the behavior can see precisely what the problem is. In a word, "One of the greatest challenges in robotics is to create machines that are able to interact with unpredictable environments in real time" (Krieger et al., 2000, p. 992). In this regard, solving a single, static barrier pattern after many trials (or generations) of trial and error is, computationally speaking, an easy problem and, at least for most primatologists, not a very exciting one. Solving any given novel barrier pattern, of many randomly presented patterns, à la one-trial learning on a learning set task (Harlow, 1949; Rumbaugh, 1997) or first-trial problem solution on tasks such as those of Hebb and Williams (described in Shore, Stanford, MacInnes, Klein, & Brown, 2001) or Köhler (1925), is a different matter. Indeed, even on the very first jump of the joystick cursor, averaged across our various test patterns, one can discriminate the performances of James's Romeo and Juliet (human adolescents or young adults) from those of nonhuman primates, and even the former fall short of perfection, although the latter are by no means dummies either.

## Method

*Subjects*

The subjects were 4 human and 16 nonhuman primates. Two female and 2 male Stony Brook undergraduate students, after having seen playbacks of the performance of chimpanzees in a class on comparative psychology, asked after class to try the same task. The students described themselves as experienced with computers and video games. All were approximately 20 years old.

The nonhuman subjects included 5 adult male rhesus monkeys (*Macaca mulatta*) ranging in age from 7 to 20 years, a 15-year-old female orangutan (*Pongo pygmaeus*), a 9-year-old male bonobo (*Pan paniscus*), and 9 chimpanzees (*Pan troglodytes*)—2 females, a 10-year-old and a 19-year old, and 7 males ranging in age from 10 to 23 years. All animals had been born and raised in the laboratory; only 2 (bonobo Kanzi and chimpanzee Panzee) had had appreciable experience with uncaged outdoor environments. Of all the subjects, the rhesus monkeys were probably the ones that had the most exposure to joysticks; however, the Language Research Center apes had served in many and varied behavioral experiments almost every week of their lives, and for all but 2 (orangutan Madu and chimpanzee Mercury), this included experiments on language learning. Four of the chimpanzees (all from the Yerkes Primate Center) were, by comparison, relatively test naive, and in particular they had had little experience with tasks involving video screens and the use of joysticks.

*Apparatus*

From a subject's point of view, the basic apparatus was a computer monitor with a color screen and a Kraft joystick by means of which one could control the movement of a cursor on the screen. The joystick returned to the center position as soon as the subject released it. The human subjects sat at an office desk and the monitor and joystick were on the desk, directly in front of them. The nonhuman subjects worked from within a cage; the joystick was outside the cage, almost flush with it and easily accessible through bars, and the monitor was approximately at eye level and just out of arm's reach.

The computer on which the tasks were originally developed was the original, Model 1, version of the IBM PC; the tasks looked and acted just the same on all subsequent monitors and computers. All programs were designed and written by Emil Menzel; most were much like others in use at the Language Research Center.

*Stimuli*

Test patterns occupied the entire video screen, which was usually about 28 cm wide and 22 cm high and always divided into 40 text cells in width and 25 text cells in height. The patterns were composed of ordinary text characters, each $8 \times 8$ pixels in size. A total of 192 patterns were generated by a simple program, according to the following factorial rules. Imagine a big letter H,

**Figure 14.2.** Umweg Patterns 1 through 96, and one chimpanzee's Trial 1 performance on each pattern. Patterns 97 through 192 were identical to Patterns 1 through 96 except that they included a hole in the center piece of the barrier. To distinguish the goal from the starting point, it is made larger than its actual size. Above each pattern, the score 0–4 indicates how greatly the optimal path deviates from a straight line toward the goal; udrl (up, down, left, right) indicate the optimal starting direction, and + or − indicate whether or not the subject had the cursor in the appropriate quadrant on its 10th jump. Scores at the top of the figure are averages over all 96 trials. corr = the directional data on Jump 10; J/MinJ = number of jumps to reach goal divided by the minimum possible.

centered on the screen, each of its four arms 9 text cells (barrier characters) long and its center bar 11 text cells long, all cells made solid and of the same color. Designate the four arms of the H as A, B, C, and D, with A on the top left, B top right, C bottom right, and D bottom left. Now generate all 16 possible patterns of the presence versus absence of each of these four arms. Next, take a single text character, say the letter S, which will define the subject's starting point. Let its position on the $x$-axis be constant and at the center of the screen, but vary its position on the $y$-axis, putting it either 5 text cells above the center bar of the H (top position) or 5 text cells below the center bar. Next, take another text character, say the letter G, which will define the goal or target. Position it 2 text cells above or below the center bar, on the opposite side of the center bar from S, and systematically vary its lateral position, putting it in the center of the screen, 4 text cells to the right of center, or 4 text cells to the left. All of the foregoing variations together yield 96 patterns, as displayed in Figure 14.2. Finally, make an extra copy of each of these patterns and modify them by creating a shortcut in the center bar of the H (i.e., replacing the three center barrier cells with space cells). Other information on Figure 14.2 is explained below.

If a subject could not discriminate cursor from goal or barrier until the cursor commenced to move, then obviously he or she would be at a loss at the very outset of each trial. We therefore tried to make such discriminations as easy as possible. The cursor was represented by a text character that looked like a face (ASCII code number 2). For the apes and humans it was colored light brown. The goal was one or more blue Gs; barriers were solid blocks of red, and empty spaces were green. The rhesus were accustomed to black backgrounds for empty space and white goals and cursor, so we returned to that once we noted on practice tasks that the new colors caused them problems. For them, the cursor was white with a green border around it, and barriers were pale blue. In formal testing as opposed to pretraining, the goal was a single G. The patterns shown in Figure 14.2 are drawn exactly to scale, except that the goal is made much larger to distinguish it from the start point.

## Test Program

The computer program that was used for testing the subjects (a) picked 1 of the 192 stimuli patterns, in the quasi-random order that it appeared on a list that had been devised in advance of testing by the experimenter; (b) put the pattern on the video screen for the subject to view; (c) controlled the operations of the joystick, as described below; (d) generated a brief click on each jump of the cursor, another sound if an impending cursor jump would land the cursor on a barrier cell or take it off screen (the impending jump was not carried out and was scored as a "bump"), and still another sound if and when the cursor landed on the goal; (e) immediately triggered the delivery of a reinforcer via a chute, if the cursor landed on the goal, assuming of course that the dispenser was connected, and for some subjects it was not; (f) paused $x$ seconds (intertrial interval); and (g) looped back to (a) to start the next trial.

Any given trial could be terminated if it lasted long enough to reach a predefined cutoff time or if the experimenter pressed an escape key on the computer keyboard. The program terminated and the session ended after the provided list of stimuli-to-be-presented ran out, but it could also be terminated on special command from the experimenter. Almost all data and other information to be reported were recorded automatically by the program; the exceptions were special notes that the experimenter typed in before or after the session or between trials.

Figure 14.3 captures the screen of one of our data analysis programs as the program was running. The figure shows, for one sample trial, the raw data that were recorded and selected additional data analyses (top-right quadrant). Note that the screen capture was made when the moving cursor was several jumps away from the goal. This program was designed to generate movielike reruns of the visual display, from the raw data, so that trials could be viewed repeatedly and at varied speeds, if so desired, with or without showing details other than the original pattern. The "map" shown in the top-left quadrant is one quarter the size of the original display. The big H is one of the barrier patterns; other details are pertinent only to data analysis and are discussed later. The trail of the cursor is plotted in single-pixel dots, most of which might not be visible on Figure 14.3; the trail for this trial also can be seen in Figure 14.2, fourth row, last column. Subjects could not see their trail or other details; only the barrier, goal, and cursor were visible during tests.

For our purposes, the most important raw datum was a string of numbers that represented the entire, jump-by-jump path of the cursor (lower-left quad-

DO PRIMATES PLAN ROUTES? 181

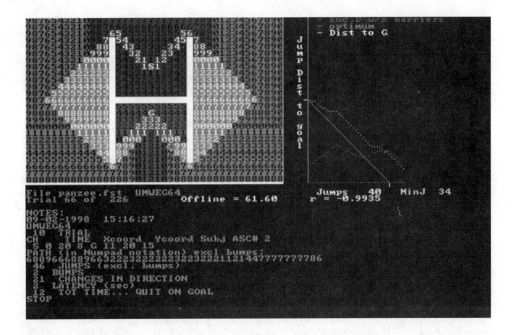

**Figure 14.3.** A screen capture of one of our data analysis programs as it was analyzing a single sample trial. The bottom-left quadrant of the figure shows the raw data that were recorded by the original program used for testing subjects. The top-left quadrant shows the umweg pattern; the subject's jump-by-jump path (very small dots, possibly not visible here); the number of jumps needed to go from start to any given cell on the map (digits, together with change in background color every 10th jump); which cells lie on an optimal path (different foreground color for digits); and one estimate of a shortest-plus-straightest path. The top-right quadrant shows graphic and numerical estimates of the relative optimality of the subject's path, as explained in the text.

rant of Figure 14.3). Each digit specifies the direction of a single jump; odd numbers are jumps in a diagonal direction, and even numbers are north, south, east, and west or, on the video screen, up, down, right, and left, respectively. Later versions of the test program included symbols to represent each tick of a clock and the direction of each aborted jump ("bumps"), in the same string; in Figure 14.3 only total counts are given for these variables.

Another feature of the test program enabled us to test how well the subjects could perform a delayed response version of the barrier tasks. Quite simply, at a set time the barriers disappeared from sight, and only the cursor and goal remained visible. In the first version of this task, the barriers were visible for a fixed number of seconds, during which time the cursor was not visible and not functional; then, simultaneously, the barriers disappeared and the cursor appeared. Later versions are described in the Results section.

The program allowed for the cursor to be moved by a variety of devices, but here all subjects used only the simplest joystick. The program ignored the force with which this joystick was moved and scored the magnitude of its movement in all-or-none fashion, according to whether it exceeded an experimenter-specified minimum value—high enough to get rid of all machine-produced "noise." More precisely, movements on the $y$- and $x$-axes were each scored as −1, zero, or +1. If the joystick's direction was nonzero, the cursor made a single jump of one text cell in the appropriate direction. After a pause of $x$ milliseconds, the program checked the joystick's direction once more. The number of jumps made by the cursor did not necessarily reflect the number of discrete joystick presses or movements that the subject made but might reflect the duration of time that the joystick was depressed. Here we set the value of the interjump interval at 0.15 seconds for all subjects—slow enough to enable the subject to make a single, discrete jump at a time if so desired and to maintain maximal control over the cursor, yet fast enough to enable the subject to complete any trial in less than 10 seconds, assuming that he or she performed in optimal fashion.

## Preliminary Training

All of the present subjects were already familiar with joysticks and the basic rule of the game, which was to move the cursor on the screen, using the joystick so as to hit a target and (in the case of the nonhumans) get a piece of food. Some subjects, however, had not had experience with the particular joystick we used or with simulated barriers, and so they were given as much time or as many sessions as we (or they, in the case of the human subjects) deemed necessary in this regard. For practice sessions we used barrier patterns from the so-called Hebb–Williams intelligence test.

## Procedure

For formal testing, the ideal for any given replication of the test was to give each subject precisely 2 trials on each of the barrier patterns shown in Figure 14.2, distributed over four sessions. Interspersed with these 192 trials would be 32 additional trials, each of which used a different pattern that had a hole in the center barrier. (Our main reason for not using all of the latter patterns is that we wished to use them as occasional catch or probe trials to assure that the subjects were indeed still taking the center barrier into account. The shortest path to the goal on all of them entailed going straight toward the goal, with no detouring, whereas the optimal initial directions on the other patterns were as shown in Figure 14.2 and always involved some detouring.) Following this, subjects were to be tested on a delayed-response version of the task, with 1 trial on each of the patterns in Figure 14.2. Several different sets of lists, each using a different, quasi-random order of the patterns, were prepared for the computer. Provided the experimenter gave each subject the right list on the right day, everything else could be, ideally speaking, completely computer-controlled and automatic.

For the human subjects, this ideal was almost met, except that on their request the number of trials per session was doubled, so that only half the number of sessions would be required. Their only reward was, as they themselves put it, "trying to beat the chimps," and by that same token they were deliberately given no further instructions on whether to minimize time, the distance traveled, or the effort expended.

The rhesus had for years been tested with fully automated procedures, with no humans in the test room, and in principle they should have posed no problems. However, there were numerous trials on which they did not respond for an hour or more or never reached the goal. Also, on some of the patterns they were given many trials in a row as further pretraining, largely because this seemed to be the custom in the rhesus lab and our own protocol had not been sufficiently explained.

Some of the Language Research Center apes were given several widely spaced replications of the basic task. We picked for analysis here their very first sessions. However, exceptions were made in the case of chimpanzee Mercury and orangutan Madu. Mercury was tested at the age of about 5 years but given only one trial per pattern at that time; then he was tested again 7 years later, in the fashion described above. Madu was mistakenly tested with an incorrect set of patterns that were very similar to those in Figure 14.2 and tested with the correct patterns 9 years later. We use only the latter data here. (Mercury's performance was far better the second time, and Madu's was slightly but significantly better. Readers who feel that it would be cheating to include their data in the intergroup comparisons may purge their data if desired.) Another exception in the ape data is that the first animals tested did not get, in their first sessions, any of the patterns with a hole in the center barrier. Bonobo Kanzi did not receive further testing, and so we have no data for him on these patterns. For chimpanzees Austin, Sherman, and Lana, data on these patterns with holes were extracted from the second replication of the basic task.

In keeping with the routine with which the apes were typically tested, fully automated testing was not always used for them. More often, a caregiver set up the test and made the joystick available and pressed a key to commence each trial only if the subject was judged to be attending to the display and motivated to work. The caregiver sat so as not to see the monitor and ordinarily delivered the food by hand, if and when the program sounded the "dinner bell." Opinions are divided as to how such human–animal interaction might affect performances; in our experience, on the present tasks they make no difference, at least after pretraining.

For all subjects, the data were reviewed before analysis and cleansed of all trials that involved obvious problems with the apparatus, disturbances in the test environment, or failure to reach the goal at all even when all stimuli were visible. We selected out the first two trials per pattern on which the subject actually reached the goal and ran a sufficient number of trials overall to assure there were at least two such trials per pattern. Without this data cleansing, all subjects, but especially the rhesus and the test-naive chimpanzees, would look worse than they now do.

## Analysis of the Test Patterns and Travel Paths

What would optimal performance on each of the test patterns look like? By tradition, maze and detour performances are most often analyzed in terms of how much time the subject took to go from start to finish and how many errors (or what sorts of errors) the subject made in the process. Time scores are, however, of use only for comparing two or more performances rather than for defining how close to optimal any given performance may be said to come. Also, deciding what is an error on any given test pattern is sometimes difficult and arbitrary, and the scoring of 100 or more patterns can become tedious (see, e.g., Figure 14.2). At the same time, behavioral data using traditional measures correlate highly with our own analyses. The relationship becomes most obvious once one recognizes that errors are typically defined by spatial criteria and according to whether, how, and where in the maze the subject's travel path deviates from a least-distance route. Here we deal directly with what is the shortest path, given barriers and given also the underlying metrics of the space and the task in question. Whether the shortest path is really the best possible path is a different problem, which we put off until the Discussion section.

The space we have thus far defined—indeed, any video display produced by a digital computer—is by definition not Euclidean or Newtonian. Perhaps the key issues are the concepts of distance and straight line. Good, relatively nontechnical descriptions of the difference between various distance metrics may be found in Cheng (2002) and Gallistel (1990), but for our purposes it is even better to go to the horse's mouth. Newton (see Mach, 1907; see also Maxwell, 1877/1991, pp. 9–12) said at the outset of his *Principia*:

> Time, space, place and motion, being words well known to everybody, I do not define. Yet it is to be remarked, that the vulgar conceive these quantities only in their relation to sensible objects. And hence certain prejudices with respect to them have arisen, to remove which it will be convenient to distinguish them into absolute and relative, true and apparent, mathematical and common, respectively. (quoted in Mach, 1907, p. 222)

Newton was taking a crack at the physics of Descartes; we are here definitely and without apology on the side of the latter, and among the vulgar. Space as we conceive it and wish further to define it is very much in relation to sensible objects—starting points, barriers, landmarks, goals—and it is commonsensical or behavioral more than mathematical.

Contrary to Euclidean geometry, the shortest distance between two points is not always a straight line, or vice versa—unless by chance one lives in a vacuum or can walk across rivers and through walls. Nor is there very often a single shortest possible path between two points, as opposed to multiple equally reasonable possibilities, especially when one is measuring the everyday world with a yardstick rather than a micrometer. Lindsay and Margenau (1936) summed up the main differences between psychological space and Newtonian space by characterizing the latter as infinite rather than finite, absolute rather than relative, public rather than private, and "homogeneous, continuous, isotropic, three-dimensional and Euclidean" (p. 61) in its geometry. Probably none of these characteristics, they pointed out, apply to psychological space. We believe that the same is also true for our video tasks.

The lack of three-dimensionality on the video screen is in one sense trivial. Yet in another sense it is not: The Pythagorean formula for (Euclidean) distance, at least in its original and simplest form ($D^2 = d_x^2 + d_y^2$), does not apply in any space or portion of space in which any barrier falls between the two points whose distance we wish to compute. If one wishes to use this formula, one must modify or extend it. In this chapter, we extend it by adding more dimensions. Put briefly, "The Pythagorean Theorem . . . is the most important theorem in mathematics . . . [and] underlies our modern notion of 'optimality', the best possible solutions to problems" (Kosko, 1993, p. 110). Most classical, parametric statistics may be derived from the Pythagorean theorem, and this includes the analysis of variance (ANOVA) and the Pearson product–moment correlation coefficient. In the present case, the optimal paths between starting points and goals, in terms of distance, may be analyzed on each axis in fairly straightforward terms in accordance with the factorial design outlined above for the design of the stimulus patterns. In effect, $d_x^2 = d_a^2 + d_b^2 + d_c^2 \ldots + d_n^2$, and much the same is true for $d_y^2$.

This, however, is getting ahead of the story. Perhaps our main point is akin to Hediger's (1964) main point: There is more than one kind of distance; moreover, distance is always in some sense a behavioral concept, if not a species-specific question, as are the questions of what constitutes a straight line and a barrier. So we focus our attention on behavioral considerations. Readers who like mathematics as well as natural history should refer to Lakoff and Núñez (2000) in addition to Hediger.

Exactly what would it take to make the motions of the video cursor, plus the joystick, plus the subject's hand (or, in some cases, foot, finger, or mouth) abide by the rules of Euclid and Newton? In the final analysis, no equipment designer or programmer knows with great precision, and few care. At some point we must all rest content with a task that looks and feels reasonably natural or intuitive to the user, and by no means are all users equal in this regard, especially if they are of diverse abilities (Church & Marston, 2003), experiential backgrounds, and species. In any event, the basic rule for the movement of the cursor in our test programs was that from any given cell of the 25 row × 40 column screen one could move to any one of the eight other immediately adjacent cells, provided that that cell was not a barrier. Most of the programs used at the Language Research Center, and many commercial video games, use the same rule. We considered allowing the cursor to move only up, down, left, and right—in geometrical language, we considered using a city block metric, in which case $D = ( |d_x| + |d_y| )$—but tests showed that this would go very much against the grain of the chimpanzees. (The other species were not tried on such a task, but we would be surprised if adult humans would have any trouble with it or if the nonhumans would not find it difficult and frustrating for some time.) We also considered having the cursor move only one pixel (1/8 cell) at a time, but this would have had no effect on the metric itself, and it also would have made the cursor movement very slow, at least on earlier computers.

We do not know if the metric implicit in our joystick software has any official name, but we call it the *jump metric*. It is not Euclidean partly because we count jumps in diagonal directions as equivalent to jumps in a vertical or horizontal direction, even though their squared Euclidean distance is $1^2 + 1^2$ units rather than $1^2 + 0^2$. Note also that all jumps take exactly the same amount of time, regardless of their direction. The simplest jump metric formula for distance, assuming no barriers, would be $D = \text{MAX}(|d_x|, |d_y|)$, or in other words, the absolute distance on the $x$-axis or the absolute distance on the $y$-axis, whichever one is greater. A more generic measure of distance is, however, even easier to remember, and it is appropriate no matter what metric one is using or how many barriers there are: How many jumps or steps does it take (at a minimum) to go from A to B?

To our knowledge, there was no good, accurate, and easy way to compute this quantity for any and all problems until the past decade or so. To be sure, the heroes and heroines of ancient myths had some clever solutions for related problems (such as Ariadne's thread, which Theseus used to get out of the Labyrinth), but true disciples of Euclid and Newton would probably not be amused and would call these solutions vulgar and common. Indeed, they might say the same for computer solutions. Nevertheless, there are now many computer programs available that will accomplish the feat in just about the way we like and need for our purposes. The first such program that we encountered (not the first one written) was one we wrote ourselves in 1990, shortly after Steve Skiena of the Stony Brook Department of Computer Science advised us that the problem of a universal barrier and maze solver looked simple to him but that we should try a depth-first search instead of the evolution-inspired genetic algorithms with which we were struggling. The problem with the latter, as Skiena put it more recently, is not that they will not work but that, first, the "pseudobiology" is largely metaphor and often misleading and simply "adds another level of complexity between you and your problem," and, second, "genetic algorithms take a very long time on nontrivial problems" and just do not work as well as alternative algorithms on "practical combinatorial problems" (Skiena, 1998, p. 131). His advice is, in our opinion, well taken and may be generalized still further. The presumed sufficiency of natural selection and its 19th-century behavioral analog, trial-and-error, to account for all of animal and human learning and behavior, comes closer to assumption than to fact; he or she who ignores emergent strategies is out of date. Still further, there is almost always more than one way to skin a cat. To assume that Mother Nature loves and perceives simplicity, consistency, and parsimony in just the way that today's theorists do might be a prime example of what William James (1890) called "the psychologist's fallacy."

The main feature of our version of a breadth-first, path-finding algorithm is that starting at point S (or at point G) on any given stimulus pattern, it explored the entire field, fanning out in all directions at the same time, first visiting each accessible cell that was one unit away, then two units away, then three units away, and so on, in an expanding circle. Thus the numeral shown in each cell of the map in the top-left quadrant of Figure 14.3 shows exactly how many jumps away from point S that cell is—provided that one also adds another 10 to the count every time one encounters the numeral 0 or a change in the background color of the display. An optimal path is any path that can start at cell S and pass (in exact sequence) through cells that are numbered 1, 2, 3 ... and so on, and land on cell G, on Jump #MinD (cells S and G are in the center of the map in the top left quadrant).

It may be seen from Figure 14.3 that on Pattern 64 (Umweg64 in the figure) there are many possible paths that would all be optimal in terms of their jump count. The distribution of all possible optimal jumps (as highlighted on Figure 14.3 by using different shades of gray) is pattern specific and metric specific and by no means always obvious, unless perchance one is an expert chess player or mathematician or studies each pattern very carefully with rule book in hand.

Coincidentally, the analytical program that produced Figure 14.3 was almost completely deterministic and used nothing whatsoever that smacks of trial and error. It used no global, as opposed to local, strategies. In other words, it could not recognize any object or pattern, such as a barrier as a whole, but operated purely one step and one cell at a time while constructing its map. Initially, we estimated the most Euclidean path among the many optimal paths by means of a genetic algorithm, but we abandoned that on discovering this much faster and more reliable, and usually more accurate, rule: Always use an even-numbered jump angle (up, down, left, or right) if you can, instead of an odd-numbered one (on a diagonal), and also do not change direction if you do not have to. The rule is not perfect, but it is surprisingly good. If Pythagoras himself did not know it, he should have.

In the interests of brevity, we skip over further details of this computer program. Any detailed account of a breadth-first search will fill in the main gaps, and Steinbock, Tóth, and Showalter (1995); Muller, Stead, and Pach (1996); and Dobkin, Gansner, Koutsofios, and North (1998), among others, do even better than that. For example, Dobkin et al. provided a better way than we just described above for finding the best Euclidean path. The article by Steinbock et al. is particularly intriguing. Even before seeing it, we and many other viewers had described movielike animations of our own program at work as looking like the spread and flow of a liquid across a barrier-filled field. Steinbock et al. literally used waves in a chemical to map out labyrinths and compute optimal paths. It is too bad that Hercules did not think of that as he flushed out the Augean stables.

We now state in general terms what optimal performance on the test patterns will look like from the standpoint of least distance. To paraphrase and expand on C. R. Menzel, Savage-Rumbaugh, and Menzel (2002, pp. 607–608): A most generic measure of the distance between Points A and B is the minimum number of idealized and well-defined steps, all presumed to be equal in length, that one would have to take to get from A to B. An optimal travel path from A to B is one that on any given step reduces the remaining distance to B by the length of that step. If one were to plot a two-dimensional graph with "distance traveled thus far" on the $x$-axis and "distance from here to Point B" on the $y$-axis—as we do in the upper right quadrant of Figure 14.3—all points along an optimal path would fall on a straight line, and this would be so regardless of whether the space in question is Euclidean, contains barriers, or is two- or three-dimensional. The normalized regression slope of this straight line would be –1. Equivalently, the ratio of total travel distance, D, to minimum possible distance, MinD, would be 1, and the Pearson product–moment correlation ($r$) between $x$ and $y$, at any given value of $x$ for which $r$ can be computed, would be −1. Hence, two highly reasonable measures of the relative optimality of any given path, in terms of least distance, are $r$ and the ratio D/MinD (in this chapter, Jumps/MinJumps). The value of $r$ can, of course, range from −1 to +1, and values greater than zero would signify going away from the goal rather than toward it. The reciprocal of the ratio D/MinD (or in other words MinD/D) is equal to the regression slope, at least as long as the regression constant is MinD.

Quantitatively and statistically speaking, these two measures, $r$ and D/MinD, or their equivalents, might arguably be said to be all one needs by way of overall summary statistics. But of course they do not capture all of the information that is in the map or the graph in Figure 14.3, any more than Figure 14.3 captures all of the information that could be obtained by watching the subjects directly. (Nor is it possible to judge exactly what the numbers mean or how many statistical degrees of freedom one may assume, without further reflection.) Watching Figure 14.3 being drawn, at the same speed with which the subject produced the original data, or even slower, is highly interesting, especially insofar as one can simultaneously monitor where the cursor is on the map and how the three curves on the graph stand (or squiggle around) in relation to one another. Every "error," no matter how slight, can be detected on the graph when the curve for distance either goes up or stays flat instead of going down. The magnitude of the error and exactly where in the field it occurred can also be seen, along with the subject's Euclidean distance to the goal (irrespective of any barriers). The easy parts of the route are those in which both distance measures go down together; the presumably hard parts are those in which the former goes down and the latter goes up. We leave it as an exercise for the student—as they say in some textbooks—to design experiments in which a record of the subject's eye movements, facial expressions, vocalizations, brain activity, and of course actual hand movements can be incorporated into a figure analogous to our Figure 14.3.

In presenting the results, we sometimes use the numbers 1, 5, 7, and 10 as if they were magic numbers. This is why: If a subject is infallible, then Jump 1 on Trial 1 should reduce the distance to the goal on any given pattern by 1 jump and never increase the distance or leave it unchanged. The nearest barrier on all patterns is always 5 jumps from the start point. The minimum number of jumps to the goal is 10 on Patterns 1 through 96 and 7 on Patterns 97 through 192.

## Results

### *Patterns 1 Through 96*

Figure 14.2 shows the performance of a single nonhuman subject on her first trial on each of Patterns 1 through 96. She was selected for display because on most of the measures that are shown in the figure, she was outdone only by the 2 male human subjects. The patterns are presented in an order that shows the systematic character of their design rather than in the sequential order in which they were presented. Panzee (indeed all subjects) showed some improvement in performance during the test, but her worst-looking paths of all were not the first ones but outliers, very likely caused by some momentary distraction. See, for example, Pattern 38, on row 3 and column 6 of Figure 14.2, which came 26th.

Above each pattern in Figure 14.2 is shown, from left to right: (a) how many degrees difference there was between a Euclidean straight line from start to goal and an optimal and most Euclidean starting direction (these scores are all divided by 45 and can range from 0 to 4); (b) the optimal starting direction on the $y$-axis (up [u] and down [d] on the video screen); (c) the optimal starting direction on the $x$-axis (l and r); and (d) whether or not (+ or –) on the $n$th jump the subject was scored as going in the correct direction (here, whether or not on the 10th jump the signs of the subject's $x$ and $y$ position relative to the starting point were the same as those in (b) and (c) above).

Our main reason for including this information is so that the relative difficulty of each pattern may be judged on a simple scale of 0 to 4 (assuming of course that all animals are Euclidean or Jamesian brutes at heart). On some patterns the correct starting direction is hard to judge. On those trials on which the correct $x$ vector is indeterminate, the correct angle was judged purely on the basis of $y$. The total count of correct in terms of angle is shown on the top of Figure 14.2, along with the mean values of the Jump/MinJump (J/MinJ) ratio and one count of errors (actual bumping into barriers).

Figure 14.4 may be thought of as the start of a cross-country marathon. It shows what each individual's first 10 jumps looked like, in terms of distance reduction, when averaged across all trials on Patterns 1 through 96. No subject was perfect or invariably $n$ jumps closer to the goal after making its $n$th jump. Only 1 subject was, on average, farther from the goal on Jump 10 than at the outset. Most were significantly closer. Individual and group differences were large and reliable. Overall, humans led the pack; test-wise chimpanzees were not far behind and almost caught up to the human mean by 10 jumps; test-wise orangutan and bonobo were only slightly behind them (but here of course $n = 1$ in each case); and the rhesus were not bad at all, for among the 4 worst subjects of all there is only 1 of them, the rest being relatively test-naive chimpanzees.

For the first two jumps in Figure 14.4, the 4 human subjects left everybody else in the dust, which strongly suggests that they were the most apt to plan their routes in advance (Collett, 1982) or to "look before they leaped." It is both interesting and important to note, however, that if the percentage of trials that each subject was headed in the correct direction were to be plotted jump by jump, some of the curves would look quite different from those in Figure 14.4. Specifically, human subjects initially did no better than the nonhuman subjects in terms of direction, but after 10 jumps they clearly improved in this regard. This, among other things, is shown in Table 14.1.

Table 14.1 presents the data of each individual on a variety of measures, averaged over all trials. Included are measures on Jumps 1, 5, 10, and all jumps combined. Many of the measures are highly intercorrelated. For example, what each subject did on Jump 1 correlates about .80 by Pearson's $r$ ($N = 20$ subjects) with other scores for Jumps 5 and 10; and most of the scores for the overall path correlate with each other well above .90. If one's major concern

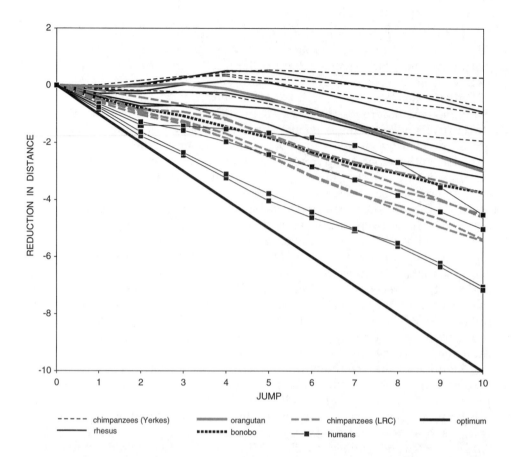

**Figure 14.4.** Each individual subject's first 10 cursor jumps on any given trial, as averaged over 192 trials (i.e., Trials 1 and 2 on Patterns 1 through 96). An optimal least-distance path is, by definition, any path that on each jump reduces the distance to the goal by one jump. Distance at the outset of each trial is transformed to zero; thus, negative scores signify that the subject is headed toward rather than away from the goal, and the larger the (negative) score the better. There is only one of each orangutan and bonobo. LRC = Language Research Center.

is only with relativistic comparisons of individuals or groups, then traditional measures of time and errors stand vindicated and need no defense. Conversely, the measures of distance minimization that we favor (the J/MinJ ratio and its corresponding $r$) seem justified on practical grounds as well as mathematical grounds.

It should be mentioned that the average $r$s shown in Table 14.1 are arithmetic averages. If Fisher's $z$ transform were used on each trial's $r$, all averages would be better (closer to –1), especially for the subjects that did the worst. To be precise, all subjects had trials on which $r$ was close to –1, especially on the problems that required minor deviation from a straight-line Euclidean path; these trials would inflate Fisher's average $r$ more than they would inflate the simple mean. The ranking of the subjects was not affected, however.

Table 14.1. Umweg Data Summary (Individual Data Means)

| Group and subject | Age (years) | Sex | Patterns 1–96 (no hole in center wall) | | | | | | | | | Patterns 97–192 (hole) | | |
|---|---|---|---|---|---|---|---|---|---|---|---|---|---|---|
| | | | j1 | r(j5) | dir | r | j/minJ | Time (seconds)[a] | Bumps | ChDir | AVX | AVY | r(j5) | r | j/minJ |
| **Rhesus** | | | | | | | | | | | | | | | |
| 103 | 9 | M | .04 | −.37 | 63 | −.95 | 1.89 | 7.69 | 3.39 | 11.46 | 8 | 4 | −.50 | −.91 | 2.25 |
| 104 | 9 | M | .33 | −.39 | 61 | −.94 | 2.04 | 9.45 | 6.51 | 13.34 | 15 | 12 | −.67 | −.90 | 2.05 |
| 224 | 13 | M | .18 | −.03 | 43 | −.89 | 2.56 | 10.59 | 11.95 | 14.89 | 8 | 4 | −.00 | −.71 | 4.54 |
| 340 | 20 | M | .08 | .21 | 44 | −.90 | 2.77 | 11.40 | 16.71 | 23.88 | 3 | 1 | −.08 | −.76 | 3.37 |
| 344 | 20 | M | .28 | −.05 | 58 | −.91 | 2.26 | 9.18 | 9.69 | 18.35 | 10 | 8 | −.30 | −.81 | 3.61 |
| **Orangutan** | | | | | | | | | | | | | | | |
| Madu | 15 | F | .04 | −.19 | 70 | −.97 | 1.88 | 7.65 | 2.62 | 12.35 | 8 | 7 | −.62 | −.93 | 2.09 |
| **Bonobo** | | | | | | | | | | | | | | | |
| Kanzi | 9 | M | .47 | −.47 | 70 | −.94 | 1.94 | 8.16 | 11.11 | 20.03 | 12 | 10 | n.d. | n.d. | n.d. |
| **Chimpanzee**[b] | | | | | | | | | | | | | | | |
| Carl | 17 | M | −.02 | .06 | 43 | −.84 | 4.27 | 15.52 | 26.71 | 35.51 | 6 | 3 | −.76 | −.63 | 4.74 |
| Clint | 23 | M | .13 | .22 | 29 | −.67 | 10.63 | 36.60 | 65.27 | 71.44 | 6 | 4 | −.08 | −.48 | 10.00 |
| Patrick | 10 | M | .09 | .13 | 39 | −.87 | 4.05 | 16.40 | 28.48 | 38.11 | 4 | 1 | −.28 | −.73 | 3.99 |
| Winston | 16 | M | .13 | −.20 | 55 | −.89 | 3.39 | 13.11 | 25.26 | 26.24 | 6 | 4 | −.75 | −.80 | 2.93 |
| **Chimpanzee**[c] | | | | | | | | | | | | | | | |
| Austin | 15 | M | .55 | −.38 | 69 | −.96 | 1.74 | 6.89 | 2.97 | 11.56 | 17 | 15 | −.49 | −.91 | 2.31 |
| Lana | 19 | F | .48 | −.57 | 69 | −.95 | 1.78 | 6.71 | 3.25 | 11.74 | 23 | 12 | +.01 | −.85 | 3.23 |
| Mercury | 12 | M | .17 | −.52 | 79 | −.97 | 1.65 | 7.17 | 2.92 | 10.82 | 9 | 12 | −.87 | −.95 | 1.66 |
| Sherman | 15 | M | .59 | −.47 | 83 | −.97 | 1.57 | 6.23 | 2.97 | 13.59 | 16 | 18 | −.68 | −.94 | 1.83 |
| Panzee | 14 | F | .32 | −.80 | 86 | −.98 | 1.53 | 7.09 | 2.36 | 9.57 | 21 | 22 | −.93 | −.92 | 1.96 |
| **Human** | | | | | | | | | | | | | | | |
| Juliet1 | YA | F | .73 | −.71 | 95 | −.98 | 1.58 | 6.23 | 0.17 | 4.06 | 16 | 21 | −.95 | −.96 | 1.34 |
| Juliet2 | YA | F | .65 | −.74 | 92 | −.97 | 1.63 | 6.94 | 1.01 | 6.14 | 20 | 16 | −.95 | −.96 | 1.35 |
| Romeo1 | YA | M | .85 | −.93 | 97 | −.98 | 1.33 | 5.79 | 0.24 | 4.99 | 19 | 26 | −.98 | −.97 | 1.23 |
| Romeo2 | YA | M | .77 | −.90 | 97 | −.99 | 1.30 | 5.34 | 0.25 | 6.50 | 26 | 31 | −.99 | −.97 | 1.26 |

*Note.* j1 = first jump's mean reduction of distance to goal (1.0 = best possible; −1.0 = worst); r(j5) = Pearson r on first 5 jumps of cursor (r on jump number vs. distance to goal); r = Pearson r over all jumps; dir = percentage of trials headed in correct direction on the 10th jump; n.d. = no data; Bump = number of times cursor touched (bumped into) barrier; ChDir = number of times cursor changed direction; AVX = total number of analysis of variance (ANOVA) components on the x-axis in Figure 14.5 for which $F \geq 3.94$; AVY = total number of ANOVA components on the y-axis in Figure 14.5 (see text) for which $F \geq 3.94$; j/minJ = number of jumps to reach goal divided by the minimum possible; M = male; F = female; YA = young adult (19–21 years).

[a]Latency excluded, geometric mean used. [b]Yerkes. [c]Language Research Center.

| Comp. | X Axis r best | data Rhes.Apes.......Hum. | Total | Y Axis r best | data Rhes.Apes.......Hum. | Total |
|---|---|---|---|---|---|---|
| L | 0.31 | .+....++...++.+.++++ | 10 | 0.00 | .................... | 0 |
| Q | 0.00 | .................... | 0 | 0.00 | .................... | 0 |
| T | 0.00 | ..+...+++...+....... | 5 | 0.33 | .++.+.++..+..+..+... | 8 |
| A | 0.38 | ++++++++..+++++++++++ | 18 | -0.25 | ..+...+...+.+++++++ | 10 |
| B | -0.38 | ++..+++...+++++++++ | 15 | -0.25 | .+...+...+++++++++ | 11 |
| C | 0.38 | .+..+.+..+++++++++++ | 14 | 0.25 | .+..+...++.+++++++ | 10 |
| D | -0.38 | ++..+++..+++++++++++ | 16 | 0.25 | ++..+.+....+++++++++ | 13 |
| LT | 0.00 | ........+......+.... | 2 | 0.00 | .................... | 0 |
| QT | 0.00 | .........+.+.....+.. | 3 | 0.06 | ........+......+.... | 2 |
| LA | 0.00 | .........+...+...... | 2 | 0.10 | ..........+......+ | 2 |
| QA | 0.00 | .+.......+.......... | 2 | 0.00 | .................... | 0 |
| LB | 0.00 | ............+.+..... | 2 | -0.10 | .......+........++ | 3 |
| QB | 0.00 | .......+............ | 1 | 0.00 | .................... | 0 |
| LC | 0.00 | ................+... | 1 | -0.10 | .......+........++ | 3 |
| QC | 0.00 | .......+.....+...... | 2 | 0.00 | .....+.............. | 1 |
| LD | 0.00 | ............+.++.... | 3 | 0.10 | ..............+..+. | 2 |
| QD | 0.00 | +................... | 1 | 0.00 | ........+........... | 1 |
| TA | -0.04 | ......+.+....+.+..... | 4 | -0.25 | .+........+++++++++ | 9 |
| TB | 0.04 | ......+....+.+....... | 2 | -0.25 | ....+.+.....++.+++++ | 9 |
| TC | 0.04 | .+..+........+..+.. | 3 | -0.25 | .+..+......+++++++++ | 11 |
| TD | -0.04 | ..+.................+ | 2 | -0.25 | ++..++......+++++++++ | 14 |
| AB | 0.00 | +........+.......... | 2 | -0.25 | .......+...++.+++++++ | 10 |
| AC | 0.00 | ....+......+.+.++.... | 5 | 0.00 | ..............++++.. | 4 |
| AD | 0.00 | .+.......+..+....... | 3 | 0.00 | ............++..++..+ | 5 |
| BC | 0.00 | ..........+....+.... | 2 | 0.00 | ........+...+..+..+ | 4 |
| BD | 0.00 | .+..+.....+.....++.+ | 6 | 0.00 | +..........+....+.. | 2 |
| CD | 0.00 | ............+....... | 1 | 0.25 | .....+........+.++++ | 6 |
| LTA | 0.00 | ....+.+............. | 2 | 0.00 | .................... | 0 |
| QTA | 0.06 | ...................+ | 1 | 0.00 | .................... | 0 |
| LTB | 0.00 | ............+....... | 1 | 0.00 | .................... | 0 |
| QTB | -0.06 | ................+... | 1 | 0.00 | ..............+.. | 1 |
| LTC | 0.00 | ............+....+.. | 2 | 0.00 | ..............+... | 1 |
| QTC | -0.06 | ..................+. | 1 | 0.00 | .................... | 0 |
| LTD | 0.00 | .................... | 0 | 0.00 | .................... | 0 |
| QTD | 0.06 | ....+.........+..... | 2 | 0.00 | .+...+..........+.. | 3 |
| LAB | 0.10 | ......+....++.+.++++ | 8 | 0.00 | .................... | 0 |
| QAB | 0.00 | .+..+..........+.. | 3 | 0.00 | .................... | 0 |
| LAC | -0.10 | ........+.+........+ | 3 | 0.00 | .................... | 0 |
| QAC | 0.00 | ........+......+.+ | 3 | 0.00 | .........+.......... | 1 |
| LAD | 0.10 | .............+.+.++ | 4 | 0.00 | ..............+.... | 1 |
| QAD | 0.00 | .....+.............. | 1 | 0.00 | ........+........... | 1 |
| LBC | 0.10 | .+......++..+.+++ | 7 | 0.00 | .................... | 0 |
| QBC | 0.00 | .................... | 0 | 0.00 | .................... | 0 |
| LBD | -0.10 | ..................+ | 1 | 0.00 | +..+............... | 2 |
| QBD | 0.00 | ..............+..... | 1 | 0.00 | +..+................ | 2 |
| LCD | 0.10 | .....+....+......+++ | 4 | 0.00 | ..............+.... | 1 |
| QCD | 0.00 | .................... | 0 | 0.00 | ................+... | 1 |
| TAB | 0.00 | .................... | 0 | -0.08 | ..............+.+++++ | 6 |
| TAC | 0.00 | ....+..+..+....... | 3 | 0.17 | ......+.+...+..++.++ | 7 |
| TAD | 0.00 | ..............++... | 2 | -0.17 | .............++++.++ | 6 |
| TBC | 0.00 | .........+...+.... | 2 | -0.17 | ..............+..+++++ | 7 |
| TBD | 0.00 | .................... | 0 | 0.17 | ...........+..++.++ | 5 |
| TCD | 0.00 | ....+............... | 1 | -0.08 | ...............+++ | 3 |
| ABC | 0.13 | .++.........++..++ | 8 | 0.00 | ........+.......... | 1 |
| ABD | -0.13 | +...+.+........+++++. | 8 | 0.00 | .................... | 0 |
| ACD | 0.13 | ..............++.++ | 4 | 0.00 | .................... | 0 |
| BCD | -0.13 | +.........+.+.++ | 5 | 0.00 | .................... | 0 |
| LTAB | 0.00 | .+................. | 1 | 0.00 | ..............+..... | 1 |
| QTAB | 0.00 | .................... | 0 | -0.06 | .+..+......+.......+ | 4 |
| LTAC | 0.00 | ...+.....+..++..... | 4 | 0.00 | .................... | 0 |
| QTAC | 0.00 | ........+........+. | 2 | -0.06 | .................... | 0 |
| LTAD | 0.00 | .................... | 0 | 0.00 | .................... | 0 |
| QTAD | 0.00 | .................... | 0 | 0.06 | ...................+ | 1 |
| LTBC | 0.00 | ..........+.+....... | 2 | 0.00 | .................... | 0 |
| QTBC | 0.00 | .................... | 0 | 0.06 | .................... | 0 |
| LTBD | 0.00 | ....+..........+... | 2 | 0.00 | .................... | 0 |
| QTBD | 0.00 | .+................+. | 2 | -0.06 | ..................+ | 1 |
| LTCD | 0.00 | .................+.. | 1 | 0.00 | ..............+..... | 1 |
| QTCD | 0.00 | .................... | 0 | -0.06 | ...................+ | 1 |
| LABC | 0.00 | ........+........+ | 1 | 0.10 | ..................++ | 2 |
| QABC | 0.00 | .............++...+ | 3 | 0.00 | ..............+..... | 1 |
| LABD | 0.00 | .............+...+ | 2 | -0.10 | ........+......+.+ | 3 |
| QABD | 0.00 | +.............+.... | 2 | 0.00 | .................... | 0 |
| LACD | 0.00 | ....+............... | 1 | -0.10 | .......+...+.....++ | 4 |
| QACD | 0.00 | .................... | 0 | 0.00 | ........+.......... | 1 |
| LBCD | 0.00 | .................... | 0 | 0.10 | .....+..........++ | 3 |
| QBCD | 0.00 | .+.................. | 1 | 0.00 | ..............+.... | 1 |
| TABC | -0.04 | .................... | 0 | 0.00 | .................... | 0 |
| TABD | 0.04 | ............+......+ | 2 | 0.00 | ................+.. | 1 |
| TACD | 0.04 | .................... | 0 | 0.00 | ................+.. | 1 |
| TBCD | -0.04 | .................... | 0 | 0.00 | .................... | 0 |
| ABCD | 0.00 | .+.................. | 1 | 0.00 | +..............+... | 2 |
| LTABC | 0.00 | ..........+....+.... | 2 | 0.00 | .................... | 0 |
| QTABC | 0.06 | ..........+..+...... | 2 | 0.00 | ..............+... | 1 |
| LTABD | 0.00 | .................... | 0 | 0.00 | .................... | 0 |
| QTABD | -0.06 | ..+...............+ | 2 | 0.00 | .................... | 0 |
| LTACD | 0.00 | .................... | 0 | 0.00 | .................... | 0 |
| QTACD | -0.06 | .................... | 0 | 0.00 | ........+.......+.+. | 3 |
| LTBCD | 0.00 | ...+..........+... | 2 | 0.00 | .................... | 0 |
| QTBCD | 0.06 | ..........+...+...+ | 3 | 0.00 | .................+. | 1 |
| LABCD | 0.31 | ..............+++++ | 5 | 0.00 | ..............+.... | 1 |
| QABCD | 0.00 | ........+......+.. | 2 | 0.00 | .................... | 0 |
| TABCD | 0.00 | .................... | 0 | -0.17 | .........+......+++++ | 6 |
| LTABCD | 0.00 | ............+....... | 1 | 0.00 | .................... | 0 |
| QTABCD | 0.00 | ............+....+. | 2 | 0.06 | .....+............+ | 2 |

To examine which features of the barrier patterns the subjects were taking into account at the outset of a trial, we focused attention on each subject's location relative to the starting point on Jump 5. We examined the $x$- and $y$-axes separately. On any given trial, $x$ and $y$ scores could each vary from $-5$ to $+5$. For each subject and for each spatial axis separately, we computed what amounts to an ANOVA. To be precise, following the logic of Cohen and Cohen (1975), we computed a Pearson $r$ between the $x$ or $y$ score and a set of coded weights. These weights were based on the variables and interactions of variables in our factorial stimulus design that related specifically to distance and direction (excluding the variable "hole in center barrier"). Then these $r$s were translated into sums of squares for ANOVA. For purposes of comparison, an analogous analysis was performed on a hypothetical best subject, namely the optimal direction data that are shown above each map in Figure 14.2 (transformed from left–right, up–down into numbers). For the $x$-axis, two trials per pattern were necessary; patterns whose correct $x$ direction could be either left or right were scored once each way. The main point of using $r$ was so that the sign as well as the magnitude of an effect could be known. For example, if the $r$ on an effect in the $x$-axis is negative, then the optimal subject should be nudged a bit to go left, the strength of the nudge depending on the absolute value of $r$ (or its square, which the statistically inclined might call the proportion of total variance of the $x$ scores explained).

Figure 14.5 shows for each individual subject, each spatial axis, and each component of variation (other than residual) in our design, broken down to single degrees of freedom, whether or not the corresponding value of $F$ in the ANOVA exceeded 3.94 (nominally $p \leq .05$, $df = 1, 96$). The point is to look at the forest of results as well as the trees; questions regarding the statistical significance of any given $F$ are of secondary, if not minor, concern. Note that each component of variation is orthogonal to (independent of) all others, as of course the $x$-axis of space is orthogonal to $y$. The 95 squared values of $r$ for the "best" subject add up to 1.00 on the $y$-axis, as indeed they should. On the $x$-axis, the "best" tally is a bit short of 1.00 because on some patterns the correct direction is indeterminate. For the real, live subjects, the tally would vary and

---

**Figure 14.5 (at left).** Umweg data analysis of variance (ANOVA) on the $x,y$ vector of the joystick cursor at Jump #5. Each column is a different subject, and each row is a component of variation. A separate ANOVA was performed on each subject; each component of variance has 1 degree of freedom. Only Patterns 1 through 96 were used in the ANOVA. Column totals for the number of "+"s for each subject are given in Table 14.1. Subjects (columns) are listed in this order: "best," 103, 104, 224, 340, 344, Madu, Kanzi, Carl, Clint, Patrick, Winston, Austin, Lana, Mercury, Sherman, Panzee, Juliet1, Juliet2, Romeo1, Romeo2. Comp. = ANOVA component; r_best = value of $r$ for an optimal path; Rhes. = rhesus ($n = 5$); Apes ....... = ape subjects ($n = 11$); Hum. = human subjects ($n = 4$); L and Q = the linear and quadratic component of the location ($x$-axis) of the target; T = the location ($y$-axis) of the target, above or below the center barrier; A, B, C, D = the various "arms" of the H pattern; + = $F(1, 96) \geq 3.94$, $p < .05$ for this component of variance.

amount to the overall multiple correlation (squared) between the subject's initial move and all of our stimulus variables. (The multiple correlations squared range from .36 to .92, and it is not hard to guess who's who.)

In the same sense, the + signs in Figure 14.5 may be added up in either the vertical or the horizontal directions. Added vertically, they show how many different components of variation a subject discriminated. The totals for the $x$-axis and the $y$-axis are shown for each subject in Table 14.1 (see columns AvX and AvY). Added horizontally, they show how powerful an effect any given stimulus component had on all subjects in general, or in other words how many subjects discriminated a particular component. These totals (shown in Figure 14.5) are highly correlated with the Pearson $r$ shown for the hypothetical "best" subject ($r = .86$ for the $x$-axis; $r = .86$ for the $y$-axis; $N = 95$ components). In other words, the stronger the influence a stimulus component had on what was the optimal path, the larger the number of subjects who were indeed affected by it. Note too the obvious group differences, and also how any given stimulus component affects the outcome on $x$ and $y$ quite differently, as of course it should. It is particularly interesting that many seemingly weak stimulus influences ($r$ values of .10 or even less) were clearly detected by the subjects, even in many seemingly complex interaction effects. Oddly enough, only 1 of the 4 human subjects and 1 of the 5 test-wise chimpanzees discriminated variable $T$, even though this involved a very substantial $r$. We have no explanation for this other than the fact that this variable did have an influence in many of the interaction effects.

Finally, consider in Figure 14.5 the apparent similarities between various pairs or groups of individuals. The humans and test-wise apes in particular tended to respond in similar fashion to the 95 variables. In many if not most cases, however, each individual resembled the hypothetical "best" subject just as clearly, if not more clearly, than he or she resembled other individuals.

Columns AvX and AvY in Table 14.1 are intercorrelated $r = .89$ ($N = 20$ subjects), which shows that individual differences are reliable across the two spatial axes. Furthermore, AvX and AvY are predictable from the Table 14.1 data on Jump 1 ($rs = .84$ and .86, respectively) and Jump 5 ($rs$ with column rJ5 in Table 14.1 = .87 and .92, respectively). The last finding is perhaps not surprising, because the ANOVAs were computed on the subjects' spatial locations relative to the start point on Jump 5; but it is significant in showing that we can still reconstruct the data after having dissected them into many components.

Finally, the reliability of the data was even higher across trials (first 96 trials, 1 per pattern, vs. second 96 trials) than across the two spatial axes. We also point out once more that the analyses rest on only 2 trials per pattern per subject and also largely on the very outset of the cursor's travels; thus, subjects that look bad are being judged by very stiff criteria, and for them one should look also at overall measures.

## Patterns 97 Through 192

On almost all trials on which a pattern contained a hole in the center part of the barrier, the subjects went through the hole. (The lowest score here was 28 out of 32 trials for chimpanzee Clint; chimpanzee Lana and rhesus 224 were among the next worst with 29 of 32.) Additional data on the relative efficiency of each subject's paths are included in Table 14.1. Data on many measures, including Jump 1, are omitted for brevity and because they tell nothing new. It is apparent from the data that group differences were virtually as clear-cut as when real detour problems (Patterns 1 through 96) were used. How is it that some subjects could do so poorly on our measures of distance minimization even though they obviously discriminated the hole and the shortcut? It was largely because on some trials they initially started off in a wrong direction (which usually would have been correct if the hole were not there) and then reversed direction. Once again, the humans were clearly the best at looking ahead.

## Delayed Response Variations on Patterns 1 Through 96

All 4 human subjects said that they found the task of navigating around the various barriers to be demanding or difficult once the barriers were rendered invisible. Certainly their performances were a bit different from those shown in Table 14.1. But the differences are small, and on every measure they still matched or surpassed the chimpanzee data in Table 14.1.

In contrast, even the most test-wise of apes, which had had extensive experience with other sorts of delayed response tests, were as poor as any barrier-naive animal. The experiment, for each ape subject, was terminated in less than one session. We did manage to shape them in later sessions by using, for example, a few of the simplest of barriers and making them flash on and off (first at fast rates and then progressively slower), by having both the cursor and the barriers continuously visible at the outset, and by letting the cursor make $n$ jumps before the barrier disappeared from sight (with $n$ at first large enough to get the cursor practically to the edge of the barrier). But even after such variations and hundreds of trials, their performances still fell far short of what they had been as shown in Table 14.1. For example, on the initial days of tasks in which the barrier stayed visible until the cursor made $n$ jumps, their paths might have been perfect until jump $n$ but then switched direction on $n + 1$ and ran into the barrier. These paths did not always head straight at the goal; more likely, they suggested that the subjects underestimated the length of the barrier and got further confused after a few bumps.

## Discussion

Space is, so some philosophers say, one of the most quarrelsome if not misguided topics in all of science. With this in mind, we save for last the most contentious, and possibly misguided, of psychology's quarrels and first consider other matters.

### What Is New Here?

Early research on the perception or learning of what is the shortest path to a goal in a situation that contains a simple barrier (especially but by no means exclusively the research of Köhler [1925], Lewin [1935], and Lorenz [1971]) was and sometimes still is characterized as being qualitative and anecdotal if not also subjective and overly mentalistic. The analyses and the data that we have presented here suggest, however, that most of our predecessors' observations and insights were right on target. It is no more difficult in principle for a practiced observer to estimate the relative optimality of an animal's travel and movement patterns by eyeball than it is for a statistician to estimate the magnitude of a correlation coefficient from a scatter plot. Nor is the analogy far-fetched. If one wants numerical and graphic rather than verbal descriptions of detour performances, they are available. Of course, the sorts of analysis we offer do not answer all of our questions. In particular, how optimal a performance appears to be, how representative it is, and exactly how it and individual differences originated are three different and complementary questions.

As stated in the introduction, we believe that the behaviors portrayed in Figure 14.1 and indeed in all of our own travel data reflect a continuum "of sorts." Our graphical and numerical techniques describe this continuum and do seem objective and quantitative and more elegant philosophically if not mathematically than counting errors. However, the qualifying phrase *of sorts* is important. What we have been attempting here to scale falls on the right-hand column of Figure 14.1; any implications for scaling on the other two columns (subjects and psychological processes) must be approached with fear and trembling.

Moreover, even the most quantitative-looking analyses must always be supplemented with some qualitative considerations. For example, James's (1890) iron particles would have scored perfectly on the first five jumps of every catch trial and would have been first to reach the goal on some trials, thus coming out "not bad" on average. For most purposes, such an average would be meaningless, if not silly. For other problems, a simple count of how many times the subjects managed to reach the goal at all will be more informative than our measures. Alternatively, one might look at the distribution of the scores and ask how often a subject has a score that is better than some "reasonable" cutoff value. For the idealized chimpanzee travel path that is portrayed in Figure 14.1, our eyeball estimate of a Pearson $r$ between distance traveled and remaining distance to target would be $-.99$ or better. That sounds like a stiff criterion. For many problems in psychology, an $r$ of $\pm.50$ is respectable, as long as it might be called better than chance. The idealized hen track in Figure 14.1 might beat that. Take your pick.

Some readers might say that everybody already knows that physical parameters such as distance are scalable and that individual if not species differences are ubiquitous. If that is so, and at that level of discussion, there is nothing new in our data or in this chapter. What does, however, still surprise us is the power of even the seemingly simplest of stimulus patterns and even a single jump in one direction, on a given trial, in a task on which five jumps can be made within 0.75 seconds, to reveal what the subjects are taking into account, if not how they are planning ahead, and how subjects differ from one another. Perhaps the most surprising finding for us was that whereas the humans could respond clearly and immediately as if in a ballistic fashion in the delayed response test, even chimpanzees that had demonstrated unprecedented learning and memory capacities in other situations (e.g., Beran, Pate, Richardson, & Rumbaugh, 2000; Brakke & Savage-Rumbaugh, 1996; C. R. Menzel, 1999; C. R. Menzel, Kelley, & Sanchez, 2002; Rumbaugh, 1977; Savage-Rumbaugh, 1986) seemed instead almost totally dependent on continuous, or near-continuous, sight of the barriers. Obviously, they had not acquired a limited set of fixed responses that were somehow triggered reflexively by a given stimulus pattern; if anyone was capable of doing this or simulating this in the present tests, it was humans. We can only wonder how monkeys and apes might be able to perform on an everyday, locomotor version of our delayed response task.

## Simplicity–Complexity

According to Newton, Nature is pleased with simplicity. We would add that humans in particular are pleased with simplicity, and if and when it seems that Nature is not, all but the Newtons among us are apt to hold it against her. If indeed anyone might have any trouble in distinguishing between humans and nonhumans on the basis of their behaviors, travel paths, and artifacts, this is a good rule to remember. It certainly applies in the case of our data.

It is apparent from inspection of our stimulus patterns, or in fact just the pattern "H," that there is a strategy by which one could solve all of our patterns without ever having to look at the barriers or ever having to contact them. (Just turn your back to the target; go to the outer wall or border of the video screen, follow the wall until you are directly to the rear of the target, then go straight for the target.) It is interesting that no subject ever used this strategy, albeit some of the chimpanzees tested with the delayed response task seemed to be headed in that direction on a few occasions, and a blind subject or an animal less visually dominant than a primate or a bird might be expected to arrive at it much sooner, especially if the animal is a wall-hugger anyway. This "universal" strategy is certainly simpler in some respects and may sometimes be quicker, shorter, and less effortful than trying to use our own maps of optimal paths. Obviously, even 100 maps are of no use if one cannot see any of them or does not know which to use. It is no wonder that Tolman (1932) dedicated his book to the wall-hugging *Mus norvegicus albinus* (see also Poucet, Thinus-Blanc, & Chapuis, 1983).

How simple were the observed paths of our subjects? That is a large topic into which we do not try to delve, for purposes of brevity; but one of many possible estimates of simplicity would be compressibility (e.g., Benedetto, Caglioti, & Loreto, 2002). Each travel path can be reduced to a string of digits or letters, as in the raw data shown in Figure 14.3. Especially if one looks at how many times each digit appears in the sequence or at our changes in direction measure, it is obvious that some strings can be compressed much farther than that. Now, then: Which subjects are going to have the smallest raw data files? Whose files will show the greatest shrinkage if you do your best to compress them? If you bet that our subjects would rank just about as they do in our figures and tables, you win a banana. Furthermore, if you go on to guess which of the stimulus patterns will have the most and least compressible files and which will produce the most and least compressible optimal paths, you get Coke and M&Ms as well.

Linear equations are even simpler (or more simplifying) than a string of discrete responses. Our subjects may be said to have estimated the shortest or straightest path between two points or to have followed the best-fitting curve around obstacles. The degree to which they did this on a single trial may, if one wishes, be reduced to a single number, such as a Pearson coefficient ($r$). The best of the live subjects, however, do not ordinarily just have high negative $r$s. Where there are many possible paths that are equally optimal, the best of the live subjects are selective. In particular, they show far fewer changes in direction than the average randomly selected optimal path and seem to go for one of the more Euclidean of the available optimal paths.

That brings us to the nonlive subject, the computer. Wasn't it really the star of the show, and smarter than anyone? Assuredly so, in the same sense that Einstein used to say that his pencil was smarter than he was. Even our "ancient" 1996 home computer can solve 100 novel barrier patterns in less than 1 minute, not to mention computing more statistical analyses in an hour than we would care to do in a lifetime with pencil and paper. The only way we could top that would be to propose a new Olympian pentathalon based on Lorenz's (1971) classic paper on psychology and phylogeny, which argued that humans are above all "specialists in non-specialization and curiosity" (p. 224).

We do not know precisely how simple or complex our barrier-solving program really is. (That is a job for computer scientists and physicists, and we doubt that many of them would claim to know, with great precision.) The program, however, solves strictly one maze at a time, independently of all others, and our subjects assuredly did not. Given our map for any given pattern, including start and goal, the program could cope with a transfer-of-training problem involving a change in either the start point or the goal, but it could not cope with random changes in both, simultaneously, nor could it find a new shortcut in a given pattern, unless of course it treated these changes as new problems, as we in fact told it to do. (The same would be true, as far as we can judge, for the robots and programs described by, e.g., Braitenberg, 1984; Deutsch, 1960; Krieger et al., 2000; Muller et al., 1996; Walker & Miglino, 1999.) It does not recognize any objects as gestalts. It does not assign any weights to any given arm of the barrier or do what amounts to an ANOVA on all patterns, taken together. The size of the problem on a 25 row × 40 column space is almost infinitesimal compared with the everyday visual world of a primate. Our shortest-path algorithm cannot handle tasks involving more than one cursor, or competition or cooperation, or multiple goals (as in the traveling salesman problem), let alone tasks with no obvious, clearly definable goals that are known to the experimenter or programmer—and the latter sorts of tasks are probably the rule rather than the exception in the real world. Compared with a chess program—which must, among other things, use a different sort of jump metric for each type of chess piece—our program is easy, but compared with a Lorenz-style pentathalon, even today's world-champion chess program is child's play.

We are not trying to eulogize people or animals or to disparage the value of computers. What we are trying to do is to note that although someone or other can probably always, in time, create a machine or train an animal to do something that someone puts forth verbally as a defining feature of, say, human intelligence, that is an exercise in human logic, not necessarily bio-logic. By now smart robots have been around for several hundred years. Each new one either excites or alarms its human designers and viewers and leads to wild speculation about the future. Still, the amazements of each decade have inexorably become the amusements and kid's toys of the next decade, whereas the more we learn about cavemen and even more remote real biological kin, the smarter and more like us they seem, at least to most of us. Our prediction of future trends is more of the same.

## Gestalts Versus Stimuli and Their Interactions

In the past, the term *gestalt* and even its synonyms, such as *whole, configuration,* or *patterning*, were considered to be controversial. Respectable American psychologists were supposed to talk instead about stimulus variables or elements and interactions thereof. With our data, readers may take their pick. Would you rather say that (a) most subjects sometimes, and some subjects most of the time, seemed to comprehend each simple barrier as a whole and to judge the shortest-looking path and plan their route from the outset, before making their first jump, even on the first occasion they saw a pattern, or (b) subjects discriminated many different variables that affected the minimum distance between start and goal plus many interactions between these variables, including higher order interactions, and thus some subjects solved simultaneously 50 or more statistically independent discriminations?

Either statement sounds plausible to us. The first sounds more like plain English and common sense, and the other will hopefully impress more professors, especially when they note that all 4 of our undergraduate subjects managed accurately to "compute" even some fourth-order interactions (LABCD on the $x$-axis and TABCD on the $y$-axis in Figure 14.5) in an ANOVA design that might raise one's hair even after graduate school. Nor were some of the nonhuman subjects far behind them. Nor is our design as detailed as it might be. It treats each arm of a barrier as a gestalt and ignores the tests that involved opening up a gap in one place. Break each arm down into its elements (cells, pixels, and whatever lies beyond that) and one really could have a statistician's ball.

As the section on simplicity–complexity might suggest, we think it far more likely that our subjects operated more on simple, intuitive snap judgments than on detailed or rational analysis. (Is our task a test of spatial perception or of spatial cognition? Are there neural processes in the primate retina or visual cortex that operate like waves to solve problems such as ours? We leave these questions to the reader, although that by no means is to say that we consider them unimportant.) We are impressed, on the one hand, with how many details the subjects did take into account in less than 1 second on a given trial and with little training. On the other hand, we are impressed equally by the fact that insofar as any given pair of subjects resembled each other (across the 95 different components of variation) they also usually resembled the hypothetical best subject. Obviously there are sources of similarity between subjects other than sources that stem from common, immediate external stimuli, but their effects are not always evident in every situation.

## Physics, Aesthetics, and Intelligence

There are several minor puzzles in our data, all of which, we think, have something in common.

- As Emilie Menzel (10 years old) put it in her first question about the data: "How come the boys did better than the girls?"
- Why did the humans on the first few cursor jumps do much better than the nonhumans in terms of distance reduction but not direction, only to do the opposite after 10 jumps?
- Why did Emil Menzel (whose data are not shown) do even better than the other boys, even though he was of retirement age; a novice at video games; and made every effort to respond quickly, intuitively, and without hesitation or reflection?
- Why do Köhler's (1925) and Hull's (1952) pictures of animals' travel paths, as they circumvent a simple barrier, look so neat, if not beautiful?
- What accounts for the following seemingly odd behaviors in the task involving shortcuts? As mentioned already, some animals on some trials initially started off in a wrong direction and then reversed direction. Sometimes, however, they had already traveled most of the way to the goal; by reversing themselves they had to travel considerably farther overall than if they had stuck to their initial direction and not bothered about trying to correct their error. Odder still, some subjects, a few times, went through the shortcut and got within a jump or two of the goal, but then went back to the start and circled the entire barrier. Humans who do the same (or who watch movielike reruns of our data) are very likely to smile, grimace, laugh, or vocalize in the process, at quite predictable points in the process.

One common denominator we have in mind here is that not everybody is necessarily trying to minimize or maximize the same thing, and not everybody necessarily focuses on a single thing or sticks to it consistently. The first author's advantage was largely that, even without deliberately trying, he focused on minimizing the total number of jumps; and of course he had written the program in the first place. Actually, because each jump took 0.15 seconds and was accompanied by a click, he could just as well have estimated the number of clock ticks, but at least according to him, he did not. The "girls' handicap" was largely that, when headed for a position that was, say, five jumps up and five jumps to the left, they were more inclined than the boys to move left and then up, rather than entirely on the diagonal—which might make good sense in terms of ease of control, with a joystick whose movements follow the rules described earlier and feel jerky until one gets used to them. That might explain finding (b) above, as well as (a), albeit not necessarily so (see, e.g., Shore et al., 2001, for reports on sex differences). After their testing was over, Juliet1 and Romeo1 watched reruns of their data, and Juliet1's comment, when they noted the difference between their strategies was, "Oh, was I supposed always to take the shortest path? My way seemed easier." Another common strategy, which many subjects used, was to sweep around a barrier in an arc that might be a bit too wide to be optimal in terms of distances but might well be easier in terms of biomechanics.

A second common denominator in the above findings is succinctly stated in Dobkin et al.'s (1998) article on graphics design:

> In contrast with techniques from other domains such as . . . robotics, where physical constraints play a major role, aesthetics play the more important role in graph layout. For graphs, we seek paths that are easy to follow and add meaning to the layout. (p. 262)

This applies not only to many of the figures that appear in journals and books but also, at least by way of analogy, to the "travel paths" and "figures" that our subjects produced with their joysticks for this chapter. Who has not discovered a new and shorter road from one town to the next, yet stuck to the familiar one, especially if it is more scenic? A route that changes direction every second or two is just plain ugly, unless of course one is a teenage hot-rodder.

Finally, as the above findings are worded, they suggest that distance minimization (or deliberate flouting of it, in some cases) has something to do with intelligence. We hope that there is some connection, and in this regard we were pleased to see the article by Shore et al. (2001) on what might still be learned from the Hebb–Williams intelligence test, a legitimate grandparent of our own stimulus patterns. The authors have developed a televised virtual maze version of the Hebb–Williams barrier patterns for use by humans, and who are better candidates for the next subjects to take such a test than monkeys and apes?

## Nature Versus Nurture

Initially, all test-naive monkeys and apes that we have observed, and plenty of humans too, produce joystick travel paths that look more like those at the bottom or middle of Figure 14.1 than like those at the top. They learn to discriminate between single barrier cells and targets in no time at all. They do not have to bump every barrier cell in a long row to get around the whole row, but the bigger the required detour the more trouble they have. Obviously, experience of some sort is necessary before one "gets to the top." Anyone who assumes that that is all there is to the so-called nature–nurture problem thereby may consider their problem to be settled. The large difference between the two groups of chimpanzees, as well as the superiority of test-wise rhesus to some of the relatively test-naive chimpanzees, is of course further evidence of the same point.

Other questions, however, almost inevitably follow. For example, could the average adult chimpanzee match or surpass the average adult human, given sufficient training? On the present tasks, the answer, in some respects, is probably yes; in other respects, no, as far as we know and would place our bets. Still, who knows what sufficient training is? We are talking like examiners, not teachers, trainers, or parent surrogates, and the world needs all of the above.

Where did the animals' strong tendency to move the cursor straight toward the target come from? This is not the same as asking how they came as infants to walk straight toward a banana, to reach their hands straight toward a banana, or even to shake a tree branch or a stick in the direction of a target; it is at least one or two levels of abstraction (or higher order conditioning, if one is a Pavlovian) up from that (E. W. Menzel, Davenport, & Rogers, 1970). Thus, for example, in the case of walking toward a banana or away from a barrier, one might invoke the simple magnification and minimization of the size of the animal's visual images as sufficient cues for minimizing or maximizing distance to the goal; but such cues are irrelevant in the present task. Here the cursor, not the subject, is doing the walking. The subject is viewing the scene as if from the heavens and trying to tell its slave what to do. Furthermore, any goal gradients that emanate from or surround the target on the video screen (see the gradients in the map in Figure 14.3), are metaphorical as far as the cursor is concerned, and their connection to the gradients that surround the food-hopper is not necessarily obvious. Studies of chimpanzee acquisition of performances on a video touch screen (Iversen & Matsuzawa, 2001) seem more directly pertinent to the tasks involved in our experiment than do everyday tasks, and they are fascinating in their own right, but they are still not the same thing.

Is it assumption or fact that all animals in general "naturally" will tend to head straight toward a distant, desirable goal once they detect it, as the quotation from James (1890) in the introduction implies? Given many seeming exceptions and complications (e.g., Hediger, 1964; Lorenz, 1971; C. R. Menzel, 1986; E. W. Menzel, 1978), we would say it is an assumption. Probably no one has been clearer on the matter than Hull (1952, p. 227). He literally called it his Theorem 61, and in effect he deduced it from Euclid. Hull was an empiricist of the old school insofar as he believed that much of vision was learned and also that the least-distance tendency might derive from what is quickest and least effortful. But he did not deny the existence of "inborn reactions." We do not know how he would account for the ontogenesis of preferences for least time and least effort or whatever in turn would explain their explanation. Although our views differ from Hull's in innumerable respects, we share Rashotte's (1987) opinion that chapter 8 of Hull remains an excellent source of ideas about spatial learning and deserves serious attention. Although our computer program for testing, as we used it, confounded distance and time, if not effort too, this was deliberate. In future experiments, it would be just as easy to unconfound these variables and to pit them against one another as it is to unconfound the Euclidean metric, a jump metric, and the city block metric.

### Behavior Versus Cognition

We had thought that disputes between cognitivists and behaviorists were by now ancient history. Mackintosh's (2002) text on the role of cognitive concepts in the domain of spatial learning, however, was about as revivalist as one can get: "Do not," his title proclaimed, "ask whether [animals] have a cognitive map, but how they find their way about." We still prefer to do both (e.g., C. R. Menzel, Savage-Rumbaugh, & Menzel, 2002) and remain unrepentant in this regard. However, Mackintosh's sentiment is by no means new or unique, and he might just as well have added *insight* to *cognitive map*, so let us throw insight in too, for good measure, because it has been with us in the domain of detour learning at least since James and Köhler. In fact, let us add *association learning* to the list, even though Mackintosh would not. Our point is fivefold. First, if anyone does not want to listen to or talk about cognitive maps, insight, or association learning, that is their prerogative. When they want to dictate what everyone else should and should not ask and talk about, we would question their authority, if not their motives. Second, we find the travel patterns of animals fascinating and intrinsically interesting, whether they occur in the woods or in simulations such as the present one. (We also would have no qualms about saying that what we are studying in this chapter are really the movements of the cursor or the joystick, not mind as such or the animal as such—in other words, we are studying an operant, not respondents, the fixed action patterns of classical ethology, introspective reports, or mind reading.) Third, in formulating this chapter and in designing our experiments, we did not worry about any of the above concepts—except insofar as our intended readers might naturally ask us what light, if any, our studies might shed on their pet concepts. Fourth, shrewd publicity is not equivalent to good science, but sometimes it helps. Tolman's (1948) *cognitive map* neologism helped to publicize ideas and data that predated that term, and Tolman, and did help to generate a fair amount of good science. The revival of the term in the 1970s had the same effect. Mackintosh has said as much himself, more than once, but typically has followed it up with a complaint. Finally, here is how we find our way out of this muddle (if it is Mackintosh's way too, we apologize for the confusion): We do not see association and cognitive mapping or insightful behavior as mutually exclusive alternatives. All of these concepts are fuzzy. All refer to emergents, particularly from the standpoint of neuroscience, evolutionary biology, and other chapters in this volume.

## Newton Versus Descartes: The Final Confrontation

We believe that the roots of many of our problems about space lie in a question that was posed by Aristotle and later considered in depth by Descartes and Newton: Imagine that a man is walking on the deck of a moving ship. Where is he, really?

Descartes and Newton did not separate their physics and geometry from their psychology, astronomy, or even theology. So the question may be interpreted as one chooses. We see it as a problem of mapping, and if it is not "cognitive" for our imaginary man, it is, in some sense, for us as outside observers.

If we were talking about, say, a statue of a man rather than a real live fellow human, we could, said Descartes, specify its location in relation to any arbitrary fixed point we chose, such as the ship's mast, or the distant shore from whence the statue came or is headed, or even the sun—for in Descartes' view, all locations and motions are relative, and there are no fixed points except insofar as we choose to view them as fixed by our own thoughts, goals, and purposes. If our subject is assumed to be as sentient and capable of feeling as we are, we do have a bit of a problem. Descartes solved that side issue by arguing that only humans (and maybe not all of them) are sentient. Newton solved the problem by doing away with relativity and, in effect, ignoring the point of view of both subject and observer.

Neither of these two solutions appeals to us very much. However, our major complaint about Descartes' position is only that we would not draw the line between objects and subjects where Descartes did, and are not entirely sure where, or by what criteria, we would draw that line. Those who yearn for certainty on that score should study theology and law as well as biology; in the domain of psychology and the emergence of mind, James (1890) called it "the deepest of all philosophical problems" (p. 8).

## References

Benedetto, D., Caglioti, E., & Loreto, V. (2002). Language trees and zipping. *Physical Review Letters, 88*, 048702–048705.

Beran, M. J., Pate, J. L., Richardson, W. K., & Rumbaugh, D. M. (2000). A chimpanzee's (*Pan troglodytes*) long-term retention of lexigrams. *Animal Learning and Behavior, 28*, 201–207.

Braitenberg, J. (1984). *Vehicles: Experiments in synthetic psychology*. Cambridge, MA: MIT Press.

Brakke, K. E., & Savage-Rumbaugh, E. S. (1996). The development of language skills in *Pan*: II. Production. *Language and Communication, 16*, 361–380.

Cheng, K. (2002). Generalization: Mechanistic and functional explanations. *Animal Cognition, 5*, 33–40.

Church, R. L., & Marston, J. R. (2003). Measuring accessibility for people with a disability. *Geographical Analysis, 35*, 83–96.

Cohen, J., & Cohen, P. (1975). *Applied multiple regression/correlation analysis for the behavioral sciences*. Hillsdale, NJ: Erlbaum.

Collett, T. S. (1982). Do toads plan routes? A study of the detour behavior of *Bufo viridis*. *Journal of Comparative Physiology, 146*, 261–271.

Collett, T. S. (2002). Spatial learning. In R. Gallistel (Ed.), *Steven's handbook of experimental psychology* (3rd ed., Vol. 3, pp. 301–364). New York: Wiley.

Deutsch, J. A. (1960). *The structural basis of behaviour*. Cambridge, England: Cambridge University Press.

Dobkin, D. P., Gansner, E. R., Koutsofios, E., & North, S. C. (1998). Implementing a general purpose edge router. In G. DiBattista (Ed.), *Lecture notes in computer science: Vol. 1353. Graph drawing* (pp. 262–271). Berlin: Springer-Verlag.

Fraenkel, G. S., & Gunn, D. L. (1961). *The orientation of animals*. Oxford, England: Clarendon Press.

Fragaszy, D., Johnson-Pynn, J., Hirsh, E., & Brakke, K. (2003). Strategic navigation of two-dimensional alley mazes: Comparing capuchin monkeys and chimpanzees. *Animal Cognition, 6*, 149–160.

Gallistel, C. R. (1990). *The organization of learning*. Cambridge, MA: MIT Press.

Harlow, H. F. (1949). The formation of learning sets. *Psychological Review, 55*, 51–65.

Hebb, D. O. (1949). *The organization of behavior*. New York: Wiley.

Hediger, H. (1964). *Wild animals in captivity*. New York: Dover.

Hull, C. L. (1952). *A behavior system*. New Haven, CT: Yale University Press.

Iversen, I. H., & Matsuzawa, T. (2001). Acquisition of navigation by chimpanzees (*Pan troglodytes*) in an automated fingermaze task. *Animal Cognition, 4*, 179–192.

James, W. (1890). *The principles of psychology* (Vol. 1). New York: Holt.

Johnson, S. (2001). *Emergence: The connected lives of ants, brains, cities and software*. New York: Simon & Schuster.
Johnson-Pynn, J., Fragaszy, D., Washburn, D., Brakke, K., Laliberte, L., & Gulledge, J. (2001). Strategic navigation of 2-D computer mazes in three genera of primates (*P. troglodytes, M. mulatta,* and *C. apella*). *American Journal of Primatology, 54*(Suppl. 1), 58–59.
Köhler, W. (1925). *The mentality of apes*. New York: Harcourt, Brace.
Kosko, B. (1993). *Fuzzy thinking: The new science of fuzzy logic*. New York: Hyperion.
Krieger, M. J. B., Billeter, J., & Keller, L. (2000, August 21). Ant-like task allocation and recruitment in cooperative robots. *Nature, 406,* 992–995.
Lakoff, G., & Núñez, R. E. (2000). *Where mathematics comes from*. New York: Basic Books.
Lewin, K. (1935). *A dynamic theory of personality*. New York: McGraw-Hill.
Lindsay, R. B., & Margenau, H. (1936). *Foundations of physics*. New York: Wiley.
Lorenz, K. (1971). *Studies in animal and human behavior* (Vol. 2). Cambridge, MA: Harvard University Press.
Lovejoy, A. O. (1936). *The great chain of being*. Cambridge, MA: Harvard University Press.
Mach, E. (1907). *The science of mechanics*. Chicago: Open Court.
Mackintosh, N. J. (2002). Do not ask whether they have a cognitive map, but how they find their way about. *Psicologica, 23,* 165–185.
Maxwell, J. C. (1991). *Matter and motion*. New York: Dover. (Original work published 1877)
Menzel, C. R. (1986). Structural aspects of arboreality in titi monkeys (*Callicebus moloch*). *American Journal of Physical Anthropology, 70,* 167–176.
Menzel, C. R. (1999). Unprompted recall and reporting of hidden objects by a chimpanzee (*Pan troglodytes*) after extended delays. *Journal of Comparative Psychology, 113,* 426–434.
Menzel, C. R., Kelley, J. W., & Sanchez, I. C. (2002). A chimpanzee's comprehension of televised spatial information. *American Journal of Primatology, 57,* 79.
Menzel, C. R., Savage-Rumbaugh, E. S., & Menzel, E. W. (1999). Organization of movement by rhesus monkeys, apes and humans in computer-presented maze tasks. *American Journal of Primatology, 49,* 80.
Menzel, C. R., Savage-Rumbaugh, E. S., & Menzel, E. W. (2002). Bonobo (*Pan paniscus*) spatial memory and communication in a 20-hectare forest. *International Journal of Primatology, 23,* 601–619.
Menzel, E. W. (1978). Cognitive mapping in chimpanzees. In S. H. Hulse, H. Fowler, & W. K. Honig (Eds.), *Cognitive processes in animal behavior* (pp. 375–422). Hillsdale, NJ: Erlbaum.
Menzel, E. W., Davenport, R. K., & Rogers, C. M. (1970). The development of tool using in wild-born and restriction-reared chimpanzees. *Folia Primatologica, 12,* 273–283.
Muller, R. U., Stead, M., & Pach, J. (1996). The hippocampus as a cognitive graph. *Journal of General Physiology, 107,* 663–694.
Poucet, B., Thinus-Blanc, C., & Chapuis, N. (1983). Route-planning in cats related to the visibility of the goal. *Animal Behavior, 31,* 594–599.
Rashotte, M. E. (1987). Behavior in relation to objects in space: Some historical perspectives. In P. Ellen & C. Thinus-Blanc (Eds.), *Cognitive processes and spatial orientation in animal and man* (Vol. 1, pp. 39–54). Dordrecht, the Netherlands: Martin Nijhoff.
Rumbaugh, D. M. (1977). *Language learning by a chimpanzee: The LANA project*. New York: Academic Press.
Rumbaugh, D. M. (1997). The psychology of Harry F. Harlow: A bridge from radical to rational behaviorism. *Philosophical Psychology, 10,* 197–210.
Rumbaugh, D. M., Washburn, D. A., Savage-Rumbaugh, E. S., & Hopkins, W. D. (1991). Language Research Center's computerized test system (LRC-CTS): Video-formatted tasks for comparative primate research. In A. Ehara, T. Kimura, O. Takenaka, & M. Iwamoto (Eds.), *Primatology today* (pp. 325–328). Amsterdam: Elsevier Science.
Savage-Rumbaugh, E. S. (1986). *Ape language: From conditioned response to symbol*. New York: Columbia University Press.
Shore, D. I., Stanford, L., MacInnes, W. J., Klein, R. M., & Brown, R. E. (2001). Of mice and men: Virtual Hebb–Williams mazes permit comparison of spatial learning across species. *Cognitive, Affective, & Behavioral Neuroscience, 1,* 83–89.
Skiena, S. S. (1998). *The algorithm design manual*. New York: Springer-Verlag.

Steinbock, O., Tóth, A., & Showalter, K. (1995, February 10). Navigating complex labyrinths: Optimal paths from chemical waves. *Science, 267,* 868–871.

Tolman, E. C. (1932). *Purposive behavior in animals and men.* New York: Century.

Tolman, E. C. (1948). Cognitive maps in rats and men. *Psychological Review, 55,* 189–228.

Walker, R., & Miglino, O. (1999). Replicating experiments in "detour behavior" with artificially evolved robots: An A-life approach to comparative psychology. In by D. Floreano, J.-D. Nicoud, & F. Mondada (Eds.), *Advances in artificial life: 5th European conference, ECAL'99, Lausanne, Switzerland, September 13–17, 1999 Proceedings* (pp. 205–214). New York: Springer.

Washburn, D. A. (1992). Analyzing the path of responding in maze-solving and other tasks. *Behavior Research Methods, Instruments, and Computers, 24,* 248–252.

# 15

# Willful Apes Revisited: The Concept of Prospective Control

## R. Thompson Putney

In 1985, I published an article with the problematic title "Do Willful Apes Know What They Are Aiming At?" (Putney, 1985). The thrust of that article, as the title suggested, was to lay a theoretical groundwork for the assertion that apes knowingly exercise intentions like humans, or at least that such an assertion is plausible. The article, referred to in this chapter as just "Willful Apes," was partly motivated by the extensive criticism arising from a noteworthy issue of *Behavior and Brain Science* in which Savage-Rumbaugh, Rumbaugh, and Boysen (1978), as well as Premack and Woodruff (1978), made such assertions with great confidence. In "Willful Apes," the foundation of a theory of intention was developed based on the Miller, Galanter, and Pribram (1960) treatment of "plan" as the central concept, supplemented by four essential points made within the philosophical theory of action presented by Davis (1979). This conceptual system is recapitulated below, embellished by connecting each of these points to phenomena and theory from human cognitive psychology. Next, the treatment of the fourth point, which deals with the core of intention found in the concept of aiming at an outcome, is expanded with some historical background. In this treatment, the aiming function is referred to as *prospective control* and is differentiated into two levels. The first is the direct control of action in the execution of plans, and the second is the supervisory control of planning of upcoming actions within the working memory system (Baddeley, 1986, 1996). Finally, the chapter ends with a discussion of recent systematic evidence that supports the applicability of this conceptual scheme to apes.

### Intentions as Plans

A convenient vehicle for discussing intentions is found in the concept of plan proposed by Miller et al. (1960). Their approach was advanced as an alternative to the then ubiquitous stimulus–response (S-R) formula founded on a much oversimplified view of behavior. Their account of the nature of plan was based on what they termed the *TOTE unit*, which stands for test, operate, test, exit, and was offered as an alternative to the reflex arc (for evidence of other basic behavior control systems, including oscillators and servomechanisms in addition to the reflex arc, see Gallistel, 1980). Plans in general were conceived as a hierarchy of such units. Contrary to the reactive nature of the S-R connection, the TOTE unit was conceptually compatible with organism-initiated action unfolding from a plan. The test phase was the critically novel part of the unit in that it contained a representation of the outcome state that the operation was working toward, so that the operation continued until the outcome was achieved, which resulted in an exit from that unit. Additional flexibility was obtained in that the "operate" phase could either be behavior or covert operations of thought, and the latter could account for unobserved processes in problem solving. Finally, the test itself was conceived as a monitoring function that compared the results of the operate phase against the goal representation. A hierarchy of TOTE units appeared necessary to conceptualize most complex behavior, with the lower levels providing necessary subroutines with their own subgoals and one or more superordinate layers with the overarching outcomes of a whole action or operational sequence. Thus, for example, to make an omelet, one needs to assemble appropriate implements in some order—a bowl, a fork or whisk, and frying pan suitable for omelets; then the eggs must be cracked and whisked and poured into an almost smoking pan greased with a little butter; and so on. Such an action sequence requires a series of subroutines, each with its test (e.g., the eggs are sufficiently whisked), or finding that the omelet is sufficiently done to fold out onto a plate and then exiting to the next subroutine, getting the plate to the table, until the consummatory phase commences with its own plan.

In Miller et al.'s (1960) concept of plan, the test and operate phases were presented as serially occurring phases of the unit. However, a useful extension of this scheme is to consider that the test and operate phases may occur in parallel in some actions, so that the monitoring of operational results is concurrent and determines the ongoing flow of performance. This would appear to be the case in continuous tasks, particularly with skilled movement. For instance, while maneuvering a car around a curve, the test phase continuously monitors the trajectory of the car to detect indications of under- or oversteering leading to compensatory adjustments as such perturbations of guidance are detected.

## Action Theory

Davis (1979) presented four points from action theory in his critical philosophical analysis that provided the necessary requirements for understanding everyday human intentions. The importance of a practical understanding of intention was highlighted in his elaboration of the concept of responsibility underlying much of our jurisprudence. The theory of action and the arguments from that literature that he summarized had come from a variety of sources within the tradition of ordinary language analysis, including Anscombe's (1957) seminal work on intention. The purpose of such analysis was to explicate our everyday understanding of language use, including the many common words referring to mental events like *beliefs*, *desires*, *intentions*, and *volitions*. Because such words occur regularly and reliably in ordinary speech to describe mental processes and their relation to actions, the kind of careful analysis that Davis (1979) presented within action theory provides a promising background for a cognitive theory of intention.

The ultimate purpose of this chapter is to apply these concepts to chimpanzees and possibly to other nonhuman species; thus, the reader might try to imagine, as each is described and related to cognitive concepts, whether the four points might pertain to familiar animals as well as to humans. This proposal to take a comparative approach to intention is based on the following view. Given that evolution is conservative, the system through which intentional action is framed is probably not novel to humans but rather has a long history within the developing vertebrate brain. Thus, the phylogenetic grade at which intentional control evolved is open to question and undoubtedly awaits a careful analysis of the underlying processes and related neural control mechanisms. One possibility is that intentional control is common at least among mammals. However, there may be a number of levels of complexity in the development of the control of action that is globally referred to as "intention." Whether or how any particular level is applicable to species more removed from humans than the great apes will require much further work in comparative cognition as well as the underlying neural control processes.

The first of Davis's (1979) points was that intentions are exercised by agents who are either aware of what they are intending or can consciously access that information when it is needed or queried by someone else. Here an agent is treated as a person with a number of requisite characteristics. These include the capacity to engage in many different actions, possessing self-awareness leading to an ability to reflect on what they are doing and the reasons why they are doing it, as well as the requirements in the other points detailed below. Such properties were the substance of the assertions about the intentional actions of chimpanzees in the controversial articles (Premack & Woodruff, 1978; Savage-Rumbaugh et al., 1978) mentioned earlier. Within cognitive psychology, substantial enlightenment on the nature of agents as persons has come from Neisser's (1988) treatment of the five aspects of self. These aspects include the ecological self most directly relevant to the operational aspects of agency (see also Neisser, 1993) but also the social, conceptual, autobiographical, and private aspects of self that under normal circumstances are all intertwined. In brief, the *ecological* self is embodied, situated in the world through various perceptual modalities, particularly vision and proprioception, and possesses from Neisser's (1988) Gibsonian perspective various learned affordances for action provided by the surfaces and objects currently perceived in the world. Thus, the floor or ground affords walking, a chair affords sitting, a cup affords drinking, and so on.

Davis's (1979) second point was that an agent as an adult person possesses a large accumulation of knowledge. Within the array of knowledge is the ability to engage in many actions, the contexts in which they may be performed, and the outcomes that they may achieve. In addition, there is knowledge of self and knowledge of the autobiographical past that make up the conceptual and autobiographical aspects of self in Neisser's (1988) account. The concept of schema has provided the basic vehicle of this knowledge and has had a long history predating current cognitive psychology. Frederick Bartlett (1932) used the term *schema* to account for the generic nature of knowledge that captures the variation in exemplars of both natural and artificial categories. He began his account by introducing the concept of the body schema discussed by Henry Head (1920), a prominent British physiologist of the time. Head characterized this central schema as a dynamic system of representation of the ongoing flow of action, changing as momentary behavioral accommodation required. Bartlett went on to adapt the concept with related experiments to a range of knowledge from perceptual categories to narrative schemas. Further empirical research into the nature of schema representation continued with Attneave's (1957) research using artificial categories made up from random shapes, and this line of work was pursued further by Posner and Keele (1968). Rosch (1974) did substantial research on natural categories in addition to artificial ones used by previous authors. This body of research demonstrated that perceptual categories manifested both central tendency and variability or typicality effects within category representation. The central tendency effects have often been referred to as *prototypes*, which are appropriate to random shapes distorted from a prototype figure (Attneave, 1957; Posner & Keele, 1968). However, according to Rosch (1974), the term is not appropriate for natural categories commonly dealt with in the real world, because they do not have true prototypes. Rather, category exemplars vary in their degree of typicality, which Rosch tapped by obtaining typicality ratings with the central tendency reflecting the most typical exemplars and variability showing the spread of decreasing typicality over less representative members.

The central tendency and variability effects nicely mirror the sort of distribution found in the topography of motor actions. Thus, a standard action like that of a player skilled in a sport, such as a tennis player serving ball, has a four-dimensional distribution depending in this case on the height the ball is tossed, followed by the trajectory of the racket with the ensuing timing of impact, and so on. Such actions show variation in response topography from one instance to another in the four physical dimensions, with a central tendency in producing the skilled action commensurate with the ability to achieve an effective outcome. The basis of an agent's knowledge of action is in motor plans that can accomplish actions like serving a tennis ball. Such plans may be encompassed within larger scenario types of schemas referred to as *scripts* by Schank and Abelson (1977), which capture the knowledge of how to accomplish various goal-directed activities requiring a set of motor plans as subroutines, sometimes with at least partial serial ordering. For instance, in the earlier omelet example, the eggs need to be broken into a bowl before they are whisked to avoid an eggshell-laced omelet. The point of this brief summary is that these systems of knowledge provide a dynamic structure for the playing out of intentions, but according to Davis (1979), the agent's knowledge alone is necessary but not sufficient to capture the meaning of intention.

Beyond just possessing knowledge, Davis's (1979) third point was that intention requires that agents engage in an action for a reason. Actions do not just come out of nowhere but have some impetus. This is typically found in the motivation provided by the anticipated outcome of the action to be engaged. More than 20 years ago, Zajonc (1980) pointed out that common laboratory tasks designed to investigate cognition are conducted under conditions of minimal motivation, resulting in a relatively cold state of cognition. Contrasted with this, Zajonc noted that everyday cognition, including intention, has some heat arising from affective processes generated by things or events that are attractive or repellant. Further, cognitions are referenced to or about these events, which in turn gives the cognitions a hedonic flavor. Thus, the events are the outcomes people engage in intentional action to acquire on the one hand, or escape or avoid on the other, and these events provide the reasons for action. These events have traditionally been referred to as *positive and negative reinforcers*. However, the related affective responses (e.g., attraction, lust, anger, fear, etc.) provide the reason in Davis's terms or the fuel for intentional action in which the reinforcing outcome is anticipated. Thus, intentions are necessarily oriented to future events with a distinct hedonic valence and attendant arousal (heat), whether they be immediate and next in a stream of behavior or extended over a longer period of time. As a consequence, the schemas, scripts, or whatever vehicle provides the knowledge base must represent possible action outcome sequences.

A facile cognitive translation of the behaviorist concepts of positive and negative reinforcement has been provided by Lindsay and Norman (1977, p. 501) in their concept of apparent causal relations. According to them, sequential relations between behavior and probable contingent motivational events are learned in circumstances in which reinforcements are applied. These relations are not merely expectancies of events following each other, because the relevant behavior must occur in order for the outcome to happen, so that the goal is predicated on the appropriate operation. Thus, much of the knowledge agents have acquired in the form of motor plans and higher order schemas must be in the form of these apparent causal relations.

This last point provides a transition to the fourth and most important of Davis's (1979) requirements on intentions. He pointed out that an agent's knowledge and reasons are not sufficient to explain the occurrence of intentional action. Beyond these, the agent must aim at an outcome to exercise intentions by directing behavior toward accomplishing that end. Aiming is, thus, the future-oriented control of the operations that can bring about the desired goal. The planning or forecasting that appears implicit within the organization of upcoming or unfolding action has come to be termed *prospective control*, or simply *prospection*, by a small number of researchers (e.g., Lee, 1993; Turvey, 1990; Wasserman, 1986) and is further explained below. This function is nicely captured in the account of plans given earlier. Although it does not appear explicitly in the Miller et al. (1960) treatment, the structure of the TOTE has the ingredients for prospective control. It is contained in the relation between the operation and the test criterion, which the operation seeks to match as the representation of the desired outcome. Thus, pursuing the match to the criterion provides the prospective forecasting that continues to drive the operations. The operations therefore have an anticipatory orientation toward the immediate and upcoming events captured in apparent causal relations, which accordingly require flexibility in aiming the trajectory of current operations. Such flexibility is attained in complex action sequences made up of components at least some of which are well practiced and skilled. The flexibility is found in the ability to vary the angle of attack with a variety of similar movements, which is a necessary part of the acquisition of motor plans. The resulting variation in the form of response manifests typicality effects similar to those found among the members of natural categories mentioned in the discussion of schemas above. So, exemplar actions vary from being close to optimal for obtaining the outcome to being on the fringe of capability (for a further discussion of flexibility, see Putney, 1985, p. 54). An example of flexibility in skilled action is found in Roberts and Ondrejko's (1994) discussion of skiing, an activity that requires rapid accommodation to immediately upcoming conditions. Thus, motor plans have the basic representational structure of schemas that range in a complex spatiotemporal domain of different dynamic shapes or response topographies as they were described in the first discussion of motor plans above.

## Two Levels of Prospective Control

Two separate groups of researchers from different traditions and perspectives have referred to the aiming function of intentions as *prospective control* (or *prospection*). The first is a group of ecological psychologists (Lee, 1993; Turvey, 1990; Von Hofsten, 1994) working within the Gibsonian tradition and specializing in the control of motion. The second consists of several researchers interested in animal working memory (Honig & Thompson, 1982; Wasserman, 1986). It is not clear whether the close similarity of their formulations came from a common origin or whether they were independently formulated, which prima facie appears to be the case. In spite of the similarity of formulation, their respective usage and the context of reference for each suggest two separate levels of prospective control in the organization of action.

My introduction to the concept of prospective control came from a talk by David Lee at a conference in 1990 on the ecological self (see Neisser, 1993). Lee's (1993) primary interest has been in the control of movement based on information in the optic flow field, which allows immediate predictive control of action. He provided examples from both humans and animals, including gannets that dive when they fish for prey from close to 100 feet and fold their wings prospectively just before entering the water and brachiating gibbons that leap across considerable distances from branch to branch in the Southeast Asian tree tops, so that each landing place must be carefully targeted to avoid missing the next branch. According to Lee (personal communication, April 11, 2001), he began using the term *prospective control* in the mid 1980s. A small group of ecological psychologists, including Turvey (1990) and Von Hofsten (1994), have joined him in this usage. They have been substantially influenced by the work of Nickolai Bernstein, a once-prominent Russian physiologist in the mid-20th century who specialized in the dynamics of movement and skill. A collection of his articles spanning the decades from the 1930s to the early 1960s was published posthumously in English as *The Co-ordination and Regulation of Movements* (Bernstein, 1967) just a year after it was published in Russian. More details on Bernstein's work appear later.

On a different level, Honig and Thompson (1982) and Wasserman (1986) both made the distinction between prospective and retrospective working or short-term memory in reviewing research paradigms with animals, such as delayed matching to sample, which could provide the basis for the distinction. The former is an anticipatory memory to engage in an action at some subsequent time; the latter is memory for an event that has already occurred. As an anticipatory memory, it represents the ability to plan subsequent goal-directed actions. Wasserman (1986) cited Konorski's *Integrative Activity of the Brain* (1967) as the source of the distinction and used the term *prospection* a number of times in his exposition, which was a thorough review of existing evidence for the distinction beginning with Hunter's delayed response task. He ended his chapter with historical references to this distinction by Lloyd Morgan and Charles Sherrington. To quote Morgan (1894), "Our past life, which we can review in memory is an extension backwards through retrospective thought. . . . Our anticipations of the future are a similar extension forwards. . . . Anticipation is prospective representation" (p. 113). Further, Honig and Thompson (1982) mentioned that Morgan discussed this distinction in one of his last publications, *Animal Mind* (1930), indicating that this conceptual distinction has some history that may yet be brought to light.

Konorski (1967) made the distinction in his book in a discussion of *transient* memory, a term he preferred to *recent* or *short-term* memory. The reference was brief, occupying about a page, and the term *prospective* was used only twice in the text. However, Konorski (1967) clearly stated the contrasting functions of prospective and retrospective memory:

> Transient memory is best manifested in planned behavior when a subject has to perform some behavioral acts programmed in advance (prospective role of recent memory), and in resistance to the perseveration of behavioral acts already performed (retrospective role of recent memory). (p. 503)

Thus, it appears that there are apparently two independent sources for the concept of prospection. The first emanates from Konorski's (1967) proposal for two different kinds of working memory (Wasserman, 1986) focused on the more global and strategic forecasting that organizes impending actions. This level would correspond to the upper stages of intentions as plans in the Miller et al. (1960) sense of a hierarchical structure of TOTE units. The second source from Lee (1993) and the other ecological psychologists deals with the lower level motor programming involved in generating a particular action. A short exploration of Bernstein's ideas on the subject should be helpful in conceptually developing this level of control.

## Bernstein on Anticipation and Foresight

Bernstein was born in 1896, 10 years after Tolman, to put him in the American historical perspective. His published work began in the 1920s and continued through the posthumous collection of essays mentioned above (Bernstein, 1967). In the foreword for the English edition, his friend Alexander Luria referred to Bernstein as "an outstanding physiologist and mathematician" (p. vii). Moreover, the latter is testified to by the differential equations in the book. Luria further pointed out that 12 years before the publication of Wiener's *Cybernetics* (1948), Bernstein had formulated basic principles of self-regulatory systems and the role of feedback in the regulation of voluntary movement. He first used the term *motor program* in 1935. An important book, *On Dexterity and Its Development*, was published in Russian in 1991 and was translated into English (Bernstein, 1996) along with accompanying essays in Latash and Turvey (1996). This book was intended for broad consumption by physicians and sports psychologists and would have been published in 1950, if Bernstein had not come under a sudden cloud resulting in the loss of his academic position and laboratories (Feigenberg & Latash, 1996). A few excerpts from this work demonstrate that he understood the basic concepts underlying prospective control without using the term.

> Consider, for example, the throwing of a javelin at a target or the hitting of a ball while playing billiards. Both these movements are very brief, nearly instantaneous. Their important feature is that after the javelin has been thrown or . . . the ball has been hit no corrections can be applied to adjust their movement. . . . Here all the corrections must be introduced on the basis of anticipation, when the movement has not yet started. (Latash & Turvey, 1996, p. 223)

Bernstein then proceeded to refer to jumping as throwing one's own body:

> Therefore everything here is also based on foresight. . . . Such foresight or anticipation as it is called in Physiology is based on a rich stock of previous experience. This experience lets you predict in advance the outcome of a throwing or striking movement. . . . Anticipation, or the generation of corrections in advance plays a very important role in motor coordination . . . it helps us calculate the point where we would hit a car crossing one's way and to modify our route correspondingly. (Latash & Turvey, 1996, p. 224)

Bernstein then went on to make similar cases for walking, tennis, wrestling, soccer, fencing, and so on. So in general, his point of view as well as his language directly reflect the aiming functions of intentions mentioned above. He further expanded the previous theme in a 1957 article (reprinted as chap. 4 in Bernstein, 1967) "in which great, sometimes decisive, importance attaches to correction of an advance or anticipatory character. This is particularly the case where during the course of any given segment of a movement, retrospective control becomes practically impossible" (Bernstein, 1967, p. 141). He then introduced the notion of movements that forestall others, which necessitate anticipations with correction in which there is movement directed not at an object but "towards an anticipated or extrapolated point of intersection with its trajectory" (Bernstein, 1967, p. 141). He gave as examples catching a moving object, passing a ball, and interposing a racket across the path of a moving ball and then summarized the above as follows:

> The existence of correction of the anticipatory type . . . directs our attention to the importance of anticipation in realizing any type of goal directed motor act. Programming, as has been demonstrated above, is determined by the

apprehension of motor problems as they arise, and represents an anticipation both of the result which is determined by its solution, and of such motor techniques as are necessary for its attainment. (Bernstein, 1967, p. 141)

In an article in 1961 Bernstein cast the process of anticipation in an even more modern vein as

the phenomenon which would be called "looking forward" in Chinese, and . . . in more scientific terms may be called *extrapolation to the future*. Indeed, planning a motor act . . . involves the recognition . . . of what must be, but is not yet the case. In a similar way in which the brain forms an image of the real external world it must possess . . . the capacity to form a representation of (or what is the essence of the matter to plan in advance) situations which are yet unrealized. . . . Only such an explanatory image of the necessary future can serve as a basis for the formulation of problems and the programming of their solutions. (Bernstein, 1967, p. 150)

Finally, these remarks were followed with a discussion of modeling from a mathematical point of view and then an unusually contemporary account of active mental modeling of the world mirroring Konorski's duality presented above: "In the brain there exist two unitary opposed categories or forms of modeling the perceptual world: the model of the past-present, or what has happened and is happening, and the model of the future" (Bernstein, 1967, p. 156).

Thus, from Bernstein's exposition as well as the preliminary remarks on the aiming function and the interacting components of plans, it appears that prospective control should include more that just anticipation or foretelling in its conceptual development. Rather, like the earlier exposition of apparent causal relations, the future-oriented forecasting found in prospective control must lead to the operation that produces the expected outcome.

## Prospective Control in the Chimpanzee

I now apply this conceptual scheme to the behavior of chimpanzees. In "Willful Apes," a few rich anecdotes were used to make the case for the role of intentions in chimpanzees without any systematic evidence to support the plausibility argument. Although the anecdotes were quite different from each other, they all had the theme of intentionality, in which systematic attempts were made to produce outcomes that strongly indicated the future-oriented expectation in working toward the desired goal. Fortunately, there is now an accumulating body of evidence for this kind of anticipatory cognitive control described above at both levels of prospection (Menzel, 1999, 2005).

I will first apply the conceptual scheme to an additional anecdote to tie the theoretical language further developed here to understanding chimpanzee behavior. In this episode, Austin, one of 2 male chimpanzees from the second series of lexigram training studies (Savage-Rumbaugh et al., 1978), was watching himself live on a TV monitor by looking at a TV camera (for an illustrated account of this episode, see Savage-Rumbaugh, 1986, p. 312). Austin had considerable experience looking in TV monitors, discriminating live from other pictures and also exploring his own face by this means. He began by opening his mouth while at the same time looking in the monitor at his exposed mouth and throat. During this part of the episode, he engaged in a plan involving various angles of attack on the problem of opening his mouth and pointing it toward the camera (operations) to see down his throat (the outcome of the ongoing monitoring). After a period of this exploration, he suddenly stopped and left the room, evidently on his way to the cupboards in the back room to retrieve a flashlight. He then returned and, snapping on the flashlight, he began the complex triangulation of light, mouth, and camera, which resulted in a number of at first less successful and then more complete glimpses far down his throat. From this, it might be inferred that he prospectively formed the plan on the spot that he could see down his throat with a flashlight, which he knew lit up dark places. So he exited to the flashlight retrieval subroutine, returned, and operated with the kind of flexible change in attempted pointing to find better angles of attack all involving the sort of prospective motor plan for varying skilled action until the desired outcomes were obtained. In general, the varied attempts to look down his throat with or without the flashlight illustrate flexible prospection at the motor control level, whereas stopping to retrieve the flashlight represents prospective planning as a working memory function.

Menzel's (1999) work with a chimpanzee called Panzee is an exemplary case of systematic evidence for prospective–intentional control of actions in *Pan troglodytes*. Panzee had extensive human enculturation in her upbringing along with some comprehension of spoken English as well as her lexigram language training (about 120 lexigrams in an array of 256 on the panel containing them). Some of her early experience involved planned trips in the woods around the Language Research Center grounds where she was reared, which provided a substantial acquisition of knowledge (schemas and scripts) to provide the basis for the behavior she continued to exhibit in Menzel's experiments. Panzee was housed in a building with several permanent cages with other chimpanzees. Her cage was connected by a short tunnel through which she could travel to a large outside enclosure in view of a considerable area of woods, and there were lexigram keyboards that she could use for communication in both inside and outside cages. In Menzel's experiments he began a trial by hiding an attractive item in the woods at various distances and orientations from the outside cage while Panzee was watching. Having seen the attractive object (e.g., M&Ms) hidden earlier, at some later time, frequently from 1 or more hours to a day or more after, and usually when it was quiet indoors, Panzee began the recruiting routine of the overall plan for retrieval of the desired item. She began by capturing the attention of one of the caretaking personnel who had been previously recruited but was ignorant of the facts of the hiding, including the item and place. This was usually done with some combination of vocalizations, pressing the lexigram keyboard with the correct symbol for the hidden object (Menzel referred to this as "reporting"), and pointing to the tunnel that she frequently began to traverse to the outdoor cage. In the meantime, the recruit left by the side door to walk around the outside enclosure, which was a lot farther than Panzee had to traverse. Once there, Panzee could view the hiding place and direct the recruit to the hidden object by positioning herself just inside the chain link fence, orienting toward the hiding place and pointing with her index finger and with her direction of gaze. She usually continued to point repetitively with "encouraging" vocalizations, sometimes retreating to the keyboard, until the recruit got closer to the hiding place and finally found the lure. Thus, this script consisted of a succession of flexible subroutines, so that with successful recruitment frequently punctuated by some conversation from the recruit, Panzee exited outside to the directing subroutine in which she prospectively directed the recruit using repeated attempts, pointing with her index finger, sometimes with accompanying vocalization. The final step in the overall retrieval plan was to return to the inside to receive the goodies.

Menzel (1999, 2005) not only replicated this scenario many times but continued to find variations on the basic experiment to test the limits of Panzee's performance. These variations have included hiding multiple items on a trial, each of which she recruited for separately, letting her use a laser pointer after recruiting to indicate the hiding place, or using a TV monitor to display the outside hiding place first to direct the recruit in one series of trials and then to both view the hiding and direct in another series. Panzee continually manifested a very high rate of success in directing the recruits to the hidden objects, frequently showing perfect performance in a given series, although there were a few cases in which her error rate increased when she was not allowed to see the objects directly, and only lexigrams or English words were used to indicate objects hidden in a box.

## Conclusion

The future-oriented, intentional character of Panzee's script was initially found in her spontaneous retrieval from long-term memory of the hidden object to be obtained, which then led to the whole recruitment plan in which she must have depended on the communicative interaction with the recruit to attain the desired goal. This use of her self-generated memory contrasts markedly with the study of human memory retrieval in which, for the experimenters' convenience, there is almost always some sort of a cue or prompt for retrieval of the to-be-remembered items, even in so-called "free-recall" tasks. The prospective nature of the plan proceeded flexibly through the recruiting and directing subroutines as the particular conditions of each unique episode unfolded. The two levels of prospective control were both manifest in the stream events. The activation and arrangement of specific forms of the subroutines were organized in planning at the working memory level. On the other hand, Panzee's selection of the appropriate lexigram from among the 256 alternatives, her well-articulated pointing, and her directive vocalizations signaling the recruit to vary the recruit's search involved considerable directed orienting with fine-tuned flexible motor control representing prospection in the volitional control of her actions.

Finally, it should be noted that the details of prospection in the motor control sense are much better developed than the prospective planning function of working memory. The influence of Bernstein's work, which has been increasingly evident among motor control specialists over the last 25 years, has resulted in well-formulated mathematically based accounts of prospective control. On the other hand, there have been references to future-oriented planning with a clear prospective flavor as early as Miller et al. (1960, p. 65) in their discussion of working memory, which may be the first reference to that term. However, conceptual development of the planning function that they made reference to as part of the featured theme of their book has not progressed much beyond their exposition, probably because intention has appeared a daunting subject to cognitive psychologists. The topic of working memory has had substantial representation following the work of Baddeley (e.g., 1986, 1996), quite apart from references to the subject from the animal cognition literature (e.g., Honig & Thompson, 1982). Further, the concept of central executive within Baddeley's system is an active area of interest, but little novel thought about prospective planning is yet apparent there, although there was a promising beginning in the work of Roberts and Ondrejko (1994) and the treatment of working memory from a neuropsychological point of view by Pennington (1994). Thus, the present perspective should provide a beginning for conceptual analysis not only within human cognitive psychology but also within comparative cognition embracing both our nearest phyletic neighbors, the chimpanzees, and, perhaps, many more of our vertebrate relatives.

## References

Anscombe, G. E. M. (1957). *Intention*. Oxford, England: Blackwell.
Attneave, F. (1957). Transfer of experience with a class-schema to identification-learning of patterns and shapes. *Journal of Experimental Psychology, 54*, 81–88.
Baddeley, A. D. (1986). *Working memory*. Oxford, England: Clarendon Press.
Baddeley, A. D. (1996). Exploring the central executive. *Quarterly Journal of Experimental Psychology, 49A*, 5–28.
Bartlett, F. (1932). *Remembering*. Cambridge, England: Cambridge University Press.
Bernstein, N. (1967). *The co-ordination and regulation of movements*. Oxford, England: Pergamon Press.
Bernstein, N. (1996). On dexterity and its development. In M. L. Latash & M. T. Turvey (Eds.), *Dexterity and its development* (pp. 3–244). Mahwah, NJ: Erlbaum.
Davis, L. H. (1979). *The theory of action*. Englewood Cliffs, NJ: Prentice-Hall.
Feigenberg, I. M., & Latash, L. P. (1996). N. A. Bernstein: The reformer of neuroscience. In M. L. Latash & M. T. Turvey (Eds.), *Dexterity and its development* (pp. 247–275). Mahwah, NJ: Erlbaum.
Gallistel, C. R. (1980). *The organization of action: A new synthesis*. Hillsdale, NJ: Erlbaum.
Head, H. (1920). *Studies in neurology* (2 vols.). London: Hodder & Stoughton.
Honig, W. K., & Thompson, R. K. R. (1982). Retrospective and prospective processing in animal working memory. In G. H. Bower (Ed.), *The psychology of learning and motivation* (pp. 239–283). New York: Academic Press.
Konorski, J. A. (1967). *Integrative activity of the brain*. Chicago: University of Chicago Press.
Latash, M. L., & Turvey, M. T. (Eds.). (1996). *Dexterity and its development*. Mahwah, NJ: Erlbaum.
Lee, D. N. (1993). Body–environment coupling. In U. Neisser (Ed.), *The perceived self: Ecological and interpersonal sources of self knowledge* (pp. 43–67). Cambridge, England: Cambridge University Press.

Lindsay, P. H., & Norman, D. A. (1977). *Human information processing*. New York: Academic Press.

Menzel, C. R. (1999). Unprompted recall and reporting of hidden objects by a chimpanzee (*Pan troglodytes*) after extended delays. *Journal of Comparative Psychology, 113*, 426–434.

Menzel, C. R. (2005) Progress in the study of chimpanzee recall and episodic memory. In H. Terrace & J. Metcalfe (Eds.), *The missing link in cognition: Origins of self-reflective consciousness* (pp. 188–224). New York: Oxford University Press.

Miller, G., Galanter, E., & Pribram, K. (1960). *Plans and the structure of behavior*. New York: Holt, Rinehart & Winston.

Morgan, C. L. (1894). *An introduction to comparative psychology* (2nd ed.). New York: Scribner's.

Morgan, C. L. (1930). *The animal mind*. New York: Longmans.

Neisser, U. (1988). Five kinds of self-knowledge. *Philosophical Psychology, 1*, 35–39.

Neisser, U. (1993). *The perceived self: Ecological and interpersonal sources of self knowledge*. Cambridge, England: Cambridge University Press.

Pennington, B. F. (1994). The working memory function of the prefrontal cortices. In M. M. Haith, J. B. Benson, R. J. Roberts, & B. F. Pennington (Eds.), *The development of future oriented processes* (pp. 243–289). Chicago: University of Chicago Press.

Posner, M. I., & Keele, S. W. (1968). On the genesis of abstract ideas. *Journal of Experimental Psychology, 83*, 304–308.

Premack, D., & Woodruff, G. (1978). Does the chimpanzee have a theory of mind? *Behavioral and Brain Sciences, 4*, 515–526.

Putney, R. T. (1985). Do willful apes know what they are aiming at? *Psychological Record, 35*, 49–62.

Roberts, R. J., & Ondrejko, M. (1994). Perception, action, and skill: Looking ahead to meet the future. In M. M. Haith, J. B. Benson, R. J. Roberts, & B. F. Pennington (Eds.), *The development of future oriented processes* (pp. 87–117). Chicago: University of Chicago Press.

Rosch, E. (1974). Cognitive representations of semantic categories. *Journal of Experimental Psychology: General, 3*, 192–233.

Savage-Rumbaugh, E. S. (1986). *Ape language: From conditioned response to symbol*. New York: Columbia University Press.

Savage-Rumbaugh, E. S., Rumbaugh, D. M., & Boysen, S. (1978). Linguistically mediated tool use and exchange by chimpanzees (*Pan troglodytes*). *Behavioral and Brain Sciences, 4*, 539–554.

Schank, R., & Abelson, R. (1977). *Scripts, goals, and understanding*. Hillsdale, NJ: Erlbaum.

Turvey, M. T. (1990). Coordination. *American Psychologist, 45*, 938–953.

Von Hofsten, C. (1994). Planning and perceiving what is going to happen next. In M. M. Haith, J. B. Benson, R. J. Roberts, & B. F. Pennington (Eds.), *The development of future oriented processes* (pp. 63–86). Chicago: University of Chicago Press.

Wasserman, E. A. (1986). Prospection and retrospection as processes of animal short term memory. In D. F. Kendrick, M. E. Rilling, & M. R. Denny (Eds.), *Theories of animal memory* (pp. 53–75). Hillsdale, NJ: Erlbaum.

Wiener, N. (1948). *Cybernetics*. New York: Wiley.

Zajonc, R. B. (1980). Feeling and thinking: Preferences need no inferences. *American Psychologist, 35*, 151–175.

# Part IV

# Language and Tools

# 16

# The Past, Present, and Possible Futures of Animal Language Research

*William A. Hillix*

The notion of animal language has inspired stories for centuries, but only in the past 100 years have lingual abilities among animals become the object of much research. In this chapter I outline the history of this research, first providing some background on the myths and facts about animals speaking, recounting the beginnings of research, and then discussing the ups and downs of language research in the past 50 years. I end the chapter with some speculations about the direction of future research in this field.

### Stage 1: Myths and Fables

Animal language has been a subject of myths and fables for thousands of years (Morris & Morris, 1966). One of the early stories was written in approximately 1000 BC, when the author or authors of the Book of Genesis reported a conversation between Eve and the serpent. That conversation allegedly accounts for Adam and Eve's ejection from Paradise. It is said that God punished the serpent for his sin by taking away his ability to speak, and that accounts for the fact that snakes do not talk. This story is a prototype for explaining why animals cannot, or do not, talk.

At about the same time, 1000 BC, an Egyptian papyrus depicted humans worshipping baboons. About 1,500 years later, in AD 500, the Egyptian scribe Horapollo Nilous wrote another papyrus explaining how the priests determined which baboons were sacred. Priests from ancient times believed that some baboons had the power of reading and writing, and they therefore tested the language abilities of baboons newly arrived in the temple. They gave the baboon a quill pen, ink, and a tablet. If the baboon passed the writing test, it was regarded as sacred; was fed wine and choice roast meats; and, according to the earlier papyrus, was even worshipped. Nilous did not report the criteria for passing the language test, but it is the earliest test of animal language abilities that I have heard about. In addition, the priests allowed the baboons to bypass the vocal channel, which was not done successfully by scientists until nearly 3,000 years later. So this story is a prototype for the second belief, that animals have a language that they do not use or one that humans do not understand. Some animal language researchers might argue that the first type of myth, that animals are forbidden to ever have any language, is also still with us in the form of stories told by linguists like Noam Chomsky (1965) and Steven Pinker (1994).

Aesop allegedly told stories of the second kind, fables about talking animals, in about 600 BC, prominent among which was the famous fox who could not reach the grapes and said they were probably sour anyway. I was disappointed to find that Aesop may be as mythological a figure as his animals; the stories may have been handed down orally from antiquity and written down by multiple writers.

In any case, the Aesop type of story is still with us: The Dr. Dolittle stories written by Hugh Lofting (e.g., Lofting, 1922/1988) and stories written by Ted Geisel (as Dr. Seuss; e.g., Geisel, 1957/1985, 1971) are two of the best-known modern examples of this genre. Humans have been fascinated by questions about animal language for at least 3,000 years and continue to be fascinated with them.

## Stage 2: The Beginnings of Empiricism

A little over 100 years ago saw the start of a new era, when mythology started to be replaced by empirical observation. Some might choose other people to represent this shift, but I am fond of a writer and adventurer named Richard Lynch Garner (Garner, 1896). In the late 19th century, he built and occupied a large cage in Gabon where he observed the passing wildlife and tried to get several apes to speak. He had little success, but he did think that he taught one of his chimpanzees to say *feu*, the French word for *fire*.

Garner (1896) believed that chimpanzees had "words" for "food," "good," "danger," "strange," and "come," which he and they understood and shared. These were native chimpanzee sounds. It is interesting that both E. Sue Savage-Rumbaugh and colleagues (Taglialatela, Savage-Rumbaugh, & Baker, 2003) and Sarah Boysen (see *Demonic Ape*, 2004) have recently demonstrated that chimpanzees really do have distinguishable vocalizations for some foods or situations, so Garner may not have been as crazy as he seemed! Garner also described natural chimpanzee gestures, including the one that Beatrix and Allen Gardner later used for "come" with their chimpanzee, Washoe; Garner aptly said that it consists merely of extending the arm, without motion of the wrist, toward the person or thing desired (Gardner, Gardner, & Van Cantfort, 1989).

Garner was a transitional figure and provided a lot of misinformation as well as information. However, on the credit side, he was the first person I know of who tried to bypass the vocal channel, recognizing the difficulty apes have in speaking. He tried to get young chimpanzees to spell out words by allowing them to rotate rods on which he had placed alphabet blocks.

Only 4 years after Garner (1896) published his book, Herr von Osten gave his first public display of his famous horse, Hans (see Candland, 1993). Hans tapped out answers to numerical questions with his right hoof and indicated readiness to answer a question by nodding his head affirmatively, thus bypassing the vocal channel. Obviously nobody expected horses to talk.

Hans indicated that he was finished by tapping once with his left hoof. Hans also tapped out answers to mathematical questions, providing correct sums, differences, products, quotients, and even squares and square roots. Hans was so clever that that adjective was added to his name, whereupon he became "Clever Hans."

However, Hans was not clever enough to fool Oskar Pfungst, a psychology student assigned by Professor Carl Stumpf to investigate Hans's abilities more thoroughly. Pfungst (1911) discovered that Hans was not clever at all if no human who knew the answer was present where Hans could see him. Pfungst believed that Hans used cues from observers who knew the answer; when Hans had tapped the correct number of times, observers tended to relax or look up at Hans or shift gaze to the other hoof. Pfungst broke new ground in that he introduced controlled observation into the field of animal language research by testing Hans with and without people present who knew the answers to the questions put to Hans. Pfungst thus introduced a tradition that for about 100 years has been an essential part of animal language research.

The use of controlled observation was the good part of the Clever Hans episode. The bad part was that people who refused to accept the possibility that animals had some language ability had an easy explanation, most often a pseudoexplanation, for any and all positive results. However, many investigators have pointed out that anyone who says that a Clever Hans effect is responsible for an animal's performance should be responsible for showing that nonlinguistic cues are responsible for the results. The parallel responsibility is that those making positive claims are responsible for showing that Clever Hans effects are not causing the behavior. That position is widely accepted, so most of the damage done by Clever Hans is in the past, and the methodological lesson is well learned.

The empirical tradition continued and gradually shaded over into intensive efforts to teach animals a human-designed language. Among the early transitional empirical studies were observations made by William Henry Furness III, Nadesha Kohts, Maria Hoyt, Winthrop and Luella Kellogg, and Catherine and Keith Hayes.

Furness acquired an orangutan and two chimpanzees in 1909 and another orangutan in 1911 (see Furness, 1916). He tried to teach all four animals to speak words, and his most apt orangutan pupil learned to say "papa" after 6 months of training. Furness then taught the orangutan to say "cup" and "th," preparatory to teaching her to say words like *the*, *this*, and *that*, but the orangutan died soon after. Furness tried for 5 years to teach a chimpanzee to say "cup," but he had no luck in doing so.

Furness was an intermediate figure between Garner's earlier short-term efforts to teach language and more intensive observations of home-reared animals. The transition to home rearing was an important step. Scientists hoped that apes reared with humans would learn to speak. Although this hope was almost completely dashed, home rearing forced investigators to recognize that something else had to be done if apes were to communicate in a human-designed language. That something else was the various prostheses now in use to circumvent the vocal channel.

In 1913, Nadesha Kohts took a male chimpanzee, Joni, into her home in Moscow (Kohts, 1935). Joni lived with Kohts from the ages of 1.5 to 4 years, until 1916. Kohts was struck by Joni's failure to vocalize or even attempt to imitate human vocalizations. Kohts reported that Joni did produce 25 sounds when moved by various emotional stimuli. However, Kohts was pessimistic about the possibility of teaching chimpanzees to speak, saying that Joni completely failed to imitate human sounds.

During the 1930s, Maria Hoyt kept a female gorilla, Toto, in her home (Hoyt, 1941). Toto communicated to her only through natural gestures; however, Hoyt thought that Toto comprehended Spanish as well as any Spanish child of her age. It was significant that Hoyt so starkly contrasted Toto's lack of vocal production with her ability to comprehend spoken Spanish. Like all adult great apes, Toto became so difficult to manage that Hoyt gave her to the circus, where she became the mate of the famous male gorilla, Gargantua.

During roughly the same period, Winthrop and Luella Kellogg raised a female chimpanzee named Gua for 9 months with their child, Donald (Kellogg & Kellogg, 1933). Their intention was to compare the development of Gua and Donald within the same environment, in part to evaluate the relative effects of nature and nurture. Gua walked, climbed, and ate with utensils as well as a human child. However, like Joni, she made no effort to imitate human speech, and the Kelloggs never reported that she said even one word. As with Joni, all or nearly all of her vocalizations appeared to be emotionally driven. Despite her shortcomings in vocalization, Gua did apparently understand 58 spoken phrases, again providing a contrast between language comprehension and language production.

Gua used nine gestures to indicate various desires, such as the desire to eat, drink, or sleep. This led Kellogg and Kellogg (1933) to suggest that apes might be able to use sign language. Samuel Pepys (1661/2000, p. 160) speculated more than 300 years ago that an ape he saw on the docks of London might be taught to use sign language, and Yerkes (1925) suggested the same thing in the 1920s.

After the Kellogg study, Keith and Cathy Hayes, in the 1940s, adopted another female chimpanzee, Viki (Hayes, 1951). It is extremely interesting, and may be significant for future research, that Viki babbled occasionally for the first 4 months and then stopped. Her vocal training began when she was 5 months old, 1 month after her babbling stopped. At 3 years she could say three words—*mama*, *papa*, and *cup*—and seemed to use *cup* fairly reliably, with little indication that she knew what the other two words meant. She also used *aaah* as a request, especially for a cigarette, and a clicking sound for a car-ride request.

In summary, then, the first two phases of human interest in animal abilities to use language produced evidence that animals comprehended more language than they could produce, that vocal language did not look promising even if animals were reared with humans, and that care must be taken to eliminate or evaluate nonlinguistic cues as causes for animal behaviors.

## Stage 3: The Era of Experimentation

Ironically, B. F. Skinner (1957), although he was mostly wrong about the essence of language, was a positive influence on animal language research, and Chomsky (1965), although he was probably more correct than Skinner, was a negative influence both directly and through adamantine acolytes like Herbert Terrace (1979), Joel Wallman (1992), and Steven Pinker (1994). Skinner was helpful because his definition of verbal behavior made it clear that language did not have to be spoken; David Premack (1976) had preprints of Skinner's book before he started research on language in apes, and Skinner's views may have influenced the Gardners as well. In addition, Skinner's emphasis on operants provided a starting point for teaching language to animals. However, researchers generally abandoned or greatly modified strictly operant procedures because they found that they were not always effective.

For whatever reason, language research with animals came alive in the 1960s and 1970s and remains a vital enterprise today. Most people are familiar with the post-1960s research, so I will only provide the most general outline. In the mid to late 1960s, Beatrix and Allen Gardner taught Washoe and other chimpanzees approximately 130 signs each; the chimpanzees as a group mastered 460 signs while in the care of the Gardners, and more later (Gardner et al., 1989). Deborah and Roger Fouts adopted the Gardners' chimpanzees and showed that they continued to use the signs they had learned, even when no humans were present (Fouts & Mills, 1997). Ann and David Premack also claimed a vocabulary of about 130 plastic symbols for their chimpanzee, Sarah, whom they adopted in 1964 (Premack, 1976). Duane Rumbaugh (1977) started his computer-assisted research teaching the chimpanzee Lana to communicate with lexigrams in 1970. Penny Patterson (see Patterson & Linden, 1981) started teaching signs to her gorilla, Koko, in 1972, not long after Rumbaugh started with Lana. Lyn Miles (1990) started teaching sign to her orangutan, Chantek, in 1978. E. Sue Savage-Rumbaugh (1986) continued and transformed the lexigram research, demonstrating that the lexigrams had true semantic function, that animals could use lexigrams to communicate with one another, and that both bonobos and chimpanzees had considerable ability to decode English sentences that they had never heard before.

Other researchers have studied animals that were not great apes. Notably, Louis Herman (1986) demonstrated that dolphins are sensitive to word order and thus have the rudiments of receptive syntactical understanding, in addition to being masterful imitators. Irene Pepperberg (1999) showed that an African Grey parrot, Alex, could analyze questions about color, shape, and material and answer the questions completely correctly about 75% of the time.

So where are we now in animal language research? We know that animals are far behind humans, especially in the production of language with present techniques. Nevertheless, they are far more able linguistically than we thought 50 years ago, and we still do not know how far they can go with improved training and language prostheses. Thus far the evidence that animals have true syntax is relatively weak, although there is evidence for order effects in their use of signs or lexigrams, and we do not know how far they can go with improved techniques. We know that animal experimentation leads to techniques that are helpful with language-deficient humans. Every method for bypassing the vocal channel that has succeeded with apes has proved useful with human children: plastic symbols, signs, and lexigrams. The technique for teaching parrots has also helped children. The only exception to the generalization that all techniques successful with animals have helped some humans is the large-scale arm signals used for communicating to marine mammals, and that is probably because it has not been tried.

## Stage 4: What Can We Expect in the Future?

I would like to suggest some avenues for future research and ways to improve current research programs in animal language. In this section, I explore the limits on acquiring syntax, new ways to circumvent the vocal channel, new approaches to animal vocal language, connections between natural animal communication and human-designed language such as American Sign Language, and the components of language cognition in multiple species. I also discuss training techniques in animals used in studies, fostering the "multimodal ape," and increasing funding for all research.

### Limits on Acquiring Syntax

One proposed line of research with animals will investigate the limits of semantic and syntactic understanding that can be reached in apes and dolphins, perhaps even in parrots, given special training and testing. One issue is whether animals can understand words that have a strictly grammatical function, such as prepositions and conjunctions. Herman and Uyeyama (1999) argued that dolphins have demonstrated comprehension of such words as well as comprehension of argument structure. Others have argued that responding to challenges about exactly what animals can and cannot do is not the most productive way to proceed. However, whether researchers are enthusiastic or not, this line of research is certain to be pursued, and its results, whether positive or negative, will be fascinating and will help to answer questions about the evolutionary underpinnings of human language.

### New Ways to Circumvent the Vocal Channel

Researchers will continue to search for more effective methods to circumvent the vocal channel or to make more effective use of it. Sign language and lexigram boards have worked well, but each has disadvantages. The use of sign requires no special equipment and is, therefore, the ultimate in portability and convenience. The rate of language production is much higher with sign than with plastic symbols or lexigrams. However, it is difficult to be certain about the form and meaning of signs as they are made, especially for the frenetic chimpanzee, and it is time consuming to record and interpret signs using film or videotape. Signs fade immediately, so there is no immediate record that provides a memory aid, as there is with plastic symbols or lexigrams that change lighting to show that they have been pressed or are echoed on a display. Computerized lexigram boards make it possible to record responses immediately and objectively, so that there is no problem connected with identifying the response. Further, key presses can produce sounds, so that words in any language (usually English in the past) can be produced by the key press.

The disadvantages of each approach can be reduced. Future research on lexigram boards could use more iconic, easily learned keys, now that it is no longer necessary to demonstrate that apes can use arbitrary symbols. Lexigram boards will also be further developed for underwater use by dolphins and modified for greater ease of use by marine animals. Lexigrams on such boards might be designed to differ in acoustic reflective properties, as well as visually, to cater to dolphins' echolocating capabilities. Signs can be developed that cater to animals' anatomic limitations and build more systematically on what is known about the animals' natural gestures. Feedback to signers could be increased with mirrors and live video.

It seems unlikely, however, that the plastic symbols used by Ann and David Premack (Premack, 1976) will undergo future development. They do have the advantage of simplicity and provide a memory aid. However, keyboards connected to computers have most of their advantages, are less unwieldy, and provide automatic data recording.

## New Approaches to Animal Vocal Language

It is possible, although perhaps less likely, that three methods for encouraging vocal production will be examined in more detail. One speculative method would require prostheses like false palates to make the ape's vocal tract more humanlike. I regard this approach as unlikely to succeed because the evidence to date indicates that the limitation of ape vocal language is in neuromuscular control, not in the vocal tract.

The second method would be to use a set of phonemes that apes can already produce to construct a limited artificial language. At least one attempt of this kind was abandoned rather quickly because the experimenters could not understand the resulting words. Although humans could, with sufficient effort, learn a language made up of such a limited sound set, apes might be incapable of producing the sounds despite the theoretical capability of their vocal tracts to produce them. However, success in this endeavor would be marvelous. The bonobo, Kanzi, tried to mimic the sounds of speech and indicated his frustration by pointing to his mouth and throat after unsuccessful attempts to imitate words. It would be wonderful to free him, or other apes, from his limitations, as some language-challenged children have been freed! E. Sue Savage-Rumbaugh has described Kanzi and Panbanisha as developing a "Creole" of English and "Bonobo" (Savage-Rumbaugh, 2004); if so, this approach starts to look feasible.

A third possible method is based on the same assumption—that the chimpanzee and bonobo vocal tracts could produce a complex language—but transfers the task of learning this language to computers. It is possible, at least in theory, to have a computer map the sounds that apes can produce into corresponding human phonemes and enunciate them through a speech synthesizer, so that a chimpanzee speaking into a microphone would be heard by a human wearing headphones as making the sounds to which he or she was accustomed. The reverse transformation, although equally possible in theory, does not seem to be necessary because great apes like Panbanisha, Kanzi, Koko, and Chantek have demonstrated that they can understand human speech (Miles, 1994; Patterson & Linden, 1981; Savage-Rumbaugh & Lewin, 1994). It would be ironic if humans needed help to understand the language but apes did not!

The approach clearly will not work if the apes simply cannot deliver enough vocal information to communicate effectively. Nevertheless, this approach will almost certainly be tried in the future, according to the axiom "anything that can be done will be done."

## Connecting Natural Animal Communication to Human-Designed Language

The study of animals' natural communication systems has often remained separate from the study of animals' acquisition of a human-designed language. Some investigators have pointed out similarities between natural gestures and the signs of American Sign Language; "gimme," for example, is similar in both. Other investigators have paid close attention to apes' natural gestures as they taught signs.

It is safe to say that future researchers will have studied their participants' natural gestures and vocalizations very carefully before embarking on a training program. They will initially work with signs or vocalizations that resemble the animals' natural communications. Further, they will be aware of the contexts in which the natural communications are used and will use the information to design better teaching techniques for the animals.

## Cognitive Components of Language in Multiple Species

Now that Alex has demonstrated what parrots can do with a small brain (see Pepperberg, 1999), it is bound to occur to future researchers that animals with brain sizes between the parrots' and the apes' may, with appropriate training, acquire rudimentary language. The results of this research will contribute to our knowledge of the evolutionary pathway that leads from stimulus–response learning through the ability to recognize semantic relationships to the ability to acquire the rudiments of syntax and finally to human language. If we can unravel this evolutionary pathway, we may finally come to understand what the language acquisition device that Chomsky (1965) postulated accomplishes for us. That is, we might understand what cognitive processes are involved, once we unravel what evolutionary and cognitive steps lead from less accomplished to more accomplished communicators.

## Improved Training Techniques

Irene Pepperberg (1999) demonstrated that systematic use of a *model-rival approach* was effective with parrots and enabled some language-deficient human children to improve their communicative abilities. The approach allows a learner to observe two accomplished users of a language interacting, "modeling" proper use of the language, with the "rival" competing with the learner for attention. The learner learns through observation and can enter the conversation at will.

Thus, it may not be necessary to bypass the vocal channel to teach animals or help children. Future researchers may be able to both use this technique and bypass the vocal channel in whatever way they find best. Combining approaches may lead to superior results. E. Sue Savage-Rumbaugh and Tetsuro Matsuzawa are among those who have approximated this procedure while bypassing the vocal channel with lexigrams (Matsuzawa, 2004).

Converging lines of evidence indicate that observational learning in the presence of extensive social interaction is critical if nonhuman animals are to learn to communicate in a human-designed language. Further, systematic exploitation of these factors, as in the model-rival approach, may lead to better results than approaches that have been standard in the past.

## The Multimodal Ape

One of the greatest opportunities that I see is the creation of a "multimodal ape." The ape would be created by combining the existing techniques in a training regimen that would begin soon after birth. This ape would be reared in an environment combining intimate contact with other apes and with humans, much as E. Sue Savage-Rumbaugh does now. The ape would have contact with a computer that reinforced different vocalizations differently and allowed the infant ape student to control its environment by emitting vocalizations. Training in gestures would begin early and would be based on all existing knowledge of the species' natural gestures so that social exchanges between the ape student and researchers could use these gestures and build on them to create additional gestures. Meanwhile, lexigrams would be available to humans and apes, but the lexigrams would be as iconic as they could be made; some words cannot, of course, be made completely iconic, but words like *above, beside, in,* and so on can be visualized to some extent. Model-rival training would be used where appropriate. Vocal, sign, and lexigram communication all would be encouraged throughout; Savage-Rumbaugh already does many of these things, but the addition of iconicity, a more systematic model-rival approach, and training in syntax might be important additions to her methods. I look forward to seeing a 10-year-old multimodal ape!

## Financial Support for Research

The last thing I will discuss is financial support for animal language research in the past, present, and future. For the past 40 years or so, the paucity of money to support research has been a scandal. Beatrix and Allen Gardner periodically ran out of money for their research and had to turn over their pioneering sign language chimpanzee, Washoe, to Roger and Deborah Fouts. It took years for the Foutses to get an outdoor facility, and two of their chimpanzees almost died of rickets before they could get out into the sunlight. Meanwhile, they had to pick through stale grocery produce to feed their chimpanzee family. Fouts and Mills (1997) have written an engaging account of the Fouts's financial problems from the time they took responsibility for the Gardners's chimpanzees up through 1996. Ann and David Premack stopped their language research about 20 years ago, perhaps partly because of funding problems, and their famous chimpanzee, Sarah, now resides with her namesake, Sarah Boysen. Herbert Terrace became so discouraged about funding that he stopped his project with Nim Chimsky after only 3 years; maybe if there had been more funding available, he would not have had to use 60 volunteers to teach Nim, and he might have continued working until he taught Nim to create sentences. Penny Patterson has struggled throughout the years to support her gorilla research; it took her years to buy Koko. Even now she is begging members of her Gorilla Foundation to support her move to Maui. Lyn Miles had to stop work with her orangutan Chantek for years because she did not own him or a laboratory in which to continue his education, and he is now in the Atlanta Zoo where she has limited access to him. The most recent disaster for those studying the minds of apes is that Ohio State University closed its primate research facility, directed by Sarah Boysen, on March 2, 2006, and sent its chimpanzees to an animal sanctuary, because Boysen's nine research proposals written to obtain funding to support the laboratory were all unsuccessful (Ohio State Research, 2006). Even Duane Rumbaugh and E. Sue Savage-Rumbaugh, who have kept their lexigram research going for over 30 years, have had to seize every opportunity and spend inordinate amounts of time to keep their research funded and their animals fed.

I now turn to my prediction for the future, that funding should and will increase. The ancient interest in talking animals is greater than ever. Intense and growing interest in the preservation of the natural environment, including earth's animal inhabitants, demonstrates that.

We might well ask why this interest has not already produced an outpouring of money. One reason may be that researchers have sniped at each other far too much, as though their funding would be threatened if someone else were funded. That is understandable, but what is not understandable is that animal language researchers do not always support each other.

An even more important problem is that the results of research on animal capabilities have not been publicized nearly enough. That was brought forcibly to my attention recently when a bright, teenage college student friend of mine called me, excited because she had stumbled across a television program about signing chimpanzees, which she had never known existed. I found that amazing, until I realized that few texts contain information about animal language research. The public seems strikingly unaware of the scope of what is happening in this field. Even the American Psychological Association (APA) has no division devoted to animal communication or to psychology of language in which scholars in both animal and human studies could participate. Hence there is no official recognition of the area as a separate field of research, and none of the pioneers, to my knowledge, have received APA's Distinguished Contribution Award.

This volume includes a great gathering of authors, but it celebrates the work of only one person. There should be an international conference celebrating the pioneers of animal language research. It would also be a media event of the first magnitude if a conference assembled as many of these pioneers and others active in animal language research as possible and arranged good advance publicity. Media people would gather like bees on honey. The goal of the conference would be precisely opposite to the intent of the debunking conference convened in 1980 by Sebeok and Umiker-Sebeok (1980). Such a conference would celebrate past accomplishments and look to the future. It would alert the public about the accomplishments and needs of animal language researchers and encourage politicians to back legislation favorable to animal language research. Perhaps a conference of this magnitude would get animal language research out of purgatory and back into the garden of Eden. It might also provide ammunition for forming an APA division devoted to the study of the psychology of language.

## References

Candland, D. K. (1993). *Feral children and clever animals.* New York: Oxford University Press.
Chomsky, N. (1965). Aspects of the theory of syntax. Cambridge, MA: MIT Press.
*Demonic ape—Transcript.* (2004, January 8). Retrieved April 3, 2006, from http://www.bbc.co.uk/science/horizon/2004/demonicapetrans.shtml
Fouts, R., & Mills, S. T. (1997). *Next of kin.* New York: Morrow.
Furness, W. H. (1916). Observations on the mentality of chimpanzees and orangutans. *Proceedings of the American Philosophical Society, 55,* 281–290.
Gardner, R. A., Gardner, B. T., & Van Cantfort, T. E. (Eds.). (1989). *Teaching sign language to chimpanzees.* Albany: State University of New York Press.
Garner, R. L. (1896). *Gorillas and chimpanzees.* London: Osgood McIlvane.
Geisel, T. (1971). *The lorax.* New York: Random House.
Geisel, T. (1985). *The cat in the hat.* New York: Random House. (Original work published 1957)
Hayes, C. (1951). *The ape in our house.* New York: Harper.
Herman, L. M. (1986). Cognition and language competencies of bottlenosed dolphins. In R. J. Schusterman, J. A. Thomas, & F. G. Wood (Eds.), *Dolphin cognition and behavior: A comparative approach* (pp. 221–251). Hillsdale, NJ: Erlbaum.
Herman, L. M., & Uyeyama, R. K. (1999). The dolphin's grammatical competency: Comments on Kako. *Animal Learning and Behavior, 27,* 18–23.
Hoyt, A. M. (1941). *Toto and I: A gorilla in the family.* Philadelphia: Lippincott.
Kellogg, W. N., & Kellogg, L. A. (1933). *The ape and the child.* New York: McGraw-Hill.
Kohts, N. (1935). *Infant ape and human child* (Vols. 1–2). Moscow: Museum Darwinianum.
Lofting, H. (1988). *The voyages of Dr. Doolittle.* New York: Bantam Doubleday Dell. (Original work published 1922)
Matsuzawa, T. (2004). Ai project: A retrospective of 25 years research on chimpanzee intelligence. In W. A. Hillix & D. M. Rumbaugh (Eds.), *Animal bodies, human minds: Ape, dolphin, and parrot language skills* (pp. 201–211). New York: Kluwer Academic/Plenum Publishers.
Miles, H. L. (1990). The cognitive foundations for reference in a signing orangutan. In S. T. Parker & K. R. Gibson (Eds.), *"Language" and intelligence in monkeys and apes: Comparative developmental perspectives* (pp. 511–539). Cambridge, England: Cambridge University Press.

Miles, H. L. (1994). Chantek: The language ability of an enculturated orangutan (*Pongo pygmaeus*). In J. Ogden, L. Perkins, & L. Sheeran (Eds.), *Proceedings of the International Conference on "Orangutans: The neglected ape"* (pp. 209–219). San Diego, CA: Zoological Society of San Diego.

Morris, R., & Morris, D. (1966). *Men and apes*. New York: McGraw-Hill.

Ohio State Research. (2006, February 21). *Ohio State to close its primate center, retire its chimpanzees*. Retrieved March 31, 2006, from http://researchnews.osu.edu/archive/chmpclos.htm

Patterson, F. G., & Linden, E. (1981). *The education of Koko*. New York: Holt, Rinehart & Winston.

Pepperberg, I. M. (1999). *The Alex studies*. Cambridge, MA: Harvard University Press.

Pepys, S. (2000). *The diary of Samuel Pepys*. Berkeley: University of California Press. (Original work published 1661)

Pfungst, O. (1911). *Clever Hans*. New York: Holt.

Pinker, S. (1994). *The language instinct*. New York: William Morrow.

Premack, D. (1976). *Intelligence in ape and man*. Hillsdale, NJ: Erlbaum.

Rumbaugh, D. M. (Ed.). (1977). *Language learning by a chimpanzee*. New York: Academic Press.

Savage-Rumbaugh, E. S. (1986). *Ape language: From conditioned response to symbol*. New York: Columbia University Press.

Savage-Rumbaugh, E. S. (2004). An overview of her work by Dr. Sue Savage-Rumbaugh. In W. A. Hillix & D. M. Rumbaugh (Eds.), *Animal bodies, human minds* (pp. 154–165). New York: Kluwer Academic/Plenum Publishers.

Savage-Rumbaugh, E. S., & Lewin, R. (1994). *Kanzi: The ape at the brink of the human mind*. New York: Wiley.

Sebeok, T. A., & Umiker-Sebeok, D. J. (Eds.). (1980). *Speaking of apes: A critical anthology of two-way communication with man*. New York: Plenum Press.

Skinner, B. F. (1957). *Verbal behavior*. New York: Appleton-Century-Crofts.

Taglialatela, J. P., Savage-Rumbaugh, S., & Baker, L. A. (2003). Vocal production by a language-competent bonobo, *Pan Paniscus*. *International Journal of Primatology, 24*, 1–17.

Terrace, H. S. (1979). *Nim*. New York: Knopf.

Wallman, J. (1992). *Aping language*. New York: Cambridge University Press.

Yerkes, R. M. (1925). *Almost human*. New York: Century.

# 17

# A Comparative Psychologist Looks at Language

*Herbert L. Roitblat*

Language has frequently been cited as the defining characteristic of human intelligence. By characterizing language, according to this view, one also characterizes intelligence. One version of this enterprise takes the approach of trying to distinguish human language from other forms of communication. Such investigators have proposed behavioral criteria that represent the competencies necessary for the existence of language. Another approach is the attempt to characterize the nature of human intelligence vis-à-vis language. In the process, these investigators have also proposed characteristics that language must have to produce human intelligence. Language use is so central to the conceptualization of what it means to be human and so tied into intellectual achievements that it may actually be the cause of a person's achievements (Bickerton, 1990).

According to Fodor (e.g., 1975) and others, the very nature of thought is its languagelike ability to manipulate symbols syntactically—that is, according to certain rules. According to this view, the very nature of human thought requires this kind of process; we could not function as language-using, intelligent beings without it. The language of thought is an example of a physical symbol system (Newell, 1980; Newell & Simon, 1976; see also Harnad, 1990). According to Newell and Simon (1976), for example, such a symbol system is both necessary and sufficient for human intelligence. By implication, any organism or system lacking such capabilities would necessarily lack intelligence. Language is assumed to be such a symbol system, and it is also assumed that we could not have language without such an internal symbolic system. Although some investigators using this approach have been willing to assume that machines can be physical symbol systems and therefore intelligent, others have been less confident of this assumption. Most would also deny that any animal, other than humans, could have such a symbol system, or at least that evidence for its existence is conspicuously absent. Hence, language and thought are inextricably intertwined; each one presupposes the other, or both presuppose some underlying mechanism with similar properties.

A physical symbol system is physical in the sense that it is implemented in some real material system (presumably including human brains, though it is not clear how neurons would actually implement a physical symbol system). It comprises a set of symbols, consisting of tokens or patterns, such as marks on paper, punches on computer tape, brain activation, or patterns of speech utterance, and a set of explicit rules. The tokens can be strung together to yield a structure or expression. The rules of the system govern how to manipulate the symbol tokens.

The key elements of a physical symbol system, such as human language, consist of two groups of features. The first group includes atomicity, systematicity, and semantic transparency. These features reflect the idea that words are independent symbols. The second group of features includes compositionality and syntax. These features reflect the idea that words can be combined according to rich rules to create more complex expressions, and complex expressions can be broken down into their elemental components.

*Atomicity* means that the system makes use of basic units that can be combined and recombined to express different ideas. In language, such atoms might be morphemes. *Systematicity* is the idea that symbols have a meaning that is independent of the context in which the symbol appears. The symbol "cat" for example is thought to have the same meaning in "the cat was hungry" as in "the cat sat on the mat." According to Pylyshyn (1989), "This sort of systematicity follows automatically from the use of structured symbolic expressions to represent knowledge and to serve as the basis for inference" (p. 62). Finally, *semantic transparency* means that the symbols can be assigned a clear and specific meaning.

*Compositionality* means that expressions consist of (perhaps ordered) atomic symbols. Complex expression can be broken down into simpler expressions, and the rules of the system describe how the symbols can be combined. *Syntax* is the collection of rules that govern how these expressions can be composed and how one expression can be related organizationally to another.

I argue that neither thought nor language has these characteristics. If we were to adopt them as necessary standards we would find that humans do not talk, and if they did talk, they could not understand one another. Rather, both language and thought are more properly characterized as fuzzy, contextual, sloppy, and productive. Rather than being the basis for intelligence, intelligence is required to use such physical symbol systems. It is no accident that logic is so difficult for students to learn. Far from being the essence of human thought, people have to simulate physical symbol systems to think logically.

## Context Independence

Systematicity, atomicity, and semantic transparency all imply that words have specific meanings that do not depend on their context. Fodor (1975) and other proponents of the language of thought hypothesis recognize that some words are ambiguous on the surface, such as *bank* or *bark*, but the underlying symbols are actually not ambiguous. They grossly underestimate the problem of ambiguity, as any Internet search engine user can attest. As an exercise, I looked up

| This | form | of | government | seems | like | the | best | one | man | can | devise. |
|------|------|----|------------|-------|------|-----|------|-----|-----|-----|---------|
| 8 | 45 | 16 | 8 | 5 | 32 | 2 | 20 | 24 | 26 | 6 | 4 |

**Figure 17.1.** A simple sentence. Numbers below each word indicate the number of meanings for that word that were found in a dictionary.

in a dictionary the number of definitions for the words in a simple sentence (Figure 17.1). This sentence was a paraphrase of a quote from John Adams. It was a paraphrase because I could not remember the exact words that are attributed to him. This, itself, is telling that the words are not the fundamental basis of my memory.

The numbers under each word show how many definitions there were for each one in a dictionary (*Webster's Encyclopedic Unabridged Dictionary of the English Language*, 1989). As a rough approximation, the average word in this sentence had 16 meanings. If you combined all of these possible meanings together, there are 4,416,602,112,000 (4.4 trillion) interpretations for this sentence, yet English speakers generally have no difficulty understanding it. They generally do not even notice that this level of ambiguity exists.

Rather than each word having a specific and unchanging meaning, each word in the sentence obtains some of its meaning from the other words in the sentence. Adherents to the atomic hypothesis might counter that the words used in the language might be ambiguous, but the person's internal representation of those meanings is not. If taken seriously, this retort is simply unfalsifiable. In any case, there is additional evidence that words do not reflect autonomous particles of meaning.

Machine translation also highlights the importance of context in understanding words. Machine translation systems are computer programs that attempt to translate utterances in one language into another. They are notoriously difficult to develop, and none has yet achieved professional levels of performance, apparently because they cannot yet account for all of the contextual constraints that characterize human language. Here is the machine translation of a real estate ad for a house in Provence, France, followed by its original French text.

> Roofs and frames remade to nine. House of Master with a small house of friends, exploited partly in rooms of hosts. Much authentic charm.
> *Toits et charpentes refaits à neuf. Maison de matre avec une petite maison d'amis, exploitée en partie en chambres d'hôtes. Beaucoup de charme authentique.*

Machine translation operates word for word with little contextual input from the other words in the sentence. The machine, in other words, cannot use the context of the sentence to disambiguate the words in the sentence.

## Other Evidence Concerning Systematicity and Atomicity

If these examples are not sufficiently convincing that words are not independent units of meaning, consider the following example of the word *mother* described by Lakoff (1987):

> I was adopted, I don't know who my real mother was.
> I am not a nurturing person, so I don't think that I could ever be a real mother to anyone.
> My real mother died when I was an embryo, and I was frozen and later implanted in the womb of a woman who gave birth to me.
> I had a genetic mother who contributed the egg that was implanted in the womb of my real mother who gave birth to me. (adapted from Lakoff, 1987, p. 75)

Lakoff's analysis of *mother* does not depend on the acceptance of dictionary definitions as evidence of a word's meaning. In some sense, all of these examples involve the same definition of *mother* as "parent," yet *mother* clearly also means something different in each of these sentences. In fact, taken together, these sentences describe mutually exclusive references. A "real" mother cannot both be and not be the woman whose egg was responsible for one's genetic makeup

The logical positivists recognized that language was much too sloppy to be the basis of a logical and positive science. In their ideal, all scientifically valid utterances would consist of observations and deductions from those observations (Carnap, 1939). It turned out, however, that even an idealized language of science fell short of this ideal.

Semantic illusions (Erickson & Mattson, 1981) also suggest that words are not atomic symbols. Try to answer the following questions as quickly as possible:

> How many animals of each kind did Moses take on the Ark?
> What is the nationality of Thomas Edison, inventor of the telephone?
> What do cows drink?

People often answer "two" to the first question, "American" to the second, and "milk" to the third. In fact, Moses had nothing to do with the Ark, Edison did not invent the telephone, and cows drink water; they give milk. Still, these sentences provide enough contextual cues to lead the answerer to erroneous responses. It cannot be the case that Moses and Edison are symbolically represented as atomic, systematic, and semantically transparent entities and still permit this kind of fuzzy reinterpretation. As atomic entities, they should head off any interpretation of the sentence that would result in erroneous answers.

Further, language is sometimes productive. People are often coining new words or using old words in new ways. If words relied on the existence of internal symbols to be expressed, then it is not clear how new words would emerge. President George W. Bush was often quoted during his campaign as creating new words. These often involved the novel combinations of various morphemes. Shakespeare also made up many new words. Some 17% of Shakespeare's vocabulary appeared for the first time in his work, including words such as *auspicious, assassination, disgraceful, dwindle, savagery,* and *honorificabilitudinitatibus.* In his own words, "They have been at a great feast of languages and stol'n the scraps" (Shakespeare, *Love's Labour's Lost*).

Rather than consisting of atomic symbols, these examples suggest that human language is fuzzy and contextual. Word meaning is "depauperate." Meaning is incompletely developed, subject to reinterpretation depending on the context of the moment. People seem to have fuzzy ideas of what they want to say and fuzzy interpretations of what they hear and read (that are more or less equivalent to what the speaker had in mind). Resemblance, rather than symbolic systematicity, seems to be the basis for much of human language and much of human thought.

> "Then you should say what you mean," the March Hare went on. "I do," Alice hastily replied; "at least—at least I mean what I say—that's the same thing, you know."
> "Not the same thing a bit," said the Hatter. "Why you might just as well say that 'I see what I eat' is the same thing as 'I eat what I see!'" (Carroll, 1865/1997, p. 71)

Rather than being inherent in the words or in the mind of the speaker and listener, meaning seems to emerge from the interaction between the text and the individual. Although one may say that the context contributes to the meaning of the words in a text, the context also seems to include the person. In the semantic illusion sentences, for example, a certain level of familiarity with the misleading words (i.e., *Moses, Edison,* and *cows*) would seem to be necessary to evoke the illusion. I would argue that it is specifically the history of contexts in which those words have previously appeared that makes for the ability to semantically process them.

## Human Language and Animal Language

Although language is often productive, it is also frequently stereotyped. Investigators have criticized animal language production as stereotyped, but similar samples of human speech might appear equally stereotyped. Consider the following clichés:

> Live and learn.
> Today is the first day of the rest of your life.
> What goes around comes around.
> Haste makes waste.
> Nobody's perfect.
> I can't change the past.
> Tomorrow is another day.
> When all is said and done.
> Nutty as a fruitcake.

These clichés and similar formulaic expressions account for a large proportion of utterances. One of the early animal language studies involving the

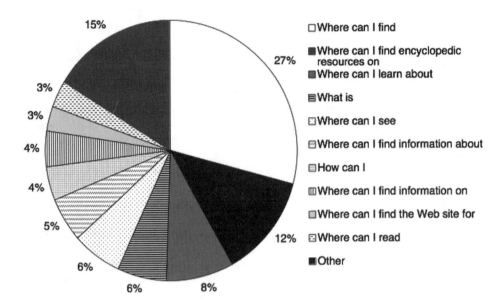

**Figure 17.2.** Typical queries for the Ask Jeeves search engine.

chimpanzee Lana and her Yerkish keyboard was criticized (Thompson & Church, 1980) because of the apparent predominance of stereotyped responses such as the following:

> Please machine give <incentive>
> Please <name> <give | move into room> <incentive>
> Please <name> <move behind room | put in machine> <incentive>
> Please Lana <want eat | want drink> <incentive>
> Please <name> <tickle | groom | swing | carry> Lana
> Please machine make <event> [1]

As it turns out, humans in a similar situation do not behave much differently. I collected queries presented to the Internet search engine "Ask Jeeves" (see http://www.ask.com). These queries were posted on the Ask Jeeves Web site and refreshed every few seconds to reflect a sample of recent queries. Exact duplicates were removed because the sampling algorithm used by the Web site apparently allows for the same query to be sampled more than once. Of the queries to the Ask Jeeves machine, 85% were 1 of 10 stereotyped queries (see Figure 17.2). If this were our sample of human speech, we would come to the conclusion that people do not speak.

---

[1] Adapted from "An Explanation of the Language of a Chimpanzee," by C. R. Thompson and R. M. Church, 1980, *Science, 208,* Table 1, p. 313. Copyright 1980 by AAAS. Adapted with permission from AAAS.

Shakespeare's use of words also showed a great deal of stereotypy. A very few words, 5% of his vocabulary, accounted for more than 80% of the words published. The same words appeared over and over again in his work. If this were our sample of human speech, we would come to the conclusion that people do not speak. In contrast, more than 40% of Shakespeare's vocabulary occurred in one work only (accounting for 1.5% of all words in the canon).

Another supposed hallmark of human language is the use of complex syntax. None of the animals studied so far in the animal language research appears to have the ability to use complex syntax in its language production. Syntactic complexity may be the exception rather than the rule. For example, the stereotypy of the human communication with the Ask Jeeves Web site is no more complex than that of the chimpanzee Lana. By at least one measure, furthermore, the complexity of human language has been diminishing for the last 200 years.

To assess syntactic complexity over a long period of time, I needed a corpus of texts that were produced in analogous contexts for this time. Starting with George Washington's second term, it has become the custom for inaugurated presidents to deliver speeches during their inauguration. The texts of these speeches are available and can be analyzed. As a measure of syntactic complexity, I chose to count the number of occurrences of the word *which* in each speech, normalized by the length of the speech. Although not without its limitations, the presence of the word *which* usually signals a subordinate clause, a construction far more complex than a simple declarative sentence. Another limitation of this method is the common confusion between *which* and *that*. Theodore Roosevelt, for example, commonly used *which* where grammarians would have told him he needed to use *that*. The proportion of *which* in his address is, therefore, abnormally high.

Despite these limitations, I expected to find that the complexity of sentences in inauguration addresses would remain at a steady level until the beginning of the 20th century, when mass communications might have put a premium on using simple declarative sentences that could be easily understood over radio or television. Instead the pattern I found was quite different. On average, syntactic complexity has been declining linearly since 1797. Figure 17.3 shows this complexity. Notable exceptions were Theodore Roosevelt (in 1905), as described earlier, and Carter (in 1977). George W. Bush is the President who used *which* the least, with only one occurrence of the word in his inauguration speech.

These views of human language use would suggest that although people are capable of great feats, most everyday conversation is mundane. Naturalistic samples of human language production do not aspire to the lofty criteria that have been attributed to it. One might suppose that an alien suddenly landing on earth and taking a speech sample at a large gathering of humans (e.g., a college football stadium on a Saturday afternoon in October) might be forgiven for coming to the conclusion that humans do not have language. With only some tongue in cheek, one might guess that the samples of animal language that have been collected all capture mundane communications that parallel those described for humans in this chapter and that nothing yet has motivated them to achieve higher communicative aspirations.

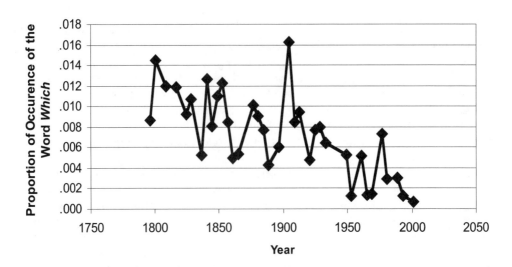

**Figure 17.3.** Complexity level (as measured by occurrence of the word *which*) of sentences in presidential inauguration addresses.

## References

Bickerton, D. (1990). *Language and species*. Chicago: University of Chicago Press.
Carnap, R. (1939). *Foundations of logic and mathematics: International encyclopaedia of unified science* (Vol. I, No. 3). Chicago: University of Chicago Press.
Carroll, L. (1997). *Alice's adventures in Wonderland and through the looking glass*. London: Penguin Group. (Original work published 1865)
Erickson, T. D., & Mattson, M. E. (1981). From words to meaning: A semantic illusion. *Journal of Verbal Learning and Behavior, 20*, 540–551.
Fodor, J. A. (1975). *The language of thought*. New York: Thomas Y. Crowell.
Harnad, S. (1990). The symbol grounding problem. *Physica D, 42*, 335–346.
Lakoff, G. (1987). *Women, fire, and dangerous things: What categories reveal about the mind*. Chicago: University of Chicago Press.
Newell, A. (1980). Physical symbol systems. *Cognitive Science, 4*, 135–183.
Newell, A., & Simon, H. A. (1976). Computer science as empirical enquiry. In J. Haugeland (Ed.), *Mind design* (pp. 35–66). Cambridge, MA: MIT Press.
Pylyshyn, Z. (1989). Computing in cognitive science. In M. I. Posner (Ed.), *Foundations of cognitive science* (pp. 49–92). Cambridge, MA: MIT Press.
Thompson, C. R., & Church, R. M. (1980, April 18). An explanation of the language of a chimpanzee. *Science, 208*, 313–314.
*Webster's encyclopedic unabridged dictionary of the English language*. (1989). New York: Random House.

# 18

# Evolution of Language and Speech From a Neuropsychological Perspective

*William D. Hopkins*

The question of language origins and their neurobiological substrates has a long history in psychology. In particular, the question of whether language and speech are uniquely human traits from a behavioral and neurobiological perspective has been the focal point of numerous theoretical and scientific investigations (see Kimura, 1993; MacNeilage & Davis, 2001; Steklis & Raleigh, 1979). The purpose of this chapter is to describe recent neuropsychological findings in great apes as they pertain to basic theories on the origin of language and speech in humans. Until recently, the typical evolutionary theory of language and speech described in many textbooks and reviews went as follows. Humans are right-handed. Most right-handed humans are left-hemisphere dominant for language and speech. Nonhuman primates and other animals do not show population-level right-handedness. Nonhuman primates do not have language or speech. Therefore, language, speech, and right-handedness are unique to humans and have their own neurobiological substrates, which are specific to the human brain. In terms of the specific brain systems involved in language and speech, Broca's and Wernicke's areas were proposed to be uniquely human. Broca's (part of the inferior frontal lobe) and Wernicke's (part of the posterior portion of the temporal lobe) areas are the two regions of the brain that are linked to speech and language production and perception, and each is more active and neuroanatomically larger in the left compared with the right cerebral hemisphere. Some have even suggested that the presence of neuroanatomical asymmetries, in any brain area, is uniquely human (Crow, 1998). Recent studies in nonhuman primates and, particularly, great apes have begun to challenge these historical and contemporary views.

---

This research was supported by National Institutes of Health Grants NS-29574, NS-36605, NS-42867, and HD-38051 to William D. Hopkins and RR-00165 to the Yerkes National Primate Research Center. The Yerkes Center is fully accredited by the American Association for Accreditation of Laboratory Animal Care. American Psychological Association guidelines for the ethical treatment of animals were adhered to during all aspects of this study.

## Neuroanatomical Asymmetries in "Language" Areas of the Brain

The planum temporale (PT) is the neocortical area bordered anteriorly by Heschl's gyrus and posteriorly by the posterior ascending ramus. Numerous studies in human subjects have reported this brain area to be larger in the left compared with the right hemisphere (Beaton, 1997). Wernicke's area is encompassed within the PT and therefore many have linked asymmetries in the PT to the evolution of speech and language comprehension in the human brain. Recent studies in cadaver specimens (Gannon, Holloway, Broadfield, & Braun, 1998) and magnetic resonance images (MRI) have shown that the PT can be anatomically defined in all the great apes (gorillas, chimpanzees, and orangutans) and is larger in the left compared with the right cerebral hemisphere (Gilissen, 2001). For example, Gannon et al. (1998) reported that 17 of 18 chimpanzee cadaver specimens showed a left-hemisphere asymmetry in the size of the PT. Measuring the PT from MRI, Hopkins and Cantalupo (2004) reported that 44 of 61 chimpanzees showed a left-hemisphere asymmetry (see also Cantalupo, Pilcher, & Hopkins, 2003; Hopkins, Marino, Rilling, & MacGregor, 1998).

Broca's area includes Brodmann's Areas 44 and 45 in the human brain (Amunts et al., 1999). Some studies have suggested that asymmetries in Broca's area of the human brain are more pronounced for Area 44 compared with Area 45. It has been known for many years that Brodmann's Areas 44 and 45 could be localized in nonhuman primate brains using cytoarchitectonic methods (Sherwood, Broadfield, Holloway, Gannon, & Hof, 2003), but until recently, there have been no studies examining the morphology and lateralization of these regions. Cantalupo and Hopkins (2001) quantified Brodmann's Area 44 from MRIs in a sample of 26 great apes and found that 20 of the apes had a left-hemisphere asymmetry, 6 apes had right-hemisphere asymmetry, and 1 ape had no bias. These findings indicate that the inferior frontal cortex, a region homologous to Broca's, is similarly asymmetrical in the great ape brain.

## Manual Gestural and Intentional Use of Vocalizations in Chimpanzees

A second recent line of research that has neuropsychological implications for theories of language origin has involved studies on the functional use of gestural and, to some extent, vocal communication in great apes, notably chimpanzees. It has been known for nearly 30 years that captive and wild chimpanzees exhibit some manual gestures (see Goodall, 1986), but more recent studies in captive chimpanzees have demonstrated that gestures used by chimpanzees are produced both intentionally and referentially (Leavens & Hopkins, 1999). For example, in chimpanzees, orangutans, and gorillas, it has been reported that when food is placed out of reach of the subjects, they will alter the frequency of their gestures on the basis of the presence or absence of a social audience (Call & Tomasello, 1994; Krause & Fouts, 1997; Leavens, Hopkins, & Bard, 1996; Poss, Kuhar, Stoinski, & Hopkins, in press). Moreover, when the apes gesture to out-of-reach food items, they alternate their gaze between the referent and the social agent, suggesting that they are monitoring the effect of their communicative behavior on the social recipient of their communicative act (Leavens et al., 1996; Leavens & Hopkins, 1998). In addition to the reports of an audience effect on gestural communication, there are at least two reports indicating that chimpanzees will alter the types of communicative behavior they engage in depending on the orientation and attentional status of human individuals around them (Hostetter, Cantero, & Hopkins, 2001; Leavens, Hostetter, Wesley, & Hopkins, 2004; but see Theall & Povenelli, 1999). In two studies, my colleagues and I examined the effect of various aspects of attentional status on the communicative behaviors of chimpanzees when they were communicating about food placed outside their home cages. We were specifically interested in the types of communicative behaviors chimpanzees engage in during these conditions, and we scored whether the chimpanzees vocalized, gestured, engaged in attention-getting behaviors (clap, spit, band their cage, or throw), pouted their lips, or did nothing. In all of these studies, we scored the first communicative behavior exhibited by the chimpanzees. In the initial experiment, humans were either oriented away from or toward the chimpanzees, and the results indicated that they were more likely to engage in either a vocalization or other nonvisual means of communication as their first communicative response when the human was oriented away from them. This is in contrast to when the human was oriented toward them, when the first communicative response was significantly more often a manual gesture or another type of visual signal (see Hostetter et al., 2001; see Figure 18.1). More recently, rather than manipulate the orientation of the human, we altered whether the human was offering a banana and looking at (a) the focal subject, (b) another chimpanzee in the same cage, or (c) another chimpanzee in an adjacent cage. When the experimenter was offering and looking to the focal subject, the subject was significantly often likely to engage in a visual form of communication, notably manually gesture or lip pout. In contrast, when the experimenter was offering to a chimpanzee in the same or adjacent cage, the focal subject was more likely to engage in a nonvisual signal, including vocalization or attention-getting behaviors (see Figure 18.2).

## Laterality, Gesture, and Facial Expression

In terms of laterality and communication in the chimpanzees, my colleagues and I have taken two distinct approaches. In one series of investigations, we examined hand use in the context of gestural communication with and without the simultaneous production of a vocalization (Hopkins & Cantero, 2003; Hopkins & Leavens, 1998; Hopkins & Wesley, 2002). Our original studies were based on previous observations on a small sample of chimpanzees in which we noted differential use of the right hand as a function of gesture type and vocal production (Leavens et al., 1996). In two subsequent follow-up studies in substantially larger samples of subjects ($N > 115$), we assessed hand use for gestures in relation to the sex and rearing (human-reared vs. chimpanzee-reared) of the subjects, the type of gesture (food begs vs. whole-hand points), and

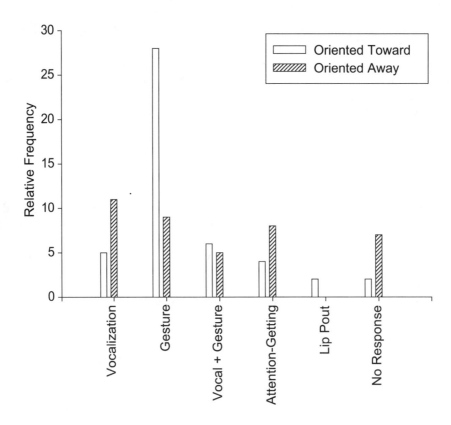

**Figure 18.1.** Relative frequency of each type of communicative behavior exhibited as the first communicative response when a human was oriented toward or away from the focal chimpanzee subject when a banana was placed outside the subject's home cage (see Hostetter et al., 2001).

whether the chimpanzees vocalized while gesturing. Four consistent findings emerged. First, population-level right-handedness was evident for gestures. Second, there was a stronger degree of right-handedness for gesturing in female compared with male chimpanzees (see Figure 18.3). Third, there was a stronger degree of right-handedness for food begs compared with whole-hand gestures (see Figure 18.4). Fourth, preferential use of the right hand for gesturing was enhanced when subjects simultaneously vocalized compared with when they did not.

The second line of investigation involved the measurement of asymmetries in the oral–facial region when chimpanzees were producing various facial expressions, including some that had an accompanying vocalization (see Fernandez-Carriba, Loeches, Morcillo, & Hopkins, 2002). In these studies, observational methods were used to obtain full-frontal images of chimpanzee facial expressions while the chimpanzees were engaged in various social interactions. The specific facial expressions we examined included silent bare teeth, scream, play, hoot, pout, and neutral. Following procedures used by others (Hook-Costigan & Rogers, 1998), the images were bifurcated down the vertical

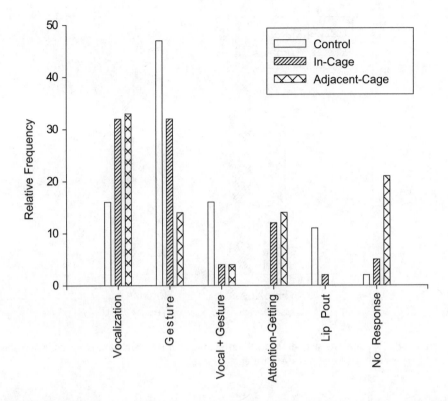

**Figure 18.2.** Relative frequency of each type of communicative behavior exhibited as the first communicative response when a human was looking at the focal subject (Control), looking at an individual living in the same cage (In-Cage), or looking at a chimpanzee housed in the adjacent cage (Adjacent-Cage; see Leavens et al., 2004).

meridian, and we measured the area of the mouth expressed on the left and right halves of the face. These results of this study indicated that the chimpanzees showed left-hemiface asymmetries for four expressions: silent bare teeth, scream, play, and hoot. No population-level bias was found for pout. Given that the left hemiface was more expressive, it indicates a right-hemisphere advantage in the production of facial expressions, a result consistent with other previous reports in nonhuman primates (Hauser, 1993; Hook-Costigan & Rogers, 1998).

## Are Great Apes Right-Handed?

Lastly, although more tangentially related to the issue of language origins, the question of whether nonhuman primates exhibit population-level handedness remains a topic of considerable empirical debate. The rationale for studying handedness in nonhuman primates is because most right-handed humans (96%) are left-hemisphere dominant for language functions. Thus, right-handedness is an indirect marker of cerebral dominance for language. The

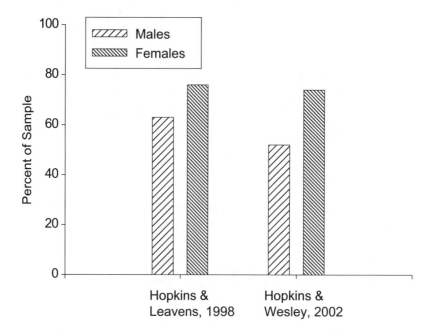

**Figure 18.3.** Relative proportion of male and female chimpanzees using the right hand to gesture in two separate studies.

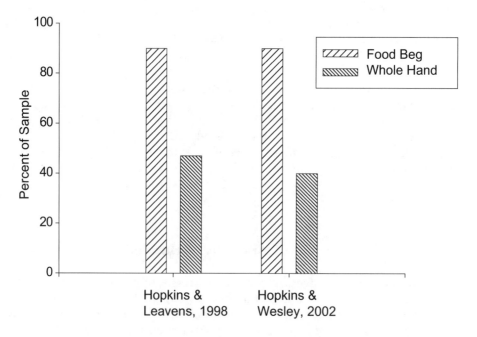

**Figure 18.4.** Relative proportion of right-hand gestures as a function of gesture type in two separate studies.

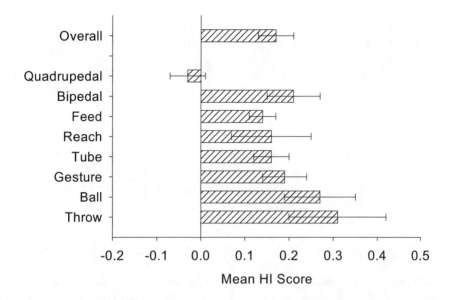

**Figure 18.5.** Mean handedness index (HI), with standard error bars, for multiple measures of hand use in chimpanzees housed at the Yerkes National Primate Research Center (see Hopkins 1993, 1994, 1995a, 1995b; Hopkins & Pearson, 2000).

appeal for studying handedness in nonhuman primates is because it is a relatively simple test to conduct in a large number of subjects.

Several recent reviews of the nonhuman primate literature on handedness have come to different conclusions regarding the degree to which any nonhuman primate species exhibits population-level handedness (for contrasting views, see Hopkins, 1999; Hopkins & Cantalupo, 2003; McGrew & Marchant, 1997; Palmer, 2002; Warren, 1980). With respect to studies in great apes, there is increasing evidence of population-level handedness, particularly in studies involving larger sample sizes of captive subjects. For example, in our laboratory, population-level right-handedness has been found for a number of measures, including throwing, reaching, bimanual feeding, and tasks that require coordinated bimanual actions. The mean handedness index for many of these measures can be seen in Figure 18.5. Although the independent measures of hand use are not strongly correlated with each other (Wesley, Fernanadez-Carriba, Hostetter, Poss, & Hopkins 2002), averaging the handedness index scores across all the measures still revealed population-level right-handedness. These results are not restricted to the chimpanzees housed at the Yerkes Center (see Hopkins, Wesley, Izard, Hook, & Schapiro, 2004) or even to chimpanzees but may also be evident in bonobos and gorillas (Hopkins & Pearson, 2000).

## Theoretical Implications and Future Directions

The collective results on neuroanatomical asymmetries, gestural communication, and handedness raise a number of theoretical issues regarding the origins of language and speech. First, the evidence of intentional and referential use of gestures by chimpanzees clearly supports theories proposing a gestural origin for the source of human language and speech (e.g., Armstrong, Stokoe, & Wilcox, 1995). The primary evidence in support of this conclusion is the lack of evidence for gestural communication in more distantly related primates and the functional use of gestures and vocalizations by apes, notably chimpanzees. Moreover, there is little evidence that more distantly related primates exhibit volitional controls of their vocal signals but rather produce them in response to emotional settings. In contrast, preliminary studies in chimpanzees suggest these apes do have volitional control of their vocalizations and can use them instrumentally to capture the attention of an otherwise inattentive person. In a larger theoretical framework, these findings link the evolution of language to the emergence of social cognition, which some have suggested is more sophisticated in apes compared with monkeys (Whiten & Byrne, 1988), and this potential relationship needs to be further explored in laboratory and field studies in primates.

The studies on neuroanatomical asymmetries as well as laterality in gestures have suggested that the neurobiological basis for intentional, referential communication was present prior to hominid evolution. These findings clearly contradict the historical and some contemporary views linking the evolution of language to lateralization in the nervous system (Crow, 1998). In fact, findings on laterality in a number of animal species, including birds, fish, reptiles, and mammals, all challenge the long-held assumptions for the uniqueness of hemispheric specialization to humans and clearly show that language was not a necessary condition for the evolution of behavioral and brain asymmetries in mammals (Bisazza, Rogers, & Vallortigara, 1998; Bradshaw & Rogers, 1993; Rogers & Andrew, 2002). What remains unclear from the findings in animals, including nonhuman primates, is the functional relevance of various behavioral asymmetries between different species. For example, birds, frogs, and chimpanzees show lateralization of function in the production of communicative signals, but the extent to which they parallel the function of human speech is unclear, particularly from a neuropsychological perspective. This discrepancy is clear when considering the nonhuman primate data. Studies have suggested that the production of species-typical vocalizations by chimpanzees, rhesus monkeys, and marmosets involves the right hemisphere (for a review, see Hopkins & Fernandez-Carriba, 2002). In contrast, the production of gestures and possibly non-species-typical vocalizations of chimpanzees are controlled by the left hemisphere. Fundamental to this discussion is whether the vocal signals indicate emotional or affective valence or whether they convey semantic or other types of information. In terms of species-typical vocal signals, there is neuropsychological evidence that they are associated with emotional expression because they are lateralized to the right hemisphere, but some have suggested that the very same vocalizations, at least in rhesus monkeys, communicate semantic information (Gouzoules, Gouzoules, & Marler, 1984), which at a simple level should be lateralized to the left hemisphere.

Lastly, the evidence of right-handedness in great apes has largely come from studies in captive populations. Evidence of population-level right-handedness in wild great apes has been less frequently reported, and it might be argued that the findings of right-handedness in captive apes are an artifact of being raised in a captive environment by right-handed humans. One problem with this interpretation is that the kinds of measures used to assess handedness in captive and wild chimpanzees vary greatly, and the differences may just as likely be due to this factor rather than the rearing conditions per se (see Hopkins & Cantalupo, 2003). Moreover, many of the captive studies have much larger sample sizes than the studies with wild apes, and therefore studies in captive apes have more statistical power and are able to detect less robust findings of population-level handedness. It is clear that greater dialogue and collaboration is needed between researchers studying captive and wild apes to clarify the ecological validity of handedness in nonhuman primates. A second important aspect of the findings on handedness is the difference in distribution of right-handedness between humans and chimpanzees. There is about an 8:1 ratio of right-to-left handedness in human populations, whereas the ratio is around 2:1 in chimpanzees. What factors explain this difference? Some have suggested that a genetic mechanism may explain this difference (Corballis, 1997), whereas others have suggested the difference may be due to cultural, life history, and measurement factors (Hopkins, Dahl, & Pilcher, 2001). Additional comparative behavioral–genetic studies of handedness are needed to pinpoint what factors account for the robust degree of handedness found in human populations compared with apes.

In summary, the recent neuroanatomical and behavioral studies on laterality and communication have clear implications for historical and contemporary theories on the origin of language. Many of these studies empirically address questions that have lingered in the literature for some time (see Steklis & Raleigh, 1979). There remain many unanswered questions, particularly with respect to the neural systems involved in higher cognitive and communicative abilities in apes. With the advent of modern functional brain-imaging techniques, the potential is real and imminent for evaluating whether similar cognitive and neurological systems are involved in the use of communicative signals by nonhuman primates. This in turn should provide valuable information on the evolution of human cognition and language from a neuropsychological perspective.

## References

Amunts, K., Schleicher, A., Bürgel, U., Mohlberg, H., Uylings, H. B., & Zilles, K. (1999). Broca's region revisited: Cytoarchitecture and intersubject variability. *Journal of Comparative Neurology, 412,* 319–341.

Armstrong, D. F., Stokoe, W., & Wilcox, S. E. (1995). *Gesture and the nature of language.* Cambridge, England: Cambridge University Press.

Beaton, A. A. (1997). The relation of planum temporale asymmetry and morphology of the corpus callosum to handedness, gender, and dyslexia: A review of the evidence. *Brain and Language, 15,* 255–322.

Bisazza, A., Rogers, L. J., & Vallortigara, G. (1998). The origin of cerebral asymmetry: A review of behavioural and brain lateralisation in fishes, reptiles, and amphibians. *Neuroscience and Biobehavioral Reviews, 22,* 411–426.

Bradshaw, J. L., & Rogers, L. (Eds.). (1993). *The evolution of lateral asymmetries, language, tool-use and intellect.* San Diego, CA: Academic Press.

Call, J., & Tomasello, M. (1994). Production and comprehension of referential pointing by orangutans (*Pongo pygmaeus*). *Journal of Comparative Psychology, 108*, 307–317.

Cantalupo, C., & Hopkins, W. D. (2001, November 29). Asymmetric Broca's area in great apes. *Nature, 414*, 505.

Cantalupo, C., Pilcher, D., & Hopkins, W. D. (2003). Are asymmetries in sylvian fissure length associated with the planum temporale? *Neuropsychologie, 41*, 1975–1981.

Corballis, M. C. (1997). The genetics and evolution of handedness. *Psychological Bulletin, 104*, 714–727.

Crow, T. J. (1998). The genetic origins of language. *Current Psychology of Cognition, 17*, 1120–1154.

Fernandez-Carriba, S., Loeches, A., Morcillo, A., & Hopkins, W. D. (2002). Asymmetry of facial expression of emotions by chimpanzees. *Neuropsychologia, 40*, 1523–1533.

Gannon, P. J., Holloway, R. L., Broadfield, D. C., & Braun, A. R. (1998, January 9). Asymmetry of chimpanzee planum temporale: Humanlike pattern of Wernicke's brain language area homolog. *Science, 279*, 220–222.

Gilissen, E. (2001). Structural symmetries and asymmetries in human and chimpanzee brains. In D. Falk & K. R. Gibson (Eds.), *Evolutionary anatomy of the primate cerebral cortex* (pp. 187–215). Cambridge, England: Cambridge University Press.

Goodall, J. (1986). *The chimpanzees of Gombe: Patterns in adaptation*. Cambridge, MA: Harvard University Press.

Gouzoules, S., Gouzoules, H., & Marler, P. (1984). Rhesus monkey (*Macaca mulatta*) screams: Representational signaling in the recruitment of agonistic aid. *Animal Behaviour, 32*, 182–193.

Hauser, M. D. (1993, July 23). Right hemisphere dominance in the production of facial expression in monkeys. *Science, 261*, 475–477.

Hook-Costigan, M. A., & Rogers, L. J. (1998). Lateralized use of the mouth in production of vocalizations by marmosets. *Neuropsychologia, 36*, 1265–1273.

Hopkins, W. D. (1993). Posture and reaching in chimpanzees (*Pan*) and orangutans (*Pongo*). *Journal of Comparative Psychology, 17*, 162–168.

Hopkins, W. D. (1994). Hand preferences for bimanual feeding in 140 captive chimpanzees (*Pan troglodytes*): Rearing and ontogenetic factors. *Developmental Psychobiology, 27*, 395–407.

Hopkins, W. D. (1995a). Hand preferences for a coordinated bimanual task in 110 chimpanzees (*Pan troglodytes*): Cross-sectional analysis. *Journal of Comparative Psychology, 105*, 178–190.

Hopkins, W. D. (1995b). Hand preferences in juvenile chimpanzees: Continuity in development. *Developmental Psychology, 31*, 619–625.

Hopkins, W. D. (1999). On the other hand: Statistical issues in the assessment and interpretation of hand preference data in nonhuman primates. *International Journal of Primatology, 20*, 851–866.

Hopkins, W. D., & Cantalupo, C. (2003). Does variation in sample size explain individual differences in hand preferences of chimpanzees (*Pan troglodytes*)? An empirical study and reply to Palmer (2002). *American Journal of Physical Anthropology, 121*, 378–381, 382–384.

Hopkins, W. D., & Cantalupo, C. (2004). Handedness in chimpanzees is associated with asymmetries of the primary motor cortex but not with homologous language areas. *Behavioral Neuroscience, 118*, 1176–1183.

Hopkins, W. D., & Cantero, M. (2003). The influence of vocalizations on preferential hand use in gestural communication by chimpanzees. *Developmental Science, 6*, 55–61.

Hopkins, W. D., Dahl, J. F., & Pilcher, D. (2001). Genetic influence on the expression of hand preferences in chimpanzees (*Pan troglodytes*): Evidence in support of the right shift theory and developmental instability. *Psychological Science, 12*, 299–303.

Hopkins, W. D., & Fernandez-Carriba, S. (2002). Laterality in communicative behaviors in nonhuman primates: A critical analysis. In L. Rogers & R. Andrews (Eds.), *Comparative vertebrate lateralization* (pp. 234–250). Oxford, England: Oxford University Press.

Hopkins, W. D., Wesley, M. J., Izard, M. K., Hook, M., & Schapiro, S. J. (2004). Chimpanzees are predominantly right-handed: Replication in three colonies of apes. *Behavioral Neuroscience, 118*, 659–663.

Hopkins, W. D., & Leavens, D. A. (1998). Hand use and gestural communication in chimpanzees (*Pan troglodytes*). *Journal of Comparative Psychology, 112*, 95–99.

Hopkins, W. D., Marino, L., Rilling, J., & MacGregor, L. (1998). Planum temporale asymmetries in great apes as revealed by magnetic resonance imaging (MRI). *NeuroReport, 9*, 2913–2918.

Hopkins, W. D., & Pearson, K. (2000). Chimpanzee (*Pan troglodytes*) handedness: Variability across multiple measures of hand use. *Journal of Comparative Psychology, 114,* 126–135.

Hopkins, W. D., & Wesley, M. J. (2002). Gestural communication in chimpanzees (*Pan troglodytes*): The effect of situational factors on gesture type and hand use. *Laterality: Asymmetry of Body, Brain and Cognition, 7,* 19–30.

Hostetter, A., Cantero, M., & Hopkins, W. D. (2001). Differential use of vocal and gestural communication in chimpanzees in response to the attentional status of a human audience. *Journal of Comparative Psychology, 115,* 337–343.

Kimura, D. (1993). *Neuromotor mechanisms in human communication.* New York: Clarendon Press.

Krause, M. A., & Fouts, R. S. (1997). Chimpanzee (*Pan troglodytes*) pointing: Hand shapes, accuracy, and the role of eye gaze. *Journal of Comparative Psychology, 111,* 330–336.

Leavens, D. A., & Hopkins, W. D. (1998). Intentional communication by chimpanzees (*Pan troglodytes*): A cross-sectional study of the use of referential gestures. *Developmental Psychology, 34,* 813–822.

Leavens, D. A., & Hopkins, W. D. (1999). The whole-hand point: The structure and function of pointing from a comparative perspective. *Journal of Comparative Psychology, 113,* 417–425.

Leavens, D. A., Hopkins, W. D., & Bard, K. A. (1996). Indexical and referential pointing in chimpanzees (*Pan troglodytes*). *Journal of Comparative Psychology, 110,* 346–353.

Leavens, D. L., Hostetter, A., Wesley, M. J., & Hopkins, W. D. (2004). Tactical use of unimodal and bimodal communication by chimpanzees (*Pan troglodytes*). *Animal Behaviour, 67,* 467–476.

MacNeilage, P. F., & Davis, B. L. (2001). Motor mechanisms in speech ontogeny: Phylogenetic, neurobiological and linguistic implications. *Current Opinion in Neurobiology, 11,* 696–700.

McGrew, W. C., & Marchant, L. F. (1997). On the other hand: Current issues in and meta-analysis of the behavioral laterality of hand function in nonhuman primates. *Yearbook of Physical Anthropology, 40,* 201–232.

Palmer, A. R. (2002). Chimpanzee right-handedness reconsidered: Evaluating the evidence with funnel plots. *American Journal of Physical Anthropology, 118,* 191–199.

Poss, S. R., Kuhar, C., Stoinski, T. S., & Hopkins, W. D. (in press). Differential use of attentional and visual communicative signaling by orangutans (*Pongo pygmaeus*) and gorillas (*Gorilla gorilla*) in response to the attentional status of a human. *American Journal of Primatology.*

Rogers, L. J., & Andrew, R. J. (2002). *Comparative vertebrate lateralization.* Oxford, England: Oxford University Press.

Sherwood, C. C., Broadfield, D. C., Holloway, R. L., Gannon, P. J., & Hof, P. R. (2003). Variability of Broca's area homologue in African great apes: Implications for language evolution. *The Anatomical Record, 271A,* 276–285.

Steklis, H. D., & Raleigh, M. J. (Eds.). (1979). *Neurobiology of social communication in primates.* New York: Academic Press.

Theall, L. A., & Povinelli, D. (1999). Do chimpanzees tailor their gestural signals to fit the attentional states of others? *Animal Cognition, 2,* 207–214.

Warren, J. M. (1980). Handedness and laterality in humans and other animals. *Physiological Psychology, 8,* 351–359.

Wesley, M. J., Fernandez-Carriba, S., Hostetter, A., Pilcher, D., Poss, S., & Hopkins, W. D. (2002). Factor analysis of multiple measures of hand use in captive chimpanzees (*Pan troglodytes*): An alternative assessment to handedness in nonhuman primates. *International Journal of Primatology, 23,* 1155–1168.

Whiten, A., & Byrne, R. W. (1988). Tactical deception in primates. *Behavioral and Brain Sciences, 11,* 233–273.

# 19

# Symbol Combination in *Pan*: Language, Action, and Culture

## Patricia Greenfield and Heidi Lyn

When a 2-year-old child says something like "hit ball" (Brown, 1973), the child is implicitly parsing an action event into two constituents, action and object. When a 2-year-old child says something like "Kendall swim" (Bowerman, 1973), the child is implicitly parsing an action event into the constituents of agent and action. Implicit there is an understanding of an agency, Kendall, as instigator of the swimming action. Such linguistic representations are therefore a window into the child's cognitive representation of action. Two related questions that we try to answer in this chapter are the following: Will apes exposed to a humanly devised symbol system parse and represent action events in the way that children do? To what extent are they representing intentional action?

Human children not only use language to parse action events but often do so in an orderly fashion, constructing preferred orders for relating two constituents. An example would be a preference for placing action before object. These ordering tendencies often reflect the language of the child's social environment; for example, verbs before object is the canonical order of English. This raises two more questions: To what extent will apes exposed to a humanly devised symbol system construct preferred orders for expressing action relations? To what extent are these orders cultural (in this case, coconstructed across species lines) rather than species specific?

Yet another question arises in connection with apes whose communicative environment is dominated by their interactions with humans: To what extent

---

We dedicate this chapter to Duane Rumbaugh, who pioneered the exploration of primate cognition with an underlying comparative perspective relevant to evolutionary theory. We are grateful for the tremendous opportunity, support, and encouragement he provided for our analyses of communication and symbol use in *Pan*. Preparation of this chapter was supported by a program project grant to E. Sue Savage-Rumbaugh from the National Institute of Child Health and Human Development, with a subcontract titled "Development and Socialization of Discourse Skills in *Pan*" from Georgia State University, Atlanta, to the University of California, Los Angeles, Patricia Greenfield, principal investigator. We thank E. Sue Savage-Rumbaugh for her key role in planning and implementing the cross-species comparative design, communicating with the apes, and directing the data collection. We also appreciate input on this chapter from members of the UCLA Foundation for Psychocultural Research Center for Culture, Brain, and Development.

does the apes' symbolic encoding of action show an awareness of a human state of mind as a prerequisite to ape action?

Whiten et al. (1999) presented evidence for chimpanzee cultural traditions. Most of the evidence presented concerned tool use, but some concerned signaling behaviors. Evidence of this nature makes more likely the possibility that one could see distinct cultural traditions emerging in captivity with regard to communicative behavior. A related question follows: To what extent does any order preference reflect perceived order of a real-world action event as opposed to a cultural norm that is independent of action structure itself?

Communication can itself be looked on as a system of action. Human children (and adults) integrate across two separate modalities: speech and gesture (or, in the case of deaf children, sign and gesture; Goldin-Meadow & Morford, 1985; Goldin-Meadow & Mylander, 1984). From a species-comparative perspective, we may therefore ask to what extent and in what way will apes exposed to a humanly devised symbol system integrate different communicative modalities.

## Method

### Cladistic Analysis

Cross-species comparisons of bonobos, chimpanzees, and humans permit basic cladistic analysis and provide evidence concerning the evolution of physical or behavioral characteristics. *Cladistics* refers to a taxonomic analysis that emphasizes the evolutionary relationships between different species. A *clade*—the basic unit of cladistic analysis—is defined as the group of species that all descended from a common ancestor unique to that clade (Byrne, 1995). Cladistic analysis separates ancestral traits, which are inherited from the ancestors of the clade, from derived traits, which are possessed by only some of the clade members (Boyd & Silk, 2000). Derived traits arose through natural selection after the divergence of the clade from the common ancestor. Ancestral traits, in contrast, have a genetic (and hence neural) foundation in the common ancestor. Bonobos, chimpanzees, and humans are considered to be a clade that diverged from the common species ancestor 5 million years ago (Stauffer, Walter, Ryder, Lyons-Weiler, & Blai-Hedges, 2001).

By examining behavior in a whole clade, we can use similarities among all three sibling species as clues to what foundations of human language may have been present in our common ancestor 5 million years ago. Such foundations would then have served as the basis from which human language evolved in the following millions of years. Differences among members of the clade would serve as clues to what may have evolved in each species since their divergence 5 million years ago in the case of humans and 2 million years ago in the case of chimpanzees and bonobos (Zihlman, 1996).

In this chapter, we examine the symbolic representation and communication of action information in the clade made up of the bonobo, chimpanzee, and human species. We also examine the role of cultural coconstruction. Although our data came from only 2 bonobos and 1 chimpanzee, the analysis is at the very least an existence proof of cross-species similarities and differences. To complete the clade, we compare our ape participants with the extant literature on human child language.

## Participants and Their Communicative Background

Our data came from 2 bonobos, Kanzi and Panbanisha, and 1 chimpanzee, Panpanzee. We compared the symbolic combinations of these apes with the body of empirical research on child language years ago.

Data collection took place at the Georgia State University Language Research Center, under the direction of Duane Rumbaugh and E. Sue Savage-Rumbaugh (Rumbaugh, 1977; Rumbaugh, Savage-Rumbaugh, & Sevcik, 1994; Savage-Rumbaugh, 1986; Savage-Rumbaugh, Shanker, & Taylor, 1998). Ongoing research with bonobos (*Pan paniscus*) and chimpanzees (*Pan troglodytes*) have allowed for some major breakthroughs in our understanding of nonhuman symbol use and cognition.

These bonobo data came from Kanzi, a male born in 1980, and his younger sibling Panbanisha, a female bonobo born in 1985. Panbanisha was raised with her agemate chimpanzee, Panpanzee, who also contributed data to this study. These apes were reared in a communication-rich environment that included English speech, gestures, and lexigrams. *Lexigrams* are symbols on a visuographic keyboard. The symbols are noniconic; that is, they do not in any way resemble their referents—the entities or actions for which they stand. They are arbitrary symbols, their meanings established through the creation of social norms in the Language Research Center. Each lexigram is highly differentiated from all others as a visual pattern. During most of the rearing of these apes, any keyboard presses would result in a computerized voice speaking the English gloss of the lexigram.

Each of these apes used the lexigram keyboard in a symbolic fashion and comprehended their caregivers' use of English at the level of a 2.5-year-old child (Brakke & Savage-Rumbaugh, 1995, 1996; Savage-Rumbaugh et al., 1993; Williams, Brakke, & Savage-Rumbaugh, 1997). At the beginning of the period of the present analysis, May 15, 1989 through October 14, 1989, Panbanisha had a productive vocabulary of about 105 lexigrams. Panpanzee had a productive referential vocabulary of about 70 lexigrams (see Brakke & Savage-Rumbaugh, 1996, Figure 1). They did not use the keyboard randomly, as they had at an earlier "babbling" period. They also had about four spontaneous yet conventionalized gestures.

## Data Collection and Data Record

The data from which these findings have emerged came from 5 months of data collection beginning when Panpanzee, the chimpanzee, and Panbanisha, the younger bonobo, were 3.5 years old. Kanzi's data, collected several years earlier, were produced when he was 5.5 years old. All 3 apes were accompanied during the day by humans, who kept a record of every communicative utterance utilizing lexigrams as well as lexigram–gesture combinations.

We focus here on the use of lexigrams and gestures in the combinatorial productions of these 3 apes and in particular, on the two-element combinations of symbol–symbol and symbol–gesture, which were the most prevalent. Presentation of the quantitative data can be found in Lyn, Greenfield, and Savage-Rumbaugh (2006).

All communicative combinations occurred in naturally occurring, everyday situations. The example below shows the information about each utterance that was recorded by a researcher at its time of occurrence:

- Ape: Panpanzee Experimenter: Jeaninne
- Date: 5/15/1989 Combination: True
- Utterance: Apple lexigram + touch gesture
- Code: 714 (request by Panpanzee)
- Context: For more apple to eat, Panpanzee again uses the keyboard to properly ask for the food followed by touching a piece of apple in a bowl. (NOTE: several other foods are also present.)

Here Panpanzee indicated apple on the lexigram board and then touched a piece of apple in a bowl. The ape was noted, as was the experimenter (caregiver) and the utterance itself. The date of the utterance was given, as well as whether that utterance was a combination of two or more elements (the term "true" after "combination" in the entry). A code was assigned to the utterance according to its perceived pragmatic force (request, comment, structured response). Finally, a description was recorded of the context in which the utterance was made. This contextual description is what allowed us to classify these combinations according to their semantic meaning.

## Corpora: Nature, Size, and Evolutionary Significance

During this period, Panbanisha had a corpus of 1,088 combinations of two or more elements; Panpanzee had a corpus of 1,000 combinations. Most of the combinations were two elements long (two lexigrams or gesture plus lexigram). After eliminating imitations and other humanly structured combinations, Kanzi's corpus of two-element utterances numbered 723, Panbanisha's numbered 642, and Panpanzee's numbered 637.

These corpora are very large in comparison with most studies of child language and offer rich material for analysis. We note, however, that unlike for children at this age, combinations constituted the minority of the corpus for our animal subjects: 10.4% for Kanzi, 16.7% for Panbanisha, and 16.0% for Panpanzee. Single-lexigram utterances were in the majority; single gestures were not recorded. Our evolutionary perspective on these rates of combining symbols is that they indicate a possible target of natural selection. That is, through natural selection in the human line, symbolic combinations may have gradually become more frequent in the language of human children in the course of the evolutionary process that took place after the divergence of the hominid line from *Pan* 5 million years ago.

> Panpanzee: **dog** (lexigram) + ***play*** (lexigram)
>          agent                 action

> Panbanisha: ***open*** (lexigram) + ***peach*** (lexigram)
>           action               object

> Panpanzee: **points to tree** (gesture) + ***play*** (lexigram)
>             goal                    action

**Figure 19.1.** Examples of two-element combinations in the communication of a chimpanzee (Panpanzee) and a bonobo (Panbanisha).

## Research Questions and Results

*Question 1: Will apes exposed to a humanly devised symbol system parse and represent action events in the way that children do?*

Published data on semantic relations of the bonobo Kanzi and the chimpanzee Washoe already indicate that the parsing into two-element relationships is very similar between children and chimpanzees (B. T. Gardner & Gardner, 1994; R. A. Gardner, Gardner, & Van Cantfort, 1989; Greenfield & Savage-Rumbaugh, 1990, 1991). Data from the bonobo Panbanisha and the chimpanzee Panpanzee confirm that the parsing of action sequences into two components is remarkably similar in all three species (Lyn et al., 2006). Examples from Panbanisha and Panpanzee's combinations are found in Figure 19.1.

In the first example (expressing a relationship between an agent and an action), Panpanzee touched the "dog" lexigram (*agent*) and then touched the "play" lexigram (*action*). She then led the caregiver over to the doghouse where Panpanzee and the dogs played together. In the second example (expressing an action on an object), Panbanisha first touched the "open" lexigram (*action*) and then touched the "peach" lexigram (*object*). Her caregiver had just broken a peach. In the third example (representing a goal–action relationship), Panpanzee pointed to a tree (*goal*) and then touched the "play" lexigram (*action*). Her caregiver said "yes" and Panpanzee climbed the tree to play. Here Panpanzee made a request, the predominant pragmatic force for all the apes; implicitly she recognized that her caregiver was an agent who must give her permission to act. Later, we show how both Panpanzee and Panbanisha could make the permission concept explicit through symbolic representation.

All three of the action relationships shown in Figure 19.1 (agent–action, action–object, and goal–action) are among the eight universal semantic relations identified by Brown (1973) in his cross-linguistic analysis of early child language. Indeed, the parsing of action events into representational categories such as action, agent, object, and goal is universal in child language. All of Brown's eight universal semantic relations are represented in our corpus and, in fact, make up the majority of relations in our corpus. These relations are (our terminology in parentheses when it differs from Brown's) agent–action; action–object; agent–object; modifier–head (attribute–X); negation–X; X–dative (X–recipient); introducer–X (demonstrative–X); and X–locative (X–location). (For attribute and negation, X can be either an action or an entity; for recipient, demonstrative, and location, X is always an entity.) All of these categories are also expressed at the one-word stage of child language, in which children relate their single words to themselves, to gestures, to other people, to objects, and to ongoing action in the present situation (Greenfield & Smith, 1976).

In conclusion, the commonality across the three species in the categories used to parse action and construct action relations is striking indeed. Although we must acknowledge that one animal does not equal a whole species, such structuring of action is highly suggestive. In addition, we are not the first researchers to identify these action structures in the symbolic productions of apes (Brown, 1970). Given that these three species are sibling species, this finding suggests the possibility that the cognitive structuring of action into categories such as agent, action, object, and goal was present before the divergence of *Pan* and *Homo* 5 million years ago. In other words, assuming that these three species are a clade, the findings suggest the possibility that the cognitive structuring of action is an ancestral trait. If this way of parsing action is indeed the foundation of human culture (Bruner, 1990) as well as a building block of human language, then this cognitive structuring of action, as manifest in our common ancestor, would have provided a foundation for the subsequent evolution of both human culture and human language. Consequently, this cognitive structuring of action may have also provided a foundation for chimpanzee culture (Whiten et al., 1999) and even chimpanzee communication in the wild, of which science is relatively ignorant.

## *Question 2: To what extent are the apes representing intentional action?*

In Bruner's (1990) cultural psychology, the categories described above are the foundation of culture, a reflection of our understanding of self and others as intentional agents with goals. Indeed, it is clear that these symbolic combinations are being used to communicate intentional action, both of self and of others. This is particularly evident in examples such as "dog play," above, where the ape expressed what the philosopher of language, John Searle (1979; see also Greenfield, 1980), called *prior intent*, a symbolic expression of a goal-state before the action takes place. In other words, behavior subsequent to the "utterance" made it clear that "dog play" announced a goal-state that Panpanzee intended to achieve.

Such expression of prior intent is present even in the single-word utterances of children (Greenfield, 1980). For example, at 22 months 21 days, a child Nicky (Greenfield & Smith, 1976) said "jump" just as he was about to jump. Here, his word expressed an action goal-state, jumping. From a linguistic perspective, agency was implicit, but from the point of view of the action, agency resided explicitly in the child who was about to jump. Similarly, the apes used their single-symbol productions to signal prior intent; for example, Panbanisha stated "open" at the keyboard and then proceeded to open a backpack.

In the last example in Figure 19.1, Panpanzee made a request, the predominant pragmatic force for all the apes (Lyn et al., 2006); implicitly she recognized that her caretaker was an agent who must give her permission to act. Later, we show how both Panpanzee and Panbanisha could make the permission concept explicit through symbolic representation. At the same time, requests intrinsically involve prior intent. That is, a request constitutes an announcement of an intended goal-state; in this case, the intended goal-state was to play in the tree.

The "apple touch" example with which we started presents another aspect of intentionality. The fact that Panpanzee subsequently ate the apple indicates that the communication was intentional rather than random. At the same time, the presence of alternative foods makes the utterance informative. Indeed, in prior work, Greenfield and Savage-Rumbaugh (1984) established that chimpanzees use their symbols informatively in the information-theory sense of *informative*. That is, they used their lexigrams to select from among available alternatives or to signal change; in contrast, they rarely used them to signal the presence of only one stimulus or to indicate a constant state of affairs (Greenfield & Savage-Rumbaugh, 1984).

Informativeness can include a kind of implicit goal-directedness or intentionality. This type of intentionality is what Searle (1979) called *intention-in-action*. In the "apple touch" example, a linguistic action contained a goal within it: to select from alternatives. This intention-in-action became a means to the partially expressed prior intent: to eat the apple. Human children, like adults, are also informative from their earliest linguistic productions (Greenfield & Smith, 1976). In conclusion, the representation of action relations by apes, like that of human children, is clearly expressing intentional action on the part of an agent. This intentionality incorporates both intention-in-action, as in the expression of informativeness, and the representation of prior intent, as in the expression of a future goal.

## Question 3: To what extent will apes exposed to a humanly devised symbol system construct preferred orders for expressing action relations?

All 3 apes constructed preferred orders for expressing action relations (Lyn et al., 2006). One instance is the affirmative–goal relationship (generally a request for an action). In one example, Panbanisha touched the "yes" lexigram (*affirmative*) and then touched the "outdoors" lexigram (*goal*). Her caretaker had just asked her what she wanted to do; she replied with this emphatic request to go outdoors.

This example illustrates the dominant order for affirmative–goal lexigram combinations. Both Panpanzee and Panbanisha showed a statistically significant preference for placing the affirmative lexigram before the goal lexigram rather than vice versa, and Panpanzee showed a stronger preference than Panbanisha (Lyn et al., 2006). This preference, although statistically significant, was not absolute. However, this is true of human children as well (e.g., Bowerman 1973; Goldin-Meadow & Mylander, 1984).

*Question 4: To what extent are these relations or orders cultural (coconstructed across species) rather than species-specific behavior?*

In this instance, by *cultural*, we mean a kind of microculture coconstructed by a group as small as a dyad; the paradigmatic case and ontogenetic origin is mother–child intersubjectivity (Trevarthen, 1980). We see this kind of coconstruction as a social byproduct of a shared environment. Panpanzee, a chimpanzee, and Panbanisha, a bonobo, had the affirmative-goal pattern in common with each other. However, the affirmative-goal relation (let alone the order) was not utilized by either Kanzi or by human caregivers; this is a cross-species (micro)cultural norm that was shared by Panpanzee and Panbanisha only. This finding seems to show the cultural capability for the construction of semantic relations within the genus *Pan*. That is to say, a common environment shared by Panpanzee and Panbanisha was more important than their species difference. In no case did the bonobo Kanzi share a semantic relation or an order preference with Panbanisha, another bonobo, but not Panpanzee, a chimpanzee. Within the limits of the *Pan* genus and symbolic combination, shared cultural environment seems more important than the innate factors associated with species. This particular shared convention, the affirmative-goal pattern, between Panpanzee and Panbanisha helps answer our next question.

*Question 5: To what extent does the apes' symbolic encoding of action show an awareness of a human state of mind (intended action) as a prerequisite to ape action?*

The apes often used an affirmative to request permission or to emphasize a request. For example: Panpanzee constructed "yes Maryanne" wanting to see Maryanne, or Panbanisha was told that there were surprises at hilltop and she replied "yes hilltop" (asking to go). This usage is not one seen in humans and reflects the social condition under which the apes were raised. A comprehension of their need to request items and to get an affirmative answer from their caregiver allowed for the creation of this type of symbolic combination. This usage also reflects an implicit belief the apes had about the state of mind of their caregivers. They were using language to request a state of permission in their caregivers as well as a lexigram response of "yes" (e.g., yes we can go see Maryanne). In cases of this type, the apes' comprehension of "permission" was reflected in the pause in action following a request until a caregiver responded.

Because theory of mind in chimpanzees is controversial, we might ask what the minimal knowledge of the caregiver and the culture would be for these examples to occur. For example, is it enough for the ape to anticipate the caregiver's behavior? To what extent is it necessary for the ape to understand the human–ape relationship? To what extent does the ape have to anticipate a changed intention? Does the ape have to know something about state of mind? At the very least, it seems that the ape must understand something about human intention-in-action. That is, the ape must know something about the human's goal-directed action. Indeed, we now know from a whole series of experiments that apes do understand intended action (Tomasello, Carpenter, Call, Behne, & Moll, 2005). The ape must also need to know something about the effect of the human's goal-directed action on his or her own action possibilities. It may not be a theory of another mind, but it is knowledge of another's intentional action and of the intersection of another creature's intentionality with one's own.

## *Question 6: To what extent will symbol order preference reflect perceived order of a real-world action event as opposed to a cultural norm that is independent of action structure itself?*

First, note that each relation can be expressed in one of two modalities: by combining two lexigrams or by combining one lexigram and one gesture. Most interesting, the ordering convention for combining two lexigrams to parse and represent an action event is not always the same as the ordering convention for combining lexigram and gesture. Specifically, in the action–goal relation for Panpanzee and Panbanisha, action is usually expressed before goal when the relationship is constructed with two lexigrams. However, goal is generally expressed before action when the relationship is constructed with lexigram and gesture (Lyn et al., 2006).

An example of the lexigram–lexigram order is as follows: Panbanisha touched the "open" lexigram (*action*) and then touched the "dog" lexigram (*goal*). She was asking her caretaker to open the door so they could visit the dogs.

An example of a lexigram–gesture combination is the following: Panpanzee touched the "string" lexigram (*goal*) and then gestured to "go" (flicking her index finger in the direction she wants to go; *action*). Her caretaker said, "Yes, you can go get a string," unsure of what string she was referring to. Panpanzee moved in the direction of the tool room, and a few feet before she got there, bent down to pick up a string from the floor

These two different constructions are analogous in human language to two different surface forms of a similar underlying semantic structure, such as active (agent first) and passive (agent last) forms of the same sentence. These orders do not uniformly conform to the real-world event order: that is, action before goal. Nor do they uniformly conform to the mental order: That is, one must have a goal in mind before undertaking the action that achieves it. Instead, the ordering preferences seem to reflect arbitrary but shared modes of expression. Social sharing is the definition of a cultural norm. The combination of social sharing with arbitrariness is the definition of linguistic convention.

Arbitrary but shared ordering preferences for combining symbols may simultaneously constitute protogrammatical conventions and important aspects of chimpanzee culture.

*Question 7: To what extent and how can apes exposed to a humanly devised symbol system integrate different communicative modalities?*

All 3 apes combined gesture and lexigram, integrating two communicative modalities. It is notable that they used this particular cross-modal strategy more than their human caregivers did; the latter tended to combine lexigrams with other lexigrams (and with speech) rather than with gestures. The frequency of the apes' cross-modal combinations was in line with a recent consensus at the 2004 Workshop on Gestural Communication in Nonhuman and Human Primates (see Liebal, 2004) concerning the importance of cross-modal integration in the evolution of human language.

Another important point is that all 3 apes adopted the "gesture-after-lexigram" preference as a strategy for accomplishing this cross-modal integration (Lyn et al., 2006). The gesture-after-lexigram strategy was first reported in a study on Kanzi in the absence of any model for this ordering pattern in his human environment (Greenfield & Savage-Rumbaugh, 1990, 1991). It may be a "natural" rule for apes, because both Panbanisha and Panpanzee used it as well; it may reflect Kanzi's enculturating influence on the younger apes; or it may be a response to the importance placed on lexigrams by the humans in this environment. Whatever the reason, this ordering convention is a way of integrating the two modalities in a systematic fashion for communicative purposes. Here are some examples of the use of the lexigram–gesture integration and use of this ordering strategy from Panbanisha's and Panpanzee's corpus. In these examples, they express two different semantic relations, entity-demonstrative and location-entity:

- *Entity-demonstrative*: Panpanzee touched the "monster" lexigram (*entity*) and then the monster mask (*demonstrative gesture*).
- *Entity-location*: Panbanisha touched the "orangutan" lexigram (*entity*) and then pointed to the colony room (*location*), where the orangutans usually were.

Thus, communicative action can in itself be complex because of the cross-modal integration that takes place. Ordering strategies are an effective way to effect such integration.

## Conclusions

We have demonstrated that apes exposed to a humanly devised symbol system parse and represent action events in the way that children do, in terms of implicit relational categories such as agent, action, and object. Like children, the apes also use their action representations to express their prior intention to carry out an action (*representation of prior intent*; Searle, 1979) as well as their intention-in-action to select from alternative possibilities. Hence, when socialized to use a symbolic tool, they spontaneously use it to parse action into familiar categories for the pragmatic purpose of expressing both prior intent and intention-in-action. In sum, a crucial underlying component of the action representations of chimpanzee, bonobo, and human child is their intentional or goal-directed nature. The neural foundation for the expression of intentionality may be in the mirror neuron system (Greenfield, in press).

Going one step further in the analysis of intentionality, we note that the use of affirmatives by Panbanisha and Panpanzee to gain permission for a desired action at minimum suggests an awareness of intended human action—the reading of intended action in another creature from another species (albeit a closely related one). It also suggests an understanding of the differential power in captivity of the ape and human—that is, the ape's dependence on the human for resources and stimulation as well as his or her understanding of ape–human power relationship as a prerequisite for ape action. Our data indicate that this explicit awareness increased over time in the community; these conventions (utilizing affirmatives) were shared by the younger apes Panbanisha and Panpanzee but were not used by Kanzi.

Like children, the apes of both species constructed preferred orders for representing particular types of action event. These order preferences reflected both cultural norms (Panpanzee and Panbanisha with affirmative constructions) and genus norms (placing gesture after lexigram). However, the preferred orders do not necessarily reflect the structure of the action itself. For example, an action-goal structure will tend to be expressed with action first when it is a two-lexigram construction but with goal first when it is a lexigram–gesture construction.

With a caveat concerning having only one or two representatives from each species, we note the presence of order preferences in each member of this evolutionary clade. Our caveat is not too strong, because a phenomenon in even one member of a species functions as an existence proof. We also note that all ordering preferences discussed in this chapter are shared across more than one ape and are therefore normative on the microlevel of a dyad or triad. The capacity for arbitrary ordering preferences for combining symbols may therefore be an ancestral trait and so may have existed in the common ancestor of *Pan* and *Homo*. Because we find it to exist in each of the three species, this capacity for arbitrary ordering may underlie the ancestral origins of the autonomous structuring of the representational system, leading ultimately to the arbitrary, yet distinct, linguistic conventions.

Apes exposed to a humanly devised symbol system integrated different communicative modalities into their communicative actions; specifically, they spontaneously integrated gesture and lexigram to express their action representations. This constitutes further evidence for the multimodal nature of human language evolution.

The apes' spontaneous integration of gesture brings up a related matter: Apes utilize gestures much more often than their human caregivers and (probably) more and later than hearing human children. The creativity of the gesture-after-lexigram organization to construct new action relations must not be underestimated; given its prevalence in all three apes, it could be an evolutionary precursor to the combinatorial creativity that is the hallmark of human language

Given that the communication of action information, both in single elements and in combinations, has been observed in the wild in gestures, vocalization, and external visual symbols used by apes (Plooij, 1978; Savage-Rumbaugh, Williams, Furuichi, & Kano, 1996; Tomasello, 1994; Whiten et al., 1999), it is possible that the cross-species commonalities we have identified in combining symbols to express intentional action have their behavioral as well as their neural roots 5 million years ago, before the divergence of *Homo* and *Pan*.

## References

Bowerman, M. (1973). *Early syntactic development: A cross-linguistic study with special reference to Finnish*. Cambridge, England: Cambridge University Press.

Boyd, R., & Silk, J. B. (2000). *How humans evolved*. New York: Norton.

Brakke, K. E., & Savage-Rumbaugh, E. S. (1995). The development of language skills in bonobo and chimpanzee: 1. Comprehension. *Language and Communication, 15*, 121–148.

Brakke, K. E., & Savage-Rumbaugh, E. S. (1996). The development of language skills in *Pan*: II. Production. *Language and Communication, 16*, 361–380.

Brown, R. (1970). The first sentences of child and chimpanzee. In R. Brown (Ed.), *Psycholinguistics* (pp. 208–231). New York: Free Press.

Brown, R. (1973). *A first language*. Cambridge, MA: Harvard University Press.

Bruner, J. (1990). *Acts of meaning: The Jerusalem–Harvard lectures*. Cambridge, MA: Harvard University Press.

Byrne, R. W. (1995). *The thinking ape: Evolutionary origins of intelligence*. New York: Oxford University Press.

Gardner, B. T., & Gardner, R. A. (1994). Development of phrases in the utterances of children and cross-fostered chimpanzees. In R. A. Gardner (Ed.), *The ethological roots of culture* (pp. 223–255). Dordrecht, the Netherlands: Kluwer Academic.

Gardner, R. A., Gardner, B. T., & Van Cantfort, T. E. (1989). *Teaching sign language to chimpanzees*. Albany: State University of New York Press.

Goldin-Meadow, S., & Morford, M. (1985). Gesture in early child language: Studies of deaf and hearing children. *Merrill-Palmer Quarterly, 31*, 145–176.

Goldin-Meadow, S., & Mylander, C. (1984). Gestural communication in deaf children: The effects and noneffects of parental input on early language development. *Monographs of the Society for Research in Child Development, 49*(3–4, Serial No. 207), 1–21.

Greenfield, P. M. (1980). Toward an operational and logical analysis of intentionality: The use of discourse in early child language. In D. Olson (Ed.), *The social foundations of language and thought: Essays in honor of J. S. Bruner* (pp. 254–279). New York: Norton.

Greenfield, P. M. (in press). Implications of mirror neurons for the ontogeny and phylogeny of cultural processes: The examples of language and tools. In M. A. Arbib (Ed.), *Action to language via the mirror neuron system*. Cambridge, England: Cambridge University Press.

Greenfield, P. M., & Savage-Rumbaugh, E. S. (1984). Perceived variability and symbol use: A common language–cognition interface in children and chimpanzees (*Pan troglodytes*). *Journal of Comparative Psychology, 98*, 201–218.

Greenfield, P. M., & Savage-Rumbaugh, E. S. (1990). Grammatical combination in *Pan paniscus*: Processes of learning and invention in the evolution and development of language. In S. T. Parker & K. R. Gibson (Eds.), *"Language" and intelligence in monkeys and apes: Comparative developmental perspectives* (pp. 540–578). New York: Cambridge University Press.

Greenfield, P. M., & Savage-Rumbaugh, E. S. (1991). Imitation, grammatical development, and the invention of protogrammar by an ape. In N. A. Krasnegor, D. M. Rumbaugh, R. L.

Schiefelbusch, & M. Studdert-Kennedy (Eds.), *Biological and behavioral determinants of language development* (pp. 235–258). Hillsdale, NJ: Erlbaum.

Greenfield, P. M., & Smith, J. H. (1976). *The structure of communication in early language development.* New York: Academic Press.

Liebal, K. (2004). Gestural Communication in Nonhuman and Human Primates. Workshop held at the Max Planck Institute for Evolutionary Anthropology, Leipzig, March 28th–30th, 2004. *Gesture, 4,* 255–263.

Lyn, H., Greenfield, P., & Savage-Rumbaugh, E. S. (2006). *Semiotic combinations in Pan: A cross-species comparison of communication in a chimpanzee and a bonobo.* Manuscript submitted for publication.

Plooij, F. X. (1978). Some basic traits of language in wild chimpanzees? In A. Lock (Ed.), *Action, gesture and symbol: The emergence of language* (pp. 111–131). London: Academic Press.

Rumbaugh, D. M. (1977). *Language learning by a chimpanzee: The Lana project—Communication and behavior.* New York: Academic Press.

Rumbaugh, D. M., Savage-Rumbaugh, E. S., & Sevcik, R. A. (1994). Biobehavioral roots of language: A comparative perspective of chimpanzee, child, and culture. In R. W. Wrangham, W. C. McGrew, F. DeWaal, & P. G. Heltne (Eds.), *Chimpanzee cultures* (pp. 319–334). Cambridge, MA: Harvard University Press.

Savage-Rumbaugh, E. S. (1986). *Ape language: From conditioned response to symbol—Animal intelligence.* New York: Columbia University Press.

Savage-Rumbaugh, E. S., Murphy, J., Sevcik, R. A., Brakke, K. E., Williams, S. L., & Rumbaugh, D. M. (1993). Language comprehension in ape and child. *Monographs of the Society for Research in Child Development, 58*(3–4, Serial No. 223).

Savage-Rumbaugh, E. S., Shanker, S. G., & Taylor, T. J. (1998). *Apes, language, and the human mind.* New York: Oxford University Press.

Savage-Rumbaugh, E. S., Williams, S. L., Furuichi, T., & Kano, T. (1996). Language perceived: *Paniscus* branches out. In W. C. McGrew, L. F. Marchant, & T. Nishida (Eds.), *Great ape societies* (pp. 173–184). New York: Cambridge University Press.

Searle, J. R. (1979, August). *Intention and action.* Paper presented at the LaJolla Conference on Cognitive Psychology, LaJolla, CA.

Stauffer, R. L., Walter, A., Ryder, O. A., Lyons-Weiler, M., & Blai-Hedges, S. (2001). Human and ape molecular clocks and constraints on paleontological hypotheses. *Journal of Heredity, 92,* 469–474.

Tomasello, M. (1994). The question of chimpanzee culture. In R. W. Wrangham, W. C. McGrew, F. B. M. deWaal, & P. G. Heltne (Eds.), *Chimpanzee cultures* (pp. 301–317). Cambridge, MA: Harvard University Press.

Tomasello, M., Carpenter, M., Call, J., Behne, T., & Moll, H. (2005). Understanding and sharing intentions: The origins of cultural cognition. *Behavioral and Brain Sciences, 28,* 675–735.

Trevarthen, C. (1980). The foundations of intersubjectivity: Development of interpersonal and cooperative understanding of infants. In D. Olson (Ed.), *The social foundations of language and thought: Essays in honor of J. S. Bruner* (pp. 316–342). New York: Norton.

Whiten, A., Goodall, J., McGrew, W. C., Nishida, T., Reynolds, V., Sugiyama, Y., et al. (1999, June 17). Cultures in chimpanzees. *Nature, 399,* 682–685.

Williams, S. L., Brakke, K. E., & Savage-Rumbaugh, E. S. (1997). Comprehension skills of language-competent and non-language-competent apes. *Language and Communication 17,* 301–317.

Zihlman, A. (1996). Reconstructions reconsidered: Chimpanzee models and human evolution. In W. C. McGrew & L. F. Marchant (Eds.), *Great ape societies* (pp. 293–304). New York: Cambridge University Press.

# 20

# Epigenesis, Mental Construction, and the Emergence of Language and Toolmaking

*Kathleen R. Gibson*

> Psychology will be based on a new foundation, that of the necessary acquirement of each mental power and capacity by gradation. Light will be thrown on the origin of man and his history.
> —Darwin (1859, p. 488)

These tantalizing sentences conclude the book that rocked the Western world. They were eventually elaborated into three strong and cohesive theses (Darwin, 1871, 1872): (a) Animal and human minds differ in degree but not in kind; (b) human mental capacities evolved in a gradual stepwise fashion from those of other animals; and (c) the differences in mental faculties between "lower" animals such as fish and "higher" animals such as apes are greater than those between apes and humans.

By the middle of the 20th century, most serious students of human origins accepted Darwin's view that our species evolved from precursor ape- or monkey-like forms. Only a few, however, embraced his concepts of the gradual evolution of human mental capacities. Instead, most scholars continued to espouse Cartesian-inspired views of qualitative distinctions between animal and human minds and hypotheses that new human capacities, such as language, toolmaking, and culture, arose suddenly and fully formed at some point in human evolution. It is not that mid-20th century scholars had disproved Darwin's behavioral theses or even argued against them. Rather, in the absence of knowledge of the behavior of wild animals, they had simply ignored them.

The onset of primate field studies in the 1950s, however, initiated a revolution in understanding of the behavioral capacities of nonhuman primates. Early in the history of modern primatology, Japanese scientists reported the invention of sweet-potato and rice-washing techniques by a juvenile female macaque and the spread, over a period of years, of these food-processing habits to most of her troop—habits that continue in these troops today (Kawai, 1965). The emergence of these behavioral traditions indicate that social learning is not, as previously supposed, a uniquely human phenomenon, and a new term coined to define socially transmitted traditions, *protoculture*, clearly implied gradations between animal and human cultural capacities. Somewhat later, wild chimpanzees at the Gombe Stream Reserve astonished the world by demonstrating that they could make simple probing tools to extract termites and ants from mounds (Goodall, 1964). This necessitated changing a long-held definition of humanity, "man the toolmaker" to "humans are the only animals who can use a tool to make a tool." Some apes, trained at Rumbaugh's Language Research Center facility at Georgia State University, however, have now met this new challenge by using stone hammers to produce sharp-edged, knifelike stone flakes, thus again lending support to concepts of gradations rather than sharp dichotomies between ape and human behaviors (Schick et al., 1999).

Washoe, a captive chimpanzee, was the first ape to challenge human linguistic uniqueness by learning to use gestural symbols of the American Sign Language of the Deaf to communicate basic wants and desires (Gardner & Gardner, 1969). Washoe, however, did not appear to combine these symbols according to grammatical rules. Human uniqueness, thus, was preserved temporarily by redefining language as syntax rather than symbolism. Again, however, some apes, including several trained by Duane Rumbaugh and E. Sue Savage-Rumbaugh, have met this challenge and demonstrated the ability to understand aspects of English grammar as well as to combine gestural and visual symbols according to systems of rules (Greenfield & Savage-Rumbaugh, 1990; Rumbaugh, 1977; Savage-Rumbaugh et al., 1993).

Meanwhile, a half century of field and laboratory research has converged to reveal far greater humanlike intelligence in great apes than had previously been supposed in additional behavioral domains, including mirror self-recognition (Parker, Mitchell, & Boccia, 1994), deception (Whiten & Byrne, 1988), imitation (Byrne & Russon, 1998), and teaching (Boesch, 1993). The findings of modern primatology thus lend credibility to Darwin's view that the differences between animal and human mentalities are primarily matters of degree rather than of kind. This is particularly so for great apes (bonobos, chimpanzees, orangutans, and gorillas), whose capacities for toolmaking, language, imitation, deception, and mirror self-recognition thus far appear to exceed those of monkeys.

The knowledge that great apes possess at least minimal capacities in varied cognitive domains previously thought to be uniquely human poses new scientific challenges. Those great apes who have exhibited the ability to acquire the rudiments of language and to use a tool to make a tool have all been captive apes, reared, all or in part, by human caretakers. Great apes reared in close proximity to humans also seem to exceed their counterparts in the wild in their abilities to imitate human tool-using behaviors and to recognize themselves in mirrors. According to the Darwinian perspective, behavioral and other traits evolved in accordance with natural selection. Yet, if wild apes do not use language, tools to make tools, or mirrors, how could such abilities have ever been selected for? The obvious answer would appear to be that these humanlike capacities reflect cognitive skills that great apes use for other purposes in their natural habitats but that are developmentally channeled in humanlike directions in apes reared by humans. How do we explain this?

Rumbaugh's (2002) response to this new scientific challenge has been to define a class of behaviors that he termed *emergents* and a new field, *rational behaviorism*, to refer to the study of emergents. Among the behaviors that qualify as emergents in Rumbaugh's scheme are unexpected skills, such as ape language and ape mirror self-recognition, and creativity, including the ability to devise novel solutions to problems. According to Rumbaugh, emergents depend on relational thinking and complex brains. This chapter suggests that modern understandings of the nature of vertebrate brain development, neural networks, and comparative great ape and human behaviors converge with and support Rumbaugh's views of emergence and relational thinking as well as Darwin's concepts of the gradual emergence of human mental capacities. Together these converging views have the potential to lead comparative and evolutionary psychology into fruitful new research directions.

## Epigenesis

Complex phenotypes develop via a dialogue between genome and environment, a phenomenon termed *epigenesis* (Waddington, 1957). The genome maps the possibilities—provides the blueprint so to speak—whereas environmental factors help determine how genes are expressed and when. For example, environmental and genetic interactions rather than genes alone determine which ant becomes a queen and which fish becomes a male (Duscheck, 2002). Similarly, in some mammalian species, such as male orangutans (Galdikas, 2003) and female naked mole rats (Faulkes, 1999), animals reach full physical or sexual maturity only in the absence of fully mature adults of the same sex.

Diverse findings verify that vertebrate brains, like other complex phenotypes, mature via epigenetic processes. For example, lateral neuroanatomical asymmetries and functional lateralization, which appear to be ubiquitous across chordates and vertebrates (Rogers & Andrew, 2002), mature through a developmental cascade of interacting genetic and environmental inputs. Early in development in chickens, for example, two genes, Shh and FGF8, are differentially expressed on the left and right sides of the embryo but not the right embryonic side (Deng & Rogers, 2002). Products of these genes stimulate expression of other genes, which initiate embryonic folding patterns that result in differential sensory input to the right and left halves of developing embryos. Chicken embryonic posture, for example, assures that the left eye rests on the body while the right eye faces the egg shell. Thus, when the mother hen moves temporarily from her resting position, the right eye, but not the left, receives a light stimulus that initiates the development of excess numbers of visual projections to the right as opposed to left brain, and ultimately leads to other hemispheric and behavioral specializations. Experimental manipulations can reverse this anatomical asymmetry and its behavioral correlates by assuring that the left rather than the right eye receives the light stimulus (Deng & Rogers, 2002). Embryonic, fetal, and neonatal exposure to other sensory stimuli have been demonstrated to similarly affect brain and behavioral lateralization in many vertebrates other than chickens, including primates (Cowell & Denenberg, 2002; Damersose & Vauclair, 2002; Güntürkün, 2002). That environmental inputs affect brain lateralization in humans is also evident from clinical situations in which damage to the left hemisphere may result in the right brain assuming language functions that are usually handled by the left brain.

Environmental stimuli also affect development of other neural systems. The receipt of species-typical sensory stimuli from tactile vibrissae, for example, plays a critical role in the development of higher brain areas in rats and mice (Simons & Land, 1987), and visual experience is essential for normal development of visual areas of the thalamus and neocortex of cats, monkeys, and humans (Hubel, Wiesel, & LeVay, 1977; von Noorden & Crawford, 1992). In humans and other primates, species-typical patterns of neural functional localization can actually be altered by unusual environmental inputs. Areas of the left brain that usually respond to auditory stimuli, thus, may respond instead to visual stimuli in congenitally deaf individuals trained in American Sign Language (Neville, 1991). Similarly, in congenitally blind humans who are trained in Braille from a young age, the primary visual cortex assumes tactile rather than visual functions (Sadato et al., 1996). Although the greatest amount of functional plasticity is found in immature brains, some plasticity continues even into adulthood. Surgical ablation of fingers in adult owl monkeys thus results in decreased neocortical representation of the ablated fingers and increased neocortical representation of adjacent fingers (Merzenich & Kaas, 1982). Similar reorganizations follow immobilization of the arms of squirrel monkeys and amputations or long-term limb immobilizations in humans (Brauth, Hall, & Dooling, 1991; Hallett, 2000; Pons, Garraghty, & Mishkin, 1988; Pons et al., 1991).

Evidence that the brain develops via epigenetic processes is thus abundant. The clear implication is that substantially different rearing conditions, such as those experienced by great apes reared in the wild versus those reared in captivity, could profoundly affect their cognitive and sensorimotor development. To the extent that caged or otherwise restrained animals experience environments that are in many respects impoverished in comparison with conditions encountered in the wild, one might expect many captive-reared apes to lag behind their species norms in certain cognitive skills, especially those related to survival in the wild and to appropriate social interactions with conspecifics. On the other hand, apes raised in captive environments that provide for rich human–ape interactions might well exceed their wild counterparts in humanlike behavioral domains, such as language. These expectations conform with findings that apes reared in human contact from an early age outperform others in ape-language experiments (Savage-Rumbaugh et al., 1993) and possibly also in mirror self-recognition abilities (Parker et al., 1994) and tendencies to imitate human acts, including human tool-making and tool use (Byrne & Russon, 1998). Epigenetic processes may, thus, account for one form of behavioral emergence as defined by Rumbaugh: the acquisition of unusual behavioral capacities under unusual environmental conditions.

## Mental Construction and the Evolution of Complex Cognitive Capacities

Although captive-reared apes often exceed wild apes in their performance in humanlike behavioral domains, no ape has yet actually matched adult human linguistic, toolmaking, mathematical, artistic, or musical capacities, and no human-reared primates other than apes have come close to the ape performance in most of these domains. This is not surprising. Species differ in brain size and in brain structural complexity and thus in the overall attainable cognitive capacity.

Rumbaugh's (2002) theory of rational behaviorism proposes that advanced cognitive skills and creative thinking depend on relational thinking, that is, on the ability to note or create relationships among stimuli or responses. Rumbaugh's emphasis on relational thinking accords with modern cognitive science theories that all learning involves strengthening relationships within and between neural networks (Fuster, 2002). Relational thinking is also implicit in the Piagetian framework with its emphasis on the construction of reality from component parts. It is explicit in Case's (1985) reanalysis of the Piagetian stages that postulates that human cognitive maturation reflects the progressive incorporation of increasing amounts of information into children's thought processes and the construction of relationships among diverse items of information to yield more complex thoughts and actions. Analyses of great ape versus human toolmaking and language by my colleagues and me (Gibson, 1983, 1990, 1996, 2002; Gibson & Jessee, 1999; Greenfield, 1991) also explicitly emphasized relational thinking in that both suggest that mental constructional capacity—that is, the ability to construct relationships among diverse stimuli, responses, or concepts to form higher order constructs—accounts for the differences in performances between humans and great apes. I further extended the concept of mental construction to suggest that mental constructional skills may also form the foundations for social intelligence, art, as well as dance, gymnastic, and other complex motor skills and that differences in mental construction may account for differences in intelligence among other vertebrate species (Gibson, 1990). That scholars from the diverse disciplines of neuroscience, anthropology, developmental psychology, and behaviorism have all converged on concepts of relational thinking as central to intelligence and creativity represents a rather remarkable bridging of scholarly traditions, some of which, such as Piagetian versus behaviorist psychology, have often been considered unbridgeable.

Mental constructional capacity as Case (1985) and I (Gibson, 1990) have envisioned it is a quantitative phenomenon related to the extent to which individual perceptions, actions, or concepts can be differentiated into component units, and to the numbers of component units that can be incorporated into any given construction. It is also hierarchical in that newly constructed units can be embedded into still higher order constructs. In the motor domain, for example, human infants younger than 6 months of age manifest "whole hand and arm control"; that is, they open and close their hands moving all fingers in unison and, when reaching, the arm and hand move as a single unit that cannot be broken into component parts (Bower, 1974). In contrast, older human children and adults can differentiate arm, wrist, and shoulder movements into component elements, and they can make precise movements of individual fingers. Differentiated hand, finger, arm, and other body postures provide for much greater motor and behavioral flexibility than is possible for infants, because these movements can then be combined and recombined simultaneously or sequentially to create varied communicative signals, tool-using behaviors, and athletic or dance routines. Some human motor constructions are hierarchical in that individual motor constructs can be subordinated into larger sequences as, for example, in the incorporation of individual dance postures or small dance sequences into larger choreographed performances. Similarly, human children and adults make very precise, discrete movements of their lips, tongue, larynx, and uvula, especially during speaking and singing endeavors. These can be combined into new simultaneous anatomical positions to create phonemes, and individual phonemic positions can then be combined and recombined into varied movement sequences to create fluid speech or song sequences.

The ability to create novel motor constructs varies greatly among animals. Humans, in particular, appear to exceed great apes and other primates in their ability to create novel vocal constructs. Hence, humans, but not other primates, can speak. Whether or not humans and apes differ in other aspects of motor constructional ability such as in the ability to construct novel dance, athletic, and tool-using postures is a topic in need of further research. It is of interest, however, that those ape-language projects that achieved the greatest success in demonstrating ape grammatical abilities required no motor skills other than the ability to point to visual icons.

At the conceptual level, language involves the construction of syllables, words, phrases, and sentences from the creation of relationships among component parts. Language is hierarchical in that individual units serve as subcomponents of higher order constructions as, for example, when words are incorporated into phrases that are then incorporated into sentences. Similarly, human toolmaking is constructional and hierarchical in nature in that separately manufactured objects such as wooden shafts, stone points, rope, and hides may be joined together to make more complex objects such as spears, axes, rafts, tools, tents, or clothing. Human linguistic and manual constructions incorporate greater amounts of information than do those of great apes, and they are often more hierarchical in that they involve greater amounts of embedding of constructs into still higher order units. Thus, although language-trained apes can use wordlike symbolic gestures and lexigrams and can regularly combine up to two or three symbols in accordance with "grammatical" rules, they fall short of even a 3-year-old human child in the numbers of "words" that they can master and in the numbers of "words" that they can combine into more meaningful units. They also do not appear to create complex utterances that involve embedding phrases into higher order linguistic units. Similarly, ape toolmaking, unlike human toolmaking, is primarily a subtractive process, usually involving the stripping of side twigs to create straight probelike shafts rather than the joining of two or more separate objects together to make complex tools.

Elsewhere, my colleagues and I (Gibson, 1990, 1996, 2002; Gibson & Jessee, 1999; Gibson, Rumbaugh, & Beran, 2001) have argued that embedding and combinatory capacities are also evident in social intelligence, art, music, and other endeavors and that, in primates, mental constructional ability correlates with absolute size of the brain as well as with size of varied brain parts. In my view, mental constructional skills would be predicted to vary among vertebrate species in accordance with overall brain size, in part, because the amount of neural tissue devoted to any sensory or motor modality determines the extent of fine sensory or motor differentiation capacity within that modality (Pubols & Pubols, 1972; Welker & Carlson, 1976; Welker et al., 1976; Welker, Johnson, & Pubols, 1964; Welker & Seidenstein, 1959). In addition, increasing brain size correlates with increasing neuronal connectivity (i.e., brain complexity) and, hence, with abilities to create relationships among neural networks (Jerison, 1982). Here again, my ideas and Rumbaugh's converge in two ways. First, Rumbaugh has suggested that relational thinking requires complex brains. Second, Rumbaugh and his collaborators have found that performance on the transfer test of mental flexibility correlates strongly with cranial capacity in captive primates (Gibson et al., 2001). That scholars from the diverse disciplines of anthropology and psychology have converged in their hypotheses that absolute brain size is a correlate of relational thinking is encouraging in that it shows potential for the development of a truly unified comparative and evolutionary psychology. That this convergence has led to an emphasis on brain size as a correlate of species differences in higher cognitive capacities also provides a potential link with the Darwinian notion that the mental differences between animals and humans are matters of degree rather than kind.

## Summary and Conclusions

Current concepts of the nature of animal versus human mental capacities derive from several philosophical and scientific traditions, some of which appear mutually contradictory. In particular, the behaviorist and Piagetian approaches to learning and intelligence have often been considered antithetical. Rumbaugh's rational behaviorism, however, converges with neo-Piagetian concepts of hierarchical mental construction in that both place heavy emphasis on relational thinking. Rational behaviorism and neo-Piagetian mental construction are both also compatible with modern neural network models of cognition and learning that emphasize strengthening connections and relationships within and between neural networks. Both also are potentially compatible with Darwinian views that mental differences between animals and humans are matters of degree rather than of kind, because brain size correlates with neural connectivity and hence with numbers of potential relationships among neural units.

Rational behaviorism as defined by Rumbaugh places heavy emphasis on emergents, that is, on the appearance of novel, unexpected behaviors. Relational thinking and mental construction explain many aspects of emergence, especially the emergence of novel solutions to problems and creativity. The emergence of unexpected humanlike cognitive skills in human-reared apes, such as language, use of a tool to make a tool, and mirror self-recognition, however, may also reflect a different process, epigenesis (i.e., gene–environment interactions during the development), a process that has been repeatedly demonstrated to account for many aspects of brain development.

## References

Boesch, C. (1993) Aspects of transmission of tool-use in wild chimpanzees. In K. R. Gibson & T. Ingold (Eds.), *Tools, language and cognition in human evolution* (pp. 171–184). Cambridge, England: Cambridge University Press.

Bower, T. G. R. (1974). *Development in infancy*. New York: Freeman.

Brauth, S. E., Hall, W. S., & Dooling, R. J. (Eds.). (1991). *Plasticity of development*. Cambridge, MA: MIT Press.

Byrne, R., & Russon, A. E. (1998). Learning by imitation: A hierarchical approach. *Behavior and Brain Sciences, 21*, 667–672.

Case, R. (1985). *Intellectual development: Birth to adulthood*. New York: Academic Press.

Cowell, P. E., & Denenberg, V. H. (2002). Development of laterality and the role of the corpus callosum in rodents and humans. In L. J. Rogers & R. J. Andrew (Eds.), *Comparative vertebrate lateralization* (pp. 274–305). Cambridge, England: Cambridge University Press.

Damerose, E., & Vauclair, J. (2002). Posture and laterality in human and non-human primates: Asymmetries in maternal handling and the infant's early motor asymmetries. In L. J. Rogers & R. J. Andrew (Eds.), *Comparative vertebrate lateralization* (pp. 306–362). Cambridge, England: Cambridge University Press.

Darwin, C. (1859). *On the origin of species by means of natural selection, or the preservation of favoured races in the struggle for life*. London: John Murray.

Darwin, C. (1871). *The descent of man and selection in relation to sex*. London: John Murray.

Darwin, C. (1872). *The expression of the emotions in man and animals*. London: John Murray.

Deng, C., & Rogers, L. (2002). Factors affecting the development of lateralization in chicks. In L. J. Rogers & R. J. Andrew (Eds.), *Comparative vertebrate lateralization* (pp. 206–246). Cambridge, England: Cambridge University Press.

Duscheck, J. (2002, October). The interpretation of genes. *Natural History, 111*, 52–59.

Faulkes, C. G. (1999). Social transmission of information in a eusocial rodent, the naked mole rat (*Heterocephalus glaber*). In H. O. Box & K. R. Gibson (Eds.), *Mammalian social learning: Comparative and ecological perspectives* (pp. 205–220). Cambridge, England: Cambridge University Press.

Fuster, J. M. (2002). *Unifying cognition*. Oxford, England: Oxford University Press.

Galdikas, B. M. (2003). *Orangutan foundation international*. Retrieved June 5, 2003, from http://www.orangutan.org/facts/orangutanfacts.php

Gardner, R. A., & Gardner, B. T. (1969, August 15). Teaching sign language to a chimpanzee. *Science, 165*, 664–672.

Gibson, K. R. (1983). Comparative neurobehavioral ontogeny and the constructionist approach to the evolution of the brain, object manipulation, and language. In E. DeGrolier (Ed.), *Glossogenetics: The origin and evolution of language* (pp. 37–62). London: Harwood Academic.

Gibson, K. R. (1990). New perspectives on instincts and intelligence: Brain size and the emergence of hierarchical mental constructional skills. In S. T. Parker & K. R. Gibson (Eds.), *"Language" and intelligence in monkeys and apes: Comparative developmental perspectives* (pp. 97–128). Cambridge, England: Cambridge University Press.

Gibson, K. R. (1996). The ontogeny and evolution of the brain, cognition, and language. In A. Lock & C. R. Peters (Eds.), *Handbook of symbolic evolution* (pp. 407–430). Oxford, England: Clarendon Press.

Gibson, K. R. (2002). Evolution of human intelligence: The roles of brain size and mental construction. *Brain, Behavior and Evolution, 59*, 10–20.

Gibson, K. R., & Jessee, S. (1999). Language evolution and the expansion of multiple neurological processing areas. In B. J. King (Ed.), *The origins of language: What non-human primates can tell us* (pp. 189–228). Santa Fe, NM: School of American Research Press.

Gibson, K. R., Rumbaugh, D., & Beran, M. (2001). Bigger is better: Primate brain size in relation to cognition. In D. Falk & K. R. Gibson (Eds.), *Evolutionary anatomy of the primate neocortex* (pp. 79–97). Cambridge, England: Cambridge University Press.

Goodall, J. (1964, March 28). Tool use and aimed throwing in a community of free-ranging chimpanzees. *Nature, 201,* 1264–1266.

Greenfield, P. M. (1991). Language, tools, and the brain. *Behavior and Brain Sciences, 14,* 531–595.

Greenfield, P. M., & Savage-Rumbaugh, E. S. (1990). Grammatical combination in *Pan paniscus*: Process of learning and invention in the evolution and development of language. In S. T. Parker & K. R. Gibson (Eds.), *"Language" and intelligence in monkeys and apes: Comparative developmental perspectives* (pp. 540–578). Cambridge, England: Cambridge University Press.

Güntürkün, O. (2002). Ontogeny of visual asymmetry in chickens. In L. J. Rogers & R. J. Andrew (Eds.), *Comparative vertebrate lateralization* (pp. 247–273). Cambridge, England: Cambridge University Press.

Hallett, M. (2000). Plasticity. In J. C. Mazziotta, A. W. Toga, & R. J. Frackowiak (Eds.), *Brain mapping: The disorders* (pp. 569–586). New York: Academic Press.

Hubel, D., Wiesel, T. N., & LeVay, S. (1977). Plasticity of ocular dominance columns in monkey striate cortex. *Philosophical Transactions of the Royal Society of London, B278,* 377–409.

Jerison, H. J. (1982). Allometry, brain size, cortical surface, and convolutedness. In E. Armstrong & D. Falk (Eds.), *Primate brain evolution, methods, and concepts* (pp. 77–84). New York: Plenum Press.

Kawai, M. (1965). Newly-acquired pre-cultural behavior of the natural troop of Japanese monkeys on Koshima Islet. *Primates, 6,* 1–30.

Merzenich, M., & Kaas, J. H. (1982). Reorganization of mammalian somatosensory cortex following peripheral nerve injury. *Trends in Neurosciences, 5,* 434–436.

Neville, H. J. (1991). Neurobiology of cognitive and language processing: Effects of early experience. In K. R. Gibson & A. C. Petersen (Eds.), *Brain maturation and cognitive development* (pp. 355–380). Hawthorne, NY: Aldine de Gruyter.

Parker, S. T., Mitchell, R. W., & Boccia, M. (Eds.). (1994). *Self-awareness in animals and humans: Developmental perspectives.* Cambridge, England: Cambridge University Press.

Pons, T. P., Garraghty, P. E., & Mishkin, M. (1988). Lesion induced plasticity in the second somatosensory cortex of adult macaques. *Proceedings of the National Academy of Sciences USA, 83,* 5279–5281.

Pons, T. P., Garraghty, P. E., Ommaya, A. K., Kaas, J. H., Taub, E., & Mishkin, M. (1991, June 28). Massive cortical reorganization after sensory deafferentiation in adult macaques. *Science, 252,* 1857–1860.

Pubols, B. H., & Pubols L. M. (1972). Neural organization of somatosensory representation in the spider monkey. *Brain, Behavior, and Evolution, 5,* 342–346.

Rogers, L. J., & Andrew, R. J. (Eds.). (2002). *Comparative vertebrate lateralization.* Cambridge, England: Cambridge University Press.

Rumbaugh, D. M. (1977). *Language learning by a chimpanzee: The Lana project.* New York: Academic Press.

Rumbaugh, D. M. (2002). Emergents and rational behaviorism. *Eye on Psi Chi, 6,* 8–14.

Sadato, N., Pascual-Leone, A., Grafman, J., Ibanez, V., Derber, M. P., Dold, G., & Hallett, M. (1996, April 11). Activation of the primary visual cortex by Braille reading in blind subjects. *Nature, 380,* 526–528.

Savage-Rumbaugh, E. S., Murphy, J., Sevcik, R., Brakke, K. E., Williams, S. L., & Rumbaugh, D. M. (1993). Language comprehension in ape and child. *Monographs in Child Development, 58,* 1–220.

Schick, K. D., Toth, N., Garufi, G., Savage-Rumbaugh, E. S., Rumbaugh, D., & Sevcik, R. (1999). Continuing investigations into the stone tool-making and tool-using capabilities of a bonobo (*Pan paniscus*). *Journal of Archaeological Science, 26,* 821–832.

Simons, D. J., & Land, P. W. (1987, April 16). Early experience of tactile stimulation influences organization of somatic sensory cortex. *Nature, 326,* 694–697.

von Noorden, G. K., & Crawford, M. L. (1992) The lateral geniculate nucleus in human strabismic amblyopia. *Investigative Ophthalmology, 33,* 2729–2732.
Waddington, C. H. (1957). *The strategy of genes.* London: Allen & Unwin.
Welker, W. I., Adrian, H. O., Lifschutz, W., Kaulen, R., Caviedes, E., & Gutman, W. (1976). Somatic sensory cortex of llama (*Lama glama*). *Brain, Behavior, and Evolution, 13,* 184–293.
Welker, W. I., & Carlson, M. (1976). Somatic sensory cortex of hyrax (*Procavia*). *Brain, Behavior, and Evolution, 13,* 294–301.
Welker, W. I., Johnson, J. I., & Pubols, B. H. (1964). Some morphological and physiological characteristics of the somatic sensory system in raccoons. *American Zoologist, 4,* 75–94.
Welker, W. I., & Seidenstein, S. (1959). Somatic sensory representation in the cerebral cortex of the raccoon (*Procyon lotor*). *Journal of Comparative Neurology, 111,* 469–501.
Whiten, A., & Byrne, R. (1988). The manipulation of attention in primate tactical deception. In R. Byrne & A. Whiten (Eds.), *Machiavellian intelligence: Social expertise and the evolution of intellect in monkeys, apes, and humans* (pp. 211–223). Oxford, England: Clarendon Press.

# 21

# Kanzi Learns to Knap Stone Tools

## E. Sue Savage-Rumbaugh, Nicholas Toth, and Kathy Schick

The idea that a living ape might become a stone knapper was first suggested by Nicholas Toth at a Wenner–Gren Conference in 1990 on the relationship of tool and language. *Knap* refers to the act of hitting, breaking apart, chipping, or flaking stones. The initial reaction of my colleagues and I at the Language Research Center to this suggestion was that this would be impossible, yet we invited paleoanthropologists Nicholas Toth and Kathy Schick to the center and asked how they might go about teaching a bonobo to knap. They had spent more than 2 decades studying the earliest archaeological evidence of human technology (Schick & Toth, 1993; Toth & Schick, 1993).

It is worth noting that as we did so, it was our unexpressed belief that (a) the bonobos would have no ready inclination to knap; (b) they most likely lacked the manual dexterity to knap with precision; (c) toolmaking must have evolved long after a simple language emerged and, given what we then saw as the simplicity of their language, toolmaking would be too demanding for them; and (d) finally, because it was so difficult for humans to master even basic knapping skills, we had little reason to anticipate that bonobos would be successful.

However, we set aside our doubts and began working with the ideas proposed by Toth and Schick. They wished to begin with an adolescent male bonobo named Kanzi because of the remarkable language skills he had acquired by observation and experience, not specific training regimens. Toth and Schick had studied contemporary human stone toolmakers in New Guinea where, around adolescence, young men train to flake stones. They acquire their skills in the context of a social group. Flaking stone is a group social activity. All learning and practice takes place in a social setting, in which the sound of the stone against stone, the comparison of work, and discussion of the product takes place. Thus, the skill of each knapper is known by the others, and the tools, when finished, become essential implements to group survival.

We mention these critical aspects of human flaking to point out that they were necessarily absent in Kanzi's experience. Kanzi began his knapping experiences at a profound cultural disadvantage. He had no cultural group for whom knapping was an essential social and survival skill. Consequently, we must temper our comparisons between Kanzi and early human knappers by reminding ourselves that Kanzi probably would behave very differently if he had been reared in a knapping culture. However, Kanzi, as an adolescent male, had at least some of the basic credentials (i.e., a history of learning from observation) to serve as a stand-in for hominid knappers.

## The Basic Experimental Design

How would Kanzi learn to knap? Toth explained that he had designed a box that could only be opened if Kanzi used a stone tool. Some prized incentive would be placed inside the box and then, from outside Kanzi's enclosure, Toth would make a stone flake and demonstrate its use. From the very beginning, Kanzi watched with interest as Toth picked up rocks, knocked off a flake, and used it to cut the string on a box that held some grapes. He then handed the grapes to Kanzi. After repeating this activity a few times, the "string box" was brought into Kanzi's enclosure.

Kanzi did not pick up rocks and start to knap, so one of us (Savage-Rumbaugh) put a rock in each of Kanzi's hands and urged him to try. He generally attempted to knap as he had observed but made the typical novice mistakes of bringing the rocks together with a horizontal motion so that their surfaces met in the middle and also using very little force. When Toth knaps, he appears to use very little force as well. In fact, to the naive observer, the way in which he strikes the rocks together so as to cause flakes to fall off into his hand seems like magic.

When knapping stone, one cannot succeed by simply hitting rocks together. It is a difficult skill to master. Just as one cannot simply pick up a violin, draw a bow back and forth across it, and produce music, one cannot simply slam rocks together and make useful tools. A successful knapper must learn to use one stone as a hammer and the other as a base or "core." The hammer stone must strike the core at the right angle and with the appropriate force to produce a single flake of the desired size and sharpness. The production of each flake determines how and where the hammer must strike to produce the next flake. The hammer blows must be produced with a more or less controlled throwing motion rather than a hitting motion. The blows must strike with precision for high-quality flakes to be produced efficiently.

Kanzi's initial efforts met with such lack of success that it appeared he was about to resign his knapping training. We encouraged him with verbal support and continued to demonstrate the knapping activity. Special incentives were placed in the box, but Kanzi refused to try after his initial failure. The impression was that he was very sensitive to failure and resented the encouragement to do something that was too difficult for him. Perhaps this is too anthropomorphic an interpretation of his reaction, but it has been reported that "every researcher with apes has learned that they will balk and simply stop working if problems encountered are in any manner beyond them" (Rumbaugh & Washburn, 2003, p. 137). This "balking" might best be understood as a refusal to engage in behavior that is too challenging, one for which there are no requisite skills for the ape subjects to use successfully.

The decision was made to enhance Kanzi's knapping efforts through social cooperation. Consequently, Toth took over the role of making the flakes for Kanzi to use to open the box. Kanzi's interest and apparent enthusiasm for the task immediately improved. Toth was so impressed with how proficiently Kanzi used the stone flake that he wondered whether Kanzi actually understood something about the properties of stone tools and whether he could determine a good flake from a bad one. That is, could Kanzi judge the sharpness and strength of a variety of flakes and choose the best one or would he essentially use any flake? Toth had not demonstrated flake selection, and Kanzi had no previous experience with stone tools. If Kanzi were to know good flakes from poor ones, this would suggest a preliminary understanding of stone geometry.

The tool site was baited, and Kanzi was given an array of flakes from which to choose. From the start he ignored the duller flakes and selected the sharper ones, generally testing them for sharpness by mouthing before using them to cut the string. Because the flakes were small, about 1/2 inch in length, it was difficult to hold them with his large fingers as he tried to cut the string. He failed to hold the flake at a downward angle and pull it toward him, as modeled by Toth. As a consequence, his cutting was neither swift nor very effective as he tried to push the flake through the string, wearing the string away fiber by fiber.

However, he quickly developed his own method of doing this with two hands instead of one. (Toth had used only one hand.) Kanzi began to hook the index finger of his left hand under the string, pull it tight, and then cut the string with the rock chip. His cutting motions were limited to one direction (toward himself) rather than back and forth, focusing his work on one part of the string until it separated. He did this patiently, using the chip even when the string was hanging only by a thread.

Toth reminded us that early hominids surely had not begun to knap in their first efforts. Accordingly, we continued to bait the tool site several times and to demonstrate flaking for Kanzi even after Toth departed at the end of the week. There was, however, a major difference in our demonstrations—knapping was not easy for us. The stone flakes did not fall as if by magic. We used more force because we could not focus the blows in the precise way that Toth did. We failed a lot, and it often took many blows to produce a flake, which all too frequently was rather dull. In some ways we were better models for Kanzi, because it was more difficult for us than it was for Toth.

We continued to encourage Kanzi to knap repeatedly and offered him various rocks with which to work. Sometimes he took the rocks and made a few half-hearted attempts at hitting them together in the mid-plane, as if he were clapping, and then gave up. Frequently, when we asked him to continue he would either hand the rocks to us or just walk away and ignore us.

Kanzi was so strong that even if he were not using the correct technique, he had sufficient strength to break the rocks, so we verbally encouraged him to hit the rocks together "harder" and tried to demonstrate this as best we could. In response to our verbal suggestion, he would produce one or more blows that sounded loud and strong but then give up again when chips did not fall. For approximately 2 weeks, no matter how tenaciously we worked to interest him or to assist him, there was no change in Kanzi's behavior.

## Overcoming Initial Obstacles

Just as we were beginning to conclude that Kanzi was not going to make any progress in this endeavor, we began to hear loud repetitive banging noises coming from the group room. We peeked in to see Kanzi, with a determined look on his face, hitting the stones together with as much force as he could muster, over and over and over again, until his arms were too tired to permit him to continue. Although because of the horizontal angle of his hands, chips were not flying off the rocks, stone powder was being created with each blow. In fact Kanzi, the first bonobo knapper, sat among a veritable cloud of stone powder produced by his own efforts. Kanzi apparently had decided to try and "open" the rocks by sheer strength alone.

We use the term *decided* because nothing had changed in the situation but Kanzi himself. It took about 15 minutes to produce his first chip, a tiny piece less than a 1/4 inch long and just a sliver in thickness. He tried to use it, but it was so small that it crumbled in his large hands before he could cut through the string. Kanzi sighed and made another flake with similar effort, though more quickly.

That day was something of a watershed for Kanzi in that he had come to realize that it was possible for him to make a flake. Kanzi had a new commitment to the task; from that point on, each time the tool site was baited he would pick up the stones and begin to knap. He no longer needed encouragement or demonstration. When he failed, we would remind him that it was possible for him to succeed. Kanzi would appear to reflect on this encouragement and then return to the task.

Kanzi now began to examine the rocks carefully, and depending on their material and angles, he would select those with the better flaking potential. His flakes continued to be very small, and many of them broke as he tried to cut the string, but he was no longer easily dissuaded and would make as many small flakes as he needed to cut the rope. Even when strongly encouraged, he did not change the plane of his blow or try to make large flakes.

Observations such as these, made across time, are neither anecdotal nor anthropomorphic. They go beyond anthropomorphism. To say that Kanzi "realized it was possible for him to make a flake" is not a simple statement about how we would interpret his behavior at any one point in time—were he a human being. Rather, it is a statement constructed across a long span of time from a multiplicity of observations. It is supported across time by observations of one sort before a unique event and observations of a different sort after that unique event. These observations stand in sharp contrast defined by one point in time. Accordingly, it justifies a statement about Kanzi's cognitive capacity to assess changes in his own competencies in this task.

Kanzi remained content with the smallest of flakes and would use them until they broke or wore away, but he spent a great deal of time and effort producing them. Often Kanzi would knap away at the stone for 10 to 15 minutes before producing the tiniest flake, only to have it wear away before he was finished cutting the string. He might have to exert this effort several times before the string was cut. Thus, understandably, Kanzi did not appear to enjoy the effort that flaking required. The task was tiring and difficult for him because he had to try to hold the stones and strike them without hitting his fingers, which he often did. He also had tried to avoid getting the flakes in his eyes.

A few months later, Kanzi began to use his left and right hands in a new and unique way. He would rest his left hand, holding a stone, against his abdomen and then strike it with the other stone held in his right hand. This enabled him to position the stones more securely and was the beginning of a bimanual differentiation of blows, with one hand holding the core and the other one the hammer. This approach also meant that he would not have to grasp the core so tightly with his fingers, the result being that he reduced the likelihood of striking his massive fingers that were wrapped securely around the core. Because his fingers are approximately twice as long as human fingers, this new method afforded his fingers substantial protection.

The innovation from "clapping" blows of the midplane to those that required one hand to act as a stabilizer of the core and the other as an impeller of the hammer made it possible for Kanzi to knap longer and to produce stronger blows. Unfortunately, because his left hand rested against his abdomen for support, the force of his blows were now partially absorbed by his own body mass. So, it took even more forceful blows to produce a chip.

As Kanzi continued to knap, he seemed somewhat more content with his newfound position and tended to work more tenaciously, but he still searched for ways to make the process simpler. The problems that Kanzi had to contend with were not ones that could be solved by observing Toth, because his anatomy was different from humans. We have shorter fingers and longer thumbs that make it easier to hold smallish stones and to knap them without hitting our fingers. On the other hand, we lack the greater strength of the bonobo and cannot easily produce chips by hitting the rocks together in the midplane.

Was Kanzi's altered use of his two hands the result of imitation, trial and error, or a reasoned attempt to solve a recurring problem defined by his physical anatomy in knapping? Although behaviorists might well argue for trial and error, we find that explanation unsatisfactory. Kanzi was never observed trying to position the core stone in various ways on his body. It seems more reasonable to believe that he was responding to the dynamic physics of the task at hand. When he used midplane blows, chips were rarely produced. If instead of hitting the rocks together mid-plane he hit one rock against another, he was more likely to produce a chip because he could achieve angular blows. It was difficult for him to do this holding the rock in one hand held in front of his body. He would sometimes hit his fingers or knock the core out of his hand. He reduced these problems by placing the core against his body. Kanzi did not randomly try a wide variety of workable and unworkable sets of possible solutions. Rather, he selected only a single stone to hold, a single hand to hold it in, and body positions indicating that he seemingly understood the physical constraints imposed by his anatomy and the operations of knapping successfully.

Notwithstanding the essential merit of his new method of knapping, it was still difficult for him to produce a large sharp flake. Too much of the striking force was absorbed by his stomach that served as the grounding for the rock being struck. Apparently, this difficulty was sufficient to induce Kanzi to seek still other methods.

## Kanzi's New Technique: Throwing

One day, after the string box was baited, Kanzi just sat and looked at it, resting his elbow on one knee and his head in one hand. We were surprised, because he always began to flake in response to an incentive being placed in the box. Kanzi did not appear uninterested on this occasion. Quite the contrary! He assumed the classic position of the "thinker" and remained frozen for some time, his eyes fixated on the rocks in front of him. Such a pose was unusual for Kanzi.

Eventually he resumed his effort and picked up one stone instead of two. He held it in his right hand and rose to a full bipedal stance instead of sitting as he normally did when preparing to knap. He then raised his arm and threw the rock onto the hard tile floor with great force, causing it to shatter into more flakes than he had produced from all of his bimanual knapping combined. Kanzi at once selected a large sharp flake and cut the string on the tool site within seconds. He had invented a way to mass produce the fabrication of stone chips with sharp edges, and they came in all shapes, sizes, and varieties.

Kanzi had come up with a technique none of us had thought of yet, and it was far more efficient than any other method and produced better flakes than the one we had encouraged him to use from more conventional knapping methods. To infer that Kanzi "thought his way" to this solution is quite reasonable. This sudden action, nested as it was within the context of Kanzi's past knapping failures and successes (e.g., his style of placing the rocks on his abdomen, his great efforts to make tiny flakes, his intense fixation on the rocks followed by a sudden change in technique), provides support for this conclusion. Moreover, if such actions were taken by a panda bear or an otter (both of whom can hold stones and sit upright), the conclusion would be the same. It is not Kanzi's physical and behavioral resemblance to us that drive these conclusions; rather it is the integration and nature of his behavior displayed across time.

We were delighted with Kanzi's innovation because it clearly demonstrated that for him, the end goal of producing a large sharp flake was well understood. With that understanding, he could invent his own method of flake production. Critically, his innovation did not result from trial and error or even from play. Rather, it was a direct response to difficulties he encountered using a human technique with a bonobo anatomy. It proved to be far more efficient, for his purposes, than what we had been encouraging him to do.

## Redirecting Kanzi to Knapping

Might apes have the potential to begin production of simple stone tools without human intervention? Although it is true that we designed a task for Kanzi that required a tool, we already knew that apes have the skill to use rocks to crack nuts and to throw missiles at predators. Inevitably in such cases, if rocks are thrown, some of them will shatter and produce sharp flakes. If there is a need for these flakes, apes clearly have the potential to produce them. As Toth suggested, the earliest stone tools in the archeological record may be more difficult to identify than previously assumed because flakes made by throwing are not as readily distinguished from natural rock fractures as are knapped flakes. Notwithstanding Kanzi's invention of a method for making sharp stone tools efficiently, Toth emphasized the need for Kanzi to utilize bimanual percussion so that the work of a living ape might be more appropriately compared with hominid flakes that were certainly not produced by throwing.

When a rock is thrown onto a hard surface and breaks open into many pieces all at once, it is not possible to reconstruct what might have been the toolmaker's intended end. However, when the flakes are produced one at time through knapping, the entire process can be directly reconstructed by putting the stone back together. The pieces of the stone become a three-dimensional puzzle, and as the original stone is reassembled, it is possible to see just what the original knapper saw: the striking platforms as they appeared to him and the selections he made at each point. Accomplished knappers can readily detect the expertise and thought process present in the mind of the original toolmaker through these methods. Thus, to compare Kanzi with hominid knappers, we decided that Kanzi would be required to knap and would not be permitted to throw the stones.

How could we achieve this goal? Our first approach was to simply ask Kanzi to knap rather than to throw, though most students of animal behavior would look askance at such a decision and concentrate instead on effective shaping procedures. However, psychologists working with human subjects typically request participation of a specific sort during an experiment. Kanzi is not human, but he could understand verbal requests. Such requests afforded Kanzi, as they do human beings, a much simpler way to achieve a particular behavior than shaping techniques.

He understood our request and accommodated our expressed desire by hitting the rocks together a few times without achieving a flake. As if to emphasize the effectiveness of his own technique, he again slammed the rock on the floor to produce an array of flakes, then looked directly at us and gestured toward his accomplishment. We tried a few times to verbally insist that Kanzi knap rather than throw, but having once demonstrated the efficiency of his technique Kanzi proceeded to ignore us. When we baited the site, he did not bother to knap or to listen to our implorings. Rather, he threw the rock onto a hard surface, made his chip, obtained his incentive, and then walked away— all in a matter of minutes.

Kanzi clearly had a strong preference for the technique that he had devised over those that we had modeled and encouraged him to use. Perhaps an empiricist would argue that we were being too anthropomorphic in our interpretation of Kanzi's behavior; however, Kanzi did not have to make his choice so clear. After all, he had several months of practice and reward for knapping and only one experience with throwing. Yet that one experience, preceded as it was by the thoughtful steady gazing at the rocks, changed his behavior unequivocally from that time forward. If reward-based experience were driving his behavior, he should have fallen back on knapping, not really understanding what he had done.

Realizing that we were at a major impasse regarding how Kanzi would make his stone tools, we attempted to devise an experimental method to encourage Kanzi to abandon his own solution. (In retrospect, and with our current knowledge of Kanzi's cognitive awareness, it might have been wiser to simply explain to Kanzi that we were performing an experiment in which he needed to knap rather than throw. This method seemed to be one that Kanzi came to prefer when we engaged his cooperation to do things that were contrary to his desires.) We determined to alter the environment to make Kanzi's technique less efficient than our own by carpeting the entire group room floor with blankets, so that any stone would bounce against the softer surface rather than flake into pieces. Kanzi entered the room and appeared to enjoy lounging on the blankets. He did not try to remove the blanket carpet and so we decided to proceed with the plan to bait the incentive box. Kanzi watched as we did so. We then presented the rocks to Kanzi, assuming that he would throw a stone on the floor only to find that it bounced off the blankets and would not shatter as he had intended. We would then encourage him to knap. Yet without any trial and error, Kanzi deduced, just as we had, that the stone would bounce. He carefully pulled a few of the carpet blankets loose from their tape to form a hole revealing the hard tile floor, threw a stone into this hole, and produced a pile of flakes.

It is clear that Kanzi had surmised, just as we did, that a stone thrown against a soft surface would not shatter. He had no previous experience throwing stones against soft surfaces nor had he observed anyone do so. Yet he seemed to be cognizant of the properties of the materials he had been working with and behaved accordingly. Some empiricists might object to this interpretation, but it is not one based on simple anthropomorphic tendencies. The anthropomorphic assumption of the experimenters planning the procedure was that Kanzi would throw the stone onto the carpet without realizing that it would bounce. Kanzi, in this case, went "beyond our anthropomorphism."

While we deliberated on our next experimental step, Kanzi's facility with throwing enabled him to make large flakes quickly and easily, which in turn permitted us to increase the thickness of the string Kanzi was required to cut. Initially, the string was 1/4 inch thick. Kanzi was easily able to cut the 1/4-inch thick string with the larger sharper flakes produced by throwing. Slowly the diameter of the string increased until it was over 1 inch in thickness. To cut such thick string, Kanzi needed very large, strong, sharp flakes. He responded by selecting his material even more carefully and throwing it much harder and completely ignored the small chips his earlier work produced—anything under 3/4 of an inch was treated as debris.

During the "throwing phase," we noticed that initially Kanzi displayed no arm bias; however, within a few days he settled on the right hand and never again used his left for this task. The videotape of this short transitional period revealed the origin of the right-arm bias. When Kanzi stood bipedally and used his left hand to throw, his right arm rather automatically moved upward and forward and across his body, mirroring in a slightly delayed manner the motion of his left arm. The same throwing motion with his right arm did not evoke the mirroring motion of the left hand, which rested in the normal position beside his body. This difference in the motion of the opposing hand during throwing was observed only when Kanzi stood fully erect while simultaneously executing a forceful throwing motion. The need for the right hand and arm to follow the motion of the left hand and arm but not the inverse indicated the existence of a neurological basis for the development of Kanzi's right-handed preference. It is not known, of course, whether a similar constraint existed among early hominids, because they were more proficient bipeds than Kanzi. The follow-through movement of the right hand and arm occurred as though it were part of the locomotor pattern of motion. If Kanzi were brachiating and moved the left hand forward to catch a branch, the right hand would need to follow in patterned precision. The fact that Kanzi was able to inhibit this primitive motor pattern when leading with the left hand and arm but not the right supports Calvin's (2002) view that it was precision throwing that placed pressure on the nervous system for extreme hemispheric specialization and for the development of the rapid sequencing capacities that underlie music, grammatical construction, dance, and many other activities thought to be exclusively human.

Warmer weather permitted us to once again attempt to exert empirical control over Kanzi's knapping methodology. The outdoor play area had a barked yard, and we planned to place stone tools there while also blocking Kanzi's access to the indoor stone floor. We believed that this procedure would force Kanzi to abandon his throwing technique and revert to bimanual percussion—unlike carpet, bark cannot be removed because there is only more bark underneath. We quickly found that we were mistaken and that we had yet again underestimated Kanzi's ingenuity and creativity.

The first time he was presented with a task that required a stone tool, he visually surveyed the entire area, looking for a hard surface against which to throw the stone. He noted the large pole that held up the chain-link cover and threw the rock at the base of this round pole. The rock glanced off. Kanzi then looked around the area again, yet it seemed that there was nothing left for him to do but to try bimanual percussion. There were some steel tables in the enclosure and a small pool but no horizontal flat hard surfaces like the tile floor indoors. This fact did not trouble Kanzi for more than a few minutes. Again he paused in thought, then calmly walked over to the rocks in the enclosure, selected a large stone from the group and positioned it directly in front of him. Then he picked up a second stone and threw it against the first! It did not shatter immediately, but within three more throws Kanzi had produced a nice large flake—without knapping! Kanzi had expanded his throwing technique to include two stones, one as substrate and one as hammer. This innovation required a much more precise aim than simply throwing onto a hard floor. Initially, Kanzi missed the target stone frequently, but after a few days of practice he became as proficient at this technique as he had been at throwing a single rock onto the floor.

Kanzi's solution of throwing one stone against another to overcome the carpet of bark yet again demonstrated his ability to come up with a functional innovation that had not been taught or even demonstrated. In many ways, this solution was more impressive than the innovation of throwing or of moving the carpet aside. Throwing required Kanzi to compute a trajectory whereby the hammer stone would hit the target stone. This likely was a technique used by our hominid ancestors to break stones too large to knap bimanually. It is clear that if Kanzi were in the forest and needed stone tools to survive, he would have a way to produce them. Toth's initial demonstrations revealed to

Kanzi that rocks could be broken and that such breakage resulted in sharp edges. In the forest, knowledge of the properties of stone could arise by other means, because one would only need to observe a rock break when it fell to see the sharp edges that were produced. Kanzi's behavior to date suggested clearly that he possessed the capacity to reason his way to stone flake production.

## Kanzi Learns to Knap Stone

We do not pretend to suggest that Kanzi's behavior has answered the question of how our ancestors began to make stone tools. We do believe, however, that Kanzi's behavior casts doubt on the commonly held assumption that hominids who existed before 2.5 million years ago did not have stone tool cultures because they lacked cognitive competence.

Kanzi's second success in foiling our experimental attempts to force bimanual percussion left us puzzled. If we were going to provide Kanzi with at least two stones, as knapping required, how could we keep him from throwing one against the other when his method produced sharp flakes more efficiently than knapping? We could, of course, train him to only use flakes produced by knapping rather than throwing by denying him access to the tool site until he had knapped. From the perspective of simply artifact comparison, it did not matter what motivated Kanzi to flake bimanually, only that he did so. However, it was also the case that to the extent we were retracing—even in some minimal and artificial sense—the emergence of stone tool production in ancestral hominids, it would be certainly inappropriate to force a less efficient technique on Kanzi by arbitrary means. Equally important was the fact that Kanzi's techniques had been self-generated, and he understood them and preferred them. If he were in the wild, he would have continued to use them as long as they were effective for the desired ends. We needed to design a situation that called for knapped flakes rather than simply sharp flakes.

We attempted to visualize a situation that might have induced a similar need in our ancestors had they, like Kanzi, hit on the idea of throwing one stone against another and found that throwing produced perfectly acceptable sharp flakes. At first it seemed that the only possible reason would be the need for a more precise tool or one with a predetermined shape. Such specific shapes could not be consistently achieved, with the geometry of the desired flake clearly in mind, except by intentional design and systematic flaking. Kanzi's geometrical needs were simple—a large sharp edge—and more rapid forceful well-aimed blows achieved that goal quite well. Of course, if stones were rare and had to be carried long distances, it would become essential to get the maximum number of flakes from each stone and to get a single flake when needed. One could not afford to waste material by simply smashing it apart.

Many archeologists have assumed that the earliest knapped flakes did not reflect any intent to produce a flake of a specific size or form. However, Kanzi's efficient throwing technique cast strong doubt on that assumption and on that of Davidson and Noble (1989, p. 312) as well, who suggested that even Acheulean hand axes were not produced by intentional effort on the part of the knapper.

Archeologists believe that the main purpose of flakes and hand axes was for skinning and butchering meat. Most stone tools are found near old lake beds where groups congregated, perhaps attracted by large hoofed stock that went there to drink. There is another property of water that might affect apparent stone tool production. If prey were killed by drowning and then brought to shore, they might be too heavy to move very far. Butchering around water would inevitably result in tools falling into water.

We began to wonder how being around water might have affected tool manufacture and the need for hand axes and decided to put Kanzi's rocks into his small wading pool, which was approximate 2.5 feet deep and 8 feet wide by 10 feet long. If he tried to throw one stone at another while the rocks were in the water, it would not work. He would either have to take the rocks out to throw them or use bimanual percussion. We thought he would take the rocks out, but even so this situation would pose an interesting problem for him.

The experiment was set up on a very hot summer day, a time when Kanzi enjoyed being in his pool. Kanzi entered the tool site area and quickly noted that the rocks were not in their usual location. He looked around and spied them in his pool. He stepped in the pool and stood bipedally in the water looking down at the rocks. He picked up one stone and looked at another on the bottom of the pool and raised his hand as if to throw. Had he done so, the water would have provided a barrier to success. He paused, leaned down and picked up a second stone and began to percuss bimanually while standing. Previously, before he began to throw, he had always sat down when bimanually percussing. When he finished a chip it fell into the water. Kanzi saw it, leaned down into the water to retrieve it, then stepped out of the pool to open the tool site.

It is clear that Kanzi could have taken his rocks out of the pool and thrown them. He had been throwing now for nearly 8 months without any bimanual percussion. His shift from throwing to bimanual percussion was as precipitous and dramatic as was his shift away from percussion to throwing. It was also hardly possible to conclude that Kanzi could not reason sufficiently to move the rocks out of the pool. Kanzi easily carried rocks long distances. When we went into the forest, he would often place rocks in his backpack and carry them to the tool site located in the forest. When playing alone in his enclosure he would frequently gather and move rocks. Moreover, we noted later that if we left rocks both inside and outside the pool, Kanzi would occasionally stand inside the pool and percuss or throw a rock obtained from the water toward one that was already on the ground.

The most critical factor that emerged from this "experiment" was that Kanzi's percussion techniques were now quite different from his earlier efforts. No longer did he "clap" the stone together in the midplane. His left hand stabilized the anvil and his right hand produced a glancing blow with the hammer stone against the anvil stone, much as a human knapper might do. Immediately and easily, Kanzi produced a sizable flake. It appeared that the "throwing period" had provided Kanzi with some new abilities that were, all at once, utilized when he began to percuss while standing in water. He could now throw with force while aiming with precision. In the water, his left hand provided the stationary platform for the anvil stone, and his right hand delivered blow of the hammer stone with considerably greater force and precision. Kanzi did not actually let go of the hammer stone; the motion was instead more of a controlled throw as opposed to the holding and hitting motion he previously had used.

Equally important, Kanzi seemed to have learned something about the geometry of knapping, because he now tended to strike toward the edges of the core rather than the center. He seemed to understand the need to knock chips off the edge rather than simply "split the rock." Apparently he had acquired these skills while throwing, even though throwing itself did not require the confluence of the abilities that he now brought to the task of knapping.

Not only was Kanzi now a much more efficient knapper, but these innovations made the task more enjoyable for him because, for the first time since he had innovatively developed the throwing technique, he began to elect to knap rather than to throw, even when the stones were no longer in the water. Sometimes he alternated between these techniques, but he slowly began to prefer knapping. The preference for knapping appeared to be a function of his increased understanding that flakes were produced by hitting the core in a precise way. Once he understood the basic principles of where to hit to produce a flake, he also realized that it was necessary to hold the core and rotate it to achieve the best striking platform for each blow. Clearly, core rotation and orientation for the next blow was not possible unless one held the core. Kanzi also began to make multiple flakes from the same core and would now make them readily on request until the core was reduced to rubble, even when no incentive was in the tool site.

Having achieved, at long last, all of the basic skills that knapping required, Kanzi settled down into a right-handed bimanual percussor and continued to improve his technique (Schick et al., 1999; Toth, Schick, Savage-Rumbaugh, Sevcik, & Rumbaugh, 1993). He began to pay increasing attention to the angle of his blows and to the striking platform for each blow. He also began to rotate the striking platform after each blow to achieve the best striking surface, and he paid close attention to the surface he selected. As his ability to flake increased, he came to ignore small- and medium-sized flakes, attending only to the larger ones, because he now wanted not just any tool but an effective and efficient tool. His increased competency, surely the product of rational processes seasoned by experience, provided him the luxury of desiring a really effective flake (Savage-Rumbaugh & Lewin, 1994).

## References

Calvin, W. H. (2002). *A brain for all seasons: Human evolution and abrupt climate change*. Chicago: University of Chicago Press.

Davidson, I., & Noble, W. (1989). The archaeology of depiction. *Current Anthropology, 30,* 125–155.

Rumbaugh, D. M., & Washburn, D. A. (2003). *The intelligence of apes and other rational beings*. New Haven, CT: Yale University Press.

Savage-Rumbaugh, E. S., & Lewin, R. (1994). *Kanzi: At the brink of the human mind*. New York: Wiley.

Schick, K. D., & Toth, N. (1993). *Making silent stones speak: Human evolution and the dawn of technology.* New York: Simon & Schuster.

Schick, K. D., Toth, N., Garufi, G., Savage-Rumbaugh, E. S., Rumbaugh, D., & Sevcik, R. (1999). Continuing investigations into the stone tool-making and tool-using: Capabilities of a bonobo *(Pan paniscus). Journal of Archaeological Science, 26,* 821–832.

Toth, N., & Schick, K. D. (1993). Early stone industries and inferences regarding language and cognition. In K. Gibson & T. Ingold (Eds.), *Tools, language and cognition in human evolution* (pp. 346–362). Cambridge, England: Cambridge University Press.

Toth, N., Schick, K. D., Savage-Rumbaugh, E. S., Sevcik, R. A., & Rumbaugh, D. M. (1993). *Pan* the tool-maker: Investigations into the stone tool-making and tool-using capabilities of a bonobo *(Pan paniscus). Journal of Archaelogical Science, 20,* 81–91.

# An Afterword—and Words of Thanks

*Duane M. Rumbaugh*

To close this wonderful volume, the reader might benefit from an account of the development of the framework for emergents and rational behaviorism. Let me first present that framework and follow it with a section that provides the reader with a more personal account of my career and words of appreciation.

## Emergents and Rational Behaviorism: Basic Rationale

The past century of research with primates led my colleagues and I to formulate a rational behaviorism that relates the first appearances of creative, seemingly insightful behavioral patterns and capabilities to conditioned and unlearned behaviors (Rumbaugh & Washburn, 2003). We have called these creative behaviors *emergents* and have proposed that they be juxtaposed with *respondents* and *operants* of classical and operant conditioning procedures. Emergents reflect creative adaptation—novel patterns of behavior and not the fixedness that is expected as a result of traditional conditioning procedures. Notwithstanding, the etiology of emergents rests on information accrued through conditioning, observation, and general experience as well. Once we have defined emergents, we seek to identify their antecedents so as to understand them.

Harlow (1949) spoke of breaking the bonds of Thorndikean stimulus–response (S-R) learning by demonstrating the orderly development of efficient learning sets by his rhesus monkeys (*Macaca mulatta*). Primates, including humans, clearly have an alternative to S-R associative learning; we call it *relational learning*, because it seemingly reflects the formulation of hypotheses and overarching principles with which new and novel problems may be solved efficiently.

Our framework necessitates a reconsideration of the concept of *reinforcement*, which implies the cultivation of rigid behavior as the end product of conditioning or learning. Our analyses have traced the significance of temporal contiguity, the role of stimulus salience, and the sharing of the response-eliciting properties of stimuli as they inhere in the procedures of sensory preconditioning, classical (Pavlovian) conditioning, and operant (Skinnerian) conditioning. We hold that stimuli that are reliably paired temporally acquire salience in their relationship and thereby come to share their individual response-eliciting properties as a positive function of the relative salience of the stimuli paired. For example, in classical conditioning, the unconditional stimulus will share relatively more of its response-eliciting properties with the conditional stimulus than vice versa because invariably it is the stronger and more salient of the two. Conditioned reinforcers and secondary reinforcers, of Skinner's and Hull's frameworks, respectively, derive their salience and effectiveness by being paired reliably with other stimuli whose salience is natural (e.g., shock and food) or acquired through conditioning or other experiences. Stimulus salience inheres in the ethologist's sign stimuli, in unconditional stimuli, and in the reinforcers of operant conditioning. In our framework, reinforcers are to be understood in terms of strong stimuli that robustly share their response-eliciting properties with other stimuli.

As discussed frequently in this volume, there is strong evidence that animal learning is not limited to specific responses to specific stimuli. Rather, animals learn about tasks on which they work, the resources of their environs, and how to obtain these resources when their motivational states make them salient. They also learn to avoid aversive situations and to cope with those that must be endured.

It remains unknown how the organism's brain extracts the reliable patterning from real-world experiences and their consequences, stores them in memory as information, and subsequently builds creatively on these representations so as to generate novel, emergent behaviors. However, ignorance of the neural mechanisms should not limit psychologists to talking only of simple behavioral mechanisms. Certainly, animals master specific response modes and use them exclusively if that is what they must do to obtain the incentive, the reward, the resource—that which traditionally has been viewed as serving as reinforcement. But what they can learn—and likely will learn—is far more comprehensive.

For at least 90 years, psychology has been limited by the inexorable but circular law of effect. We have argued that reinforcement of behavior is not the end game. Rather, it is information that will be stored and possibly used in the future to provide the foundations of symbolic functioning, communication, and even basic language skills in great apes, dolphins, and parrots (Hillix & Rumbaugh, 2004). Our framework is intended to provide a unified understanding of behaviors from the most simple and predictable to the most creative and improbable.

## Personal Views and Notes

The Festschrift of October 2003 was a thoroughly enjoyable occasion, replete with a cadre of colleagues whose presentations were nothing short of excellent. Now, with this volume in hand, I am particularly impressed with the breadth of science and the diversity of questions and methods that it offers to the reader. The authors' array of philosophies and strategies should serve richly to enhance an integrated and positive view of psychology—a point of emphasis advanced by the American Psychological Association. They give us pause to reflect on the best values of the academy: the rights of professors to freedom of thought and research and to the formulation of new perspectives on basic issues that simply will never slip quietly away into the dark until answered! Studies of comparative learning and the more comprehensive psychology of animals in comparative perspective have brought us a long way over the past century.

In retrospect, my own opportunity to become a comparative psychologist was rather happenstance as well as amusing. I offer the following account, not only for the record but also for the pleasure of the reader.

The setting in which my opportunity developed was as unusual as the dynamics by which it developed. In 1954 I was appointed for 1 year at San Diego State College (now University). I received my appointment by telegram in the summer of 1954; I was to replace a faculty member on sabbatical in England. The basic architecture of the campus was Spanish and graced with covered walkways and beautiful palm trees.

The periphery of this otherwise beautiful, indeed idyllic, campus had become marred by a host of temporary buildings erected to accommodate the burgeoning enrollments generated by World War II veterans. Psychology's departmental office in 1954 was housed in one of these modest structures—a Butler building of military vintage. The building's front door consisted of two halves that when shut were secured by a padlock, attached to the door with a chain. A few wooden steps and a wooden platform unceremoniously graced the entry. Inside, perhaps 800 square feet afforded a classroom for 24 students, two small storerooms jammed with aged equipment, and a 6-foot-square office for the chairman. The office had a chair for one person other than the chairman. It was in that office and setting that an important career-structuring conference was held with Chairman Wolcott Treat.

By the spring of 1955, I had been offered a continuing appointment if an appropriate role could be defined within the department for my interests and credentials as a general–experimental psychologist with a PhD from the University of Colorado. What would I teach, what would I do as an instructor to earn my modest salary of about $2,700—hardly enough to live on, even in 1954–1955? Chairman Treat called me in to talk this question through to mutual satisfaction.

Wolcott smoked cigarettes—surely several packs a day. His office was always heavily laced with smoke. In those days, just about everyone smoked, but Wolcott had originality and style as he smoked. He always used a cigarette holder with a filter and generally held it slightly upward with his jaws tightly closed, much as President Franklin Roosevelt did during the days of the World War II crisis. He defined an interesting profile, indeed, as we set about the task at hand in his 6-foot-square office.

Wolcott pulled out the college's catalog, with its gray back and black-print title on its front, and turned to the department's offerings. What would I teach? Here's where the agony, the despair, and the ultimate delight began.

"Well, now," said Wolcott emphatically, "you can teach introductory psychology, but then we all do that! But what will be your specialty?" He read the titles of various courses as questions: "Mental health? Physiological psychology?

Statistics? Industrial? Social? Learning? History?" Each of these was struck out as he commented something to the effect that no, that course "belongs" to so and so and there was no need for more sections.

We were nearing the end of the course listings when, with substantial relief and exuberance, he quickly took his cigarette holder from his mouth and announced, "Well, now, here's Psych 170! Comparative psychology! 'The evolution of behavior from amoeba to man.' We haven't offered that course in a long time!" As he did this, ashes fell onto the catalog and his desk. He paused to brush them onto the floor. Then leaning slightly forward in his swivel-based wooden armchair and looking at me expectantly, he asked, "How would you like to be our comparative psychologist?" Without hesitation I said, "I'd love to be your comparative psychologist!" He leaned back in his chair and said, "Good! That will be fine."

So, I not only had a position, I had one as a comparative psychologist fall into my lap. I doubt if at the time there was another such position available in the whole country. Of course, there were the challenges—the lack of an animal laboratory, equipment, or anything much else—but the opportunities were not lost on me as I set about my new assignment as the San Diego State College Psychology Department's comparative psychologist.

The absence of an animal laboratory directed my attention to the San Diego Zoo. Having done learning research with rats while at the University of Colorado, I gained entry to the zoo through its Research Council. I was given space in the Animal Hospital for a maze or for whatever I wanted and could get. Rats for my studies came from the breeding colony maintained to provide food for the reptiles. So, I could get white rats of any age level, use them, and then return them to their destiny.

My interest in studying rats' learning of mazes, as tests of Clark L. Hull's learning theory, soon extinguished. Despite the fact that it was a reinforcement theory, it was not very reinforcing to me professionally. So, one day as the nearby beautiful vocalizations of exotic birds from South America and siamangs from the Orient wafted through an open window in my laboratory room at the zoo, I concluded that there must be other, more exciting venues available to a comparative psychologist. Chimpanzees and other primates of the zoo's rich primate collection captured my attention. At San Diego State I was given a Butler building, recently vacated by the Home Economics Department, as an animal lab. Across the ensuing years, my students and I had a great time there with homing pigeons, assorted monkeys from the zoo, kittens, chickens, and even rats.

In 1957, the Suez crisis broke and I faced duty as a reservist with the navy. Like any true red-blooded American academic, I decided that I would rather serve my country's interests as a stateside officer scientist than as an enlisted man on the deck of a carrier in the oceans of the Middle East. I obtained a direct commission in the U.S. Navy as a medical service corps officer. In 1958, I spent the summer at the U.S. Naval Medical Research Institute and was ever so fortunate to work under the direction of Lieutenant Robert B. Voas, who later became the training officer for the first seven astronauts. That summer we demonstrated the feasibility of using squirrel monkeys in a Jupiter-C missile (one of Hitler's Vengeance weapons used against England).

The National Aeronautics and Space Administration was founded later that same year, 1958. The monkeys were launched in a suborbital flight in January 1959.

Meanwhile, I had returned to the academy in San Diego and addressed the beautiful array of great apes at the San Diego Zoo from a more informed perspective. I had benefited immeasurably from a presentation by William A. Mason at the American Psychological Association's Convention in Washington, DC, in August 1958. His paper was "The Primary Role of Primates in Comparative Psychology."

The complex learning skills of great apes in comparative perspective became my focal topic. With the generous cooperation of the zoo officials, I had test facilities installed for studies of learning-set skills in six species of great and lesser apes. Support was obtained from the National Science Foundation for my research in 1960. Since that time I have enjoyed unbroken grant and contract support from a variety of funding agencies.

In 1969, I left San Diego State to accept a position at the Yerkes Regional Primate Research Center, Emory University, Atlanta, Georgia. There, I organized the LANA (chimpanzee) language project and the development of the first computer-monitored keyboard for language research with apes and language-challenged humans. Research based on that seminal project continues to this day. In 1971, I became chair of psychology at Georgia State University, also in Atlanta, and with E. Sue Savage-Rumbaugh founded the Language Research Center with that university in 1981. In 2004, I was given the opportunity to work with the Great Ape Trust in Iowa, whose mission is to conserve the great apes through education and research. It is timely to serve primarily the interests of the great apes' futures wherever they are.

It is safe to say that without the kind counsel of Wolcott Treat in 1955, I might never have had the opportunity for long-term studies of the intelligence of apes and monkeys. So, I thank Chairman Wolcott Treat of San Diego State College for a highly improbable, wonderful assignment for a young PhD at the age of 25! I also thank Carl Haywood, one of my first students in comparative psychology, for his contribution to the Festschrift. Carl was my first laboratory assistant in comparative psychology and now is an international leader in developmental psychology.

My own career, now spanning 50 years, has been rich and fulfilling. I trust that my enthusiasm for comparative psychology and primates in particular has been conveyed appropriately, from the 1960s with the making of *Survey of the Primates* (Rumbaugh, Riesen, & Lee, 1971), a film produced with the San Diego Zoo, to the publication of recent books this past year with colleagues. I hope that my enthusiasm will serve to help ensure a vital future for psychology.

I hold that this volume is not to honor any one person and career as much as it is to say collectively that we care very much about our science, that we will continue to build it for future strength, and that we strive for a better future for all life. It is my hope that this volume will serve to integrate perspectives of comparative psychology and psychology generally. I hope that it will help end the pestilence, the drought, the ravaging of our discipline from within and without and that it will give way to the clearing of skies, the integration of the otherwise seemingly disparate perspectives of learning and behavior, and the revitalization of academic freedom and constructive thought. I know that in this volume and in our other books we have worked to synthesize some of the basic perspectives and data of comparative psychology.

In closing, I want to express my enduring gratitude to everyone who has contributed to my research programs and life and for the reliable support of the National Institute of Child Health and Human Development and other granting agencies—notably, the National Science Foundation and the National Aeronautics and Space Administration—across the past 4 decades. Without their support and that of several universities along the way, the bridges offered here and elsewhere by the authors in this volume would not have been possible. I want particularly to note the remarkable colleagueship and friendship that I find in David A. Washburn. To him, I am forever indebted for proposing and organizing such a wonderful Festschrift and book. He is, indeed, a wonderful builder of bridges to tomorrow!

I thank one and all for learning of our views, for helping our work, and for understanding our love for the broad reaches of psychology. I also thank Georgia State University and the American Psychological Association for sponsoring both the Festschrift and this volume.

## References

Harlow, H. F. (1949). The formation of learning sets. *Psychological Review, 56,* 51–63.

Hillix, W. A., & Rumbaugh, D. M. (2004). *Animal bodies, human minds: Ape, dolphin, and parrot language skills.* New York: Kluwer/Academic Press.

Rumbaugh, D. M., Riesen, A. H. (Writers), & Lee, R. E. (Creative Director). (1971). *Survey of the primates* [Motion picture]. New York: Academic Press. (Available from the Psychological Cinema Register, Pennsylvania State University, University Park, PA 16802)

Rumbaugh, D. M., & Washburn, D. A. (2003). *The intelligence of apes and other rational beings.* New Haven, CT: Yale University Press.

# Author Index

*Numbers in italics refer to listings in the references.*

Abelson, R., 210, *219*
Adams, M. D., 19, *26*, *28*
Adrian, H. O., *278*
Allendoerfer, K. R., 64, *78*
Allport, G. W., 50, *57*
Amit, D., 151, *159*
Amunts, K., 244, *251*
Anastasi, A., 125, *132*
Ancrenaz, M., *28*
Anderson, J. R., 162, *170*
Andrew, R. J, 250, *253*, 271, *277*
Angleitner, A., *59*
Anscombe, G. E. M., 208, *218*
Armstrong, D. F., 250, *251*
Arnold, R. C., 10, *15*
Aronson, L. R., 82, *94*
Astur, R. S., 42, *45*
Attneave, F., 111, *122*, 210, *218*
Aureli, F., 55, *58*
Awazu, S., 162, *170*
Axhelm, M. K., 55, *58*

Baddeley, A. D., 207, 218, *218*
Bak, P., 101, *107*
Baker, K. C., 55, *58*
Baker, L. A., 21, *27*, 224, *234*
Balda, R. P., 144, 146, *158*, *159*
Baldwin, P. J., 35, *36*
Ballard, D. H., 41, *45*
Bard, K. A., 244, *253*
Barnes-Holmes, D., 104, *107*
Baroncelli, S., 53, *58*
Barrow-Green, J., 101, *107*
Barth, H., 164, *170*
Bartlett, F., 209, *218*
Basmanian, J. V., 21, *28*
Bateson, P. P. G., 54, *59*
Beall, A., 38, *45*
Beaton, A. A., 244, *251*
Beck, B. B., 30, 35, *36*, 162, *172*
Beem, N. O., 41, *44*
Behne, T., 263, *267*
Bekoff, M., 18, *25*
Bell, G. L., 139, *142*
Benbow, C. P., 127, *132*
Benedetto, D., 197, *204*
Beran, M. J., 13, *15*, 20, *26*, 64, 76, 77–78, 162, 164, 165, 168, 169, *170*, 197, *204*, 275, *277*
Beran, M. M., 162, 169, *170*

Berkley, M. A., 91, *96*
Bernieri, F. J., 49, *58*
Bernstein, N., 212–215, *218*
Berntson, G. G., 148, *159*, 162–165, *170*, *171*
Bickerton, D., 235, *242*
Billeter, J., 177, *205*
Binet, A., 125, *132*
Biro, D., 162–164, *170*
Bisazza, A., 250, *251*
Bitterman, M. E., 147, *159*
Blai-Hedges, S., 256, *267*
Blascovich, J. J., 38, *45*
Boakes, R., 143, *159*
Boccia, M., 270, *277*
Boesch, C., 22, 23, *25*, 270, *276*
Boker, S. M., 82, *94*
Bond, A. B., 146, *159*
Borgen, G., *28*
Boule, P., 23, *25*
Bower, T. G. R., 274, *276*
Bowerman, M., 255, 262, *266*
Boyd, R., 256, *266*
Boysen, S. T., 148, 149, *159*, 162–165, *170*, *171*, 207, *219*
Bradford, S. A., 162, 163, *171*
Bradshaw, J. L., 250, *251*
Braitenberg, J., 177, 198, *204*
Brakke, K. E., 42, *44*, 97, 176, 197, *204*, *205*, 257, *266–267*, *277*
Bramble, D. M., 20, *25*
Brannon, E. M., 151, *160*, 162, *171*
Braun, A. R., 244, *252*
Brauth, S. E., 272, *276*
Breland, K., 114, *122*
Breland, M., 114, *122*
Brim, O. G., 94, *95*
Britten, R. J., 19, *25*
Broadfield, D. C., 244, *252*, *253*
Bronfenbrenner, U., 94, *95*
Brooks, P. H., 129, 130, *133*
Brown, J. D., 49, *58*
Brown, R., 255, 259, 260, *266*
Brown, R. E., 177, *205*
Bruner, J., 260, *266*
Bryant, P. E., 146, *159*
Buirski, P., 48, *58*
Bunge, M., 86, *95*
Bunsey, M. D., 143, *159*
Burdea, G., 40, 41, *44*
Bürgel, U., *251*
Burghardt, G. M., 11, *15*

Burke, W. P., 130, *133*
Burns, R. A., 162, 163, *171*
Burt, S. T., 162, *171*
Buss, D. M., 57, *58*
Butynski, T. M., 31, *36*
Byrne, D., 88, 92, *95*
Byrne, R. W., 17, *25*, 38, *46*, 250, *253*, 256, *266*, 270, 272, *276*, 278

Cacioppo, J. T., 162, *170*
Caglioti, E., 197, *204*
Cairns, R. B., 94, *95*
Call, J., 162, *171*, 244, *252*, 263, *267*
Calvert, S., 55, *60*
Calvin, W. H., 287, *290*
Campbell, A., 55, *58*
Campbell, D. T., 88, *95*
Candland, D. K., 224, *233*
Cannon, E., 164, *171*
Cantalupo, C., 244, 249, 251, *252*
Cantero, M., 245, *252*, *253*
Cantor, N., 55, *58*
Capaldi, E. J., 162, 163, *171*
Capitanio, J. P., 53, *58*
Cardaciotto, L., 70, *79*
Carey, S., 162, *171*
Carlson, M., 275, *278*
Carnap, R., 238, *242*
Carpenter, M., 263, *267*
Carroll, L., 239, *242*
Carroll, S. B., 19, *25*
Case, R., 273, *276*
Casse-Perrot, C., 67, *77*
Castro, C. A., 68, *77*
Cavalieri, J. M., 17, *25*
Caviedes, E., *278*
Chalmers, M., 148, *159*
Chaplin, W. F., 49, *58*
Chappell, J., 106, *107*, *108*
Chapuis, N., 197, *205*
Chen, S., 146, 148, 151, 152, 154, *159*, *160*
Cheney, D. L., 144, *159*
Cheng, K., 184, *204*
Chomsky, N., 224, 227, 230, *233*
Chorover, S. L., 81, *95*
Christiansen, M. H., 151, *159*
Church, R. L., 185, *204*
Church, R. M., 38, *44*, 240, *242*
Cicchetti, D., 94, *95*
Clanton, N. R., 49, *58*
Claverie, J. M., 19, *25*
Cohen, D. J., 94, *95*
Cohen, J., 19, *25*, 193, *204*
Cohen, P., 193, *204*
Coiffet, P., 40, 41, *44*
Cole, M., 22, *25*

Collett, T. S., 175, 189, *204*
Colombo, M., 146, 148, 158, *159*
Conservation International, 24n, *25*
Conway, C. M., 151, *159*
Cook, R. G., 68, *79*
Coppens, Y., 35, *36*
Corballis, M. C., 251, *252*
Corbey, R., 19, *26*
Costa, P. T., Jr., 56, *58*, *59*
Costello, J., 94, *95*
Couvillon, P. A., 147, *159*
Covey, P., 101, *107*
Cowell, P. E., 272, *276*
Craig, A. B., 40, *45*
Crawford, M. L., 272, *278*
Creelman, C. D., 72, *77*
Crelin, E. S., 21, *26*
Critchfield, T., 106, *107*
Cronbach, J. P., 53, *58*
Crow, T. J., 243, 250, *252*
Czelusniak, J., *26*

Dahl, J. F., 251, *252*
D'Amato, M. R., 146, 148, *159*
Damerose, E., 272, *276*
Damon, W., 82, *95*
Danks, J. H., 143, *159*
Darwin, C., 18, *26*, 269, *276*
Davenport, R. K., 202, *205*
Davidson, I., 288, *290*
Davies, P., 81, 87, *95*
Davis, B. L., 243, *253*
Davis, H., 48, *58*, 146, *159*, 162–163, 168, *171*
Davis, L. H., 207–211, *218*
Deacon, T. W., 91, *95*
Dean, A., 93, *95*
DeLillo, C., 158, *159*
Delius, J. D., 146, *160*, 162, 164, *173*
DeLoache, J. S., 139, *142*
*Demonic Ape—Transcript*, 224, *233*
Denenberg, V. H., 272, *276*
Deng, C., 271, *276*
Derber, M. P., *277*
Deruelle, C., 67, *77*
de Schonen, S., 67, *77*
Desiderio, F., 8, *15*
Deutsch, J. A., 177, 198, *204*
de Waal, F. B. M., 17, 22, *26*, *28*, 31, *36*, 54, *59*
Dewey, J., 37, *44*
Diamond, J. M., 17, *26*
Dicara, L., 137, *142*
Diener, E., 54–56, *58*, *59*
Digman, J. M., 50, *58*
Dillingham, B., 23, *28*
Dobbs, V., 130, *133*

Dobkin, D. P., 187, 201, *204*
Dold, G., *277*
Donnai, D., 67, *77*
Dooley, G. B., 162, 165, *171*
Dooling, R. J., 272, *276*
Doyle, A. C., 117, *122*
DuBrul, E. L., 21, *26*
Duchon, A. P., *45*
Duke, M., 54, *59*
Durgin, F. H., 41, *44*, *45*
Duscheck, J., 271, *276*

Eaton, G. G., 55, *58*
Egnor, R., *78*
Eichenbaum, H., 146, *160*
Ekstrand, B., 139, *142*
Elder, G. H., 94, *95*
Elliott, J., 132, *134*
Emmerton, J., 162, 164, *171*, *173*
Emmons, R. A., 55, *58*
Enns, R. M., 57, *60*
Erb, L., *78*
Erickson, T. D., 238, *242*

Fagot, J., 39, *44*, *46*
Fajen, B. R., 41, *44*
Farmer, K. H., 51, *59*
Faulkes, C. G., 271, *276*
Feibleman, J. K., 82, *95*
Feigenberg, I. M., 214, *218*
Fernandez-Carriba, S., 246, 249, 250, *252*, *253*
Ferster, C. B., 162, *171*
Feuerstein, R., 129, 130, *133*
Fields, C., 19, *26*
Figueredo, A. J., 49, 51, 53, 54, 56, *59*
Filion, C., 113, *122*
Findley, J. D., 162, *171*
Finlay, T. W., 11, *15*
Flavell, J. H., 64, *77*
Fodor, J. A., 235, 236, *242*
Fountain, S. B., 143, *159*
Fouts, R. S., 18, *26*, 227, 232, *233*, 244, *253*
Fowlkes, D., 162, *172*
Fox, L. H., 127, *134*
Fraenkel, G. S., 175, 176, *204*
Fragaszy, D. M., 38, 42, *44*, *45*, 113, *122*, 176, *204*, *205*
Fraser, D., 29, *36*
Frost, N., 158, *159*
Fujita, K., 162, *170*
Fujiyama, A., *26*, 30, *36*
Funder, D. C., 49, *58*, *59*
Furness, W. H., 225, *233*
Furuichi, T., 266, *267*

Fuster, J. M., 273, *276*
Fuwa, K., *36*

Gaggioli, A., 41, 42, *44*
Galanter, E., 207, *219*
Galdikas, B. M., *28*, 271, *276*
Galimberti, C., 38, *45*
Galizio, M., 106, *107*
Gallistel, C. R., 164, 165, *171*, *173*, 175, 184, *204*, 208, *218*
Gannon, P. J., 20, *26*, 244, *252*, *253*
Gansner, E. R., 187, *204*
Garcia, P., 162, *171*
Gardner, B. T., 224, *233*, 259, *266*, 270, *276*
Gardner, H., 127, *133*
Gardner, R. A., 224, 227, *233*, 259, *266*, 270, *276*
Garner, R. L., 224, *233*
Garraghty, P. E., 272, *277*
Garufi, G., *277*, *291*
Gasman, D., 18, *26*
Geertz, C., 23, *26*
Geisel, T., 224, *233*
Gelade, G., 116, *123*
Gelman, R., *17*, 164, 165, *171*
Gershon, J., 54, *59*
Gibson, E., 117, *122*
Gibson, J. J., 116–118, *122*
Gibson, K. R., 20, *26*, 273, 275, *276*, *277*
Gilissen, E., 244, *252*
Gill, T., 162, 165, *171*
Gillan, D. J., 146, 148, 149, 152, *159*
Goettl, M. E., 162, *171*
Gold, K. C., 13, *15*, 48, *58*
Goldberg, L. R., 47, 50, 51, *58*, *59*
Goldin-Meadow, S., 256, 262, *266*
Goldstein, J., 87, 88, 90, 94, *95*
Goodall, J., 17, *26*, 28, 48, *58*, 244, *252*, 267, 270, *277*
Goodman, M., 17, *26*
Goodwin, B., 81, 82, *95*
Gordon, D., 84, *95*
Gosling, S. D., 48, 49, *58*
Gottlieb, G., 82, *95*
Gouzoules, H., 250, *252*
Gouzoules, S., 250, *252*
Grafman, J., *277*
Grant, J., 67, *77*
Greenberg, G., 83, 85, 92, 93, *95*
Greenfield, P. M., 258, 259–261, 264, 265, *266*, *267*, 270, 273, *277*
Gribbin, J., 81, *95*
Griffin, D. R., 67, *77*, 144, *159*
Grigorenko, E. L., 132, *134*
Gron, G., 41, *44*
Guagnano, G. A., 162, *172*

Gulledge, J. P., 167, *171*, *205*
Gunn, D. L., 175, 176, *204*
Güntürkün, O., 272, *277*
Gutman, W., *278*
Guttmannova, K., 64, *78*

Haag, R., 103, *108*
Haeckel, E., 18, *26*
Haggbloom, S. J., 81, *95*
Hall, J. A., 49, *58*
Hall, W. S., 272, *276*
Hallett, M., 272, *277*
Hallgrimsson, B., 21, *28*
Ham, M., 139, *142*
Hampton, R. R., 68, *77*
Hannan, M. B., 162, *170*
Haraway, M. M., 83, 85, *95*
Harlow, H. F., 38, *44*, 63, *77*, 140, *142*, 175, 177, *204*
Harnad, S., 235, *242*
Harris, M., 23, *26*
Harrison, M. C., 41, 43, *44*
Haskell, M., 55, *60*
Hattori, M., 19, *26*
Hauser, L. B., 162, *171*
Hauser, M. D., 162, *171*, *172*, 247, *252*
Hayes, C., 145, *159*, 226, *233*
Hayes, K. J., 145, *159*
Hayes, S., 104, 106, *107*
Hayhoe, M. M., 41, *45*
Haywood, H. C., 125, 128, 129, 130, 132, *133–134*
Head, H., 210, *218*
Hebb, D. O., 175, *204*
Hediger, H., 11, *15*, 175, 185, 202, *204*
Hegel, M. T., 162, *172*
Heimbuch, R. C., 21, *26*
Heltne, P. G., 17, *28*
Herman, L. M., 227, 228, *233*
Hershberger, W. A., 120, 121, *122*
Herzog, H. A., Jr., 11, *15*
Heyes, C. M., 48, *59*, 143, *159*
Hicks, L. H., 162, *171*
Highfield, R., 101, *107*
Hillix, W. A., 63, *78*, 100, *108*, 110, *122*, 143, *160*
Hirata, S., 30, *36*
Hirsh, E., 42, *44*, 176, *204*
Ho, M. -W., 83, *96*, *97*
Hochstein, S., 151, *159*
Hof, P. R., 20, *26*, 244, *253*
Hoffman, M. B., 129, 130, *133*
Hogan, R., 57, *59*
Holloway, R. L., 244, *252*, *253*
Honig, W. K., 212, 213, 218, *218*
Hook, M., 249, *252*

Hook-Costigan, M. A., 246, 247, *252*
Hopkins, W. D., 39, *44*, *45*, 66, *78*, 162, 164, *172*, 177, *205*, 244–246, 248–251, *252*, *253*
Horn, J. L., 128, *134*
Hostetter, A., 245, 246, 249, *253*
Hoyt, A. M., 226, *233*
Hubel, D., 272, *277*
Huber, L., 143, *159*
Huettel, S. A., 41, *45*
Hull, C. L., 138, *142*, 175, 200, 202, *204*
Hulme, C., 67, *77*
Hulse, S. H., 144, 158, *159*
Hunt, J. McV., 128, 130, *133*, *134*
Huntley-Fenner, G., 164, *171*
Hutchins, M., 11, *16*, *36*
Hyatt, C. W., 39, *44*

Ibanez, V., *277*
Ickes, W., 48, *59*
Idani, G., 29, 35, *36*
Iliff-Sizmore, S. A., 55, *58*
Inman, A., 68, *77*
Itani, J., 31, *36*
Itoh, T., *36*
Iversen, I. H., 39, *44*, 202, *204*
Izard, M. K., 249, *252*

James, W., 176, 186, 196, 202, 204, *204*
Jane Goodall Institute, 54, *59*
Jerison, H. J., 91, *95*, 275, *277*
Jessee, S., 273, 275, *277*
John, O. P., 48, *58*
Johnson, J. I., 275, *278*
Johnson, M., 37, *44*
Johnson, S., 175, *205*
Johnson-Pynn, J., 42, *44*, 176, *204*, *205*
Jones, F. W., 21, *26*
Jones, V. K., *95*

Kaas, J. H., 272, *277*
Kacelnik, A., 106, *107*, *108*
Kagan, J., 94, *95*
Kahneman, D., 121, *122*
Kamil, A. C., 144, 146, 158, *159*
Kanamori, M., 35, *36*
Kano, T., 35, *36*, 266, *267*
Kansky, M. T., 146, *160*
Kantor, J. R., 82, *95*
Kanwisher, N., 164, *170*
Karmiloff-Smith, A., 67, *77*
Karoma, N. J., 25, *26*
Kauffman, S. A., 81, 82, 85, *95*, *96*
Kaufman, E. L., 165, *172*

Kaufman, J., 41, *45*
Kaulen, R., *278*
Kawai, M., 269, *277*
Kawai, N., 39, *45*
Kearns, M. J., 41, *44, 45*
Keating, D. P., 127, *134*
Keele, S. W., 210, *219*
Keller, L., 177, *205*
Kellerman, H., 48, *58*
Kellert, S. R., 11, *15*
Kelley, J. W., 197, *205*
Kellogg, L. A., 226, *233*
Kellogg, W. N., 226, *233*
Kelly, S. T., 55, *58*
Kendrick, D. F., 68, *79*
Kennedy, J. S., 48, *59*
Kenrick, D. T., 49, *59*
Kheck, N. M., 20, *26*
Kimura, D., 243, *253*
King, J. E., 49, 51, 53–57, *59, 60*, 71, 76, 77
Klein, R. M., 177, *205*
Kluckhohn, C., 22, *26*
Knott, C. D., *28*
Ko, S. J., 49, *58*
Koehler, O., 161, *172*
Köhler, W., 37, 38, *45*, 47, *59*, 76, 77, 175–177, 196, 200, *205*
Kohts, N., 226, *233*
Konorski, J. A., 213, *218*
Koriat, A., 64, *77*
Kornell, N., 70, *79*
Kosko, B., 185, *205*
Koutsofios, E., 187, *204*
Kozma, A., 56, *59*
Kraemer, P. J., 68, *78*
Krause, M. A., 244, *253*
Krieger, M. J. B., 177, 198, *205*
Kroeber, A. L., 22, *26*
Kuhar, C., 244, *253*
Kunimatsu, Y, 21, *27*
Kuo, Z. Y., 82, *96*
Kuper, A., 22, *26*

Laing, E., 67, *77*
Laitman, J. T., 21, *26*
Lakoff, G., 185, *205*, 238, *242*
Laliberte, L., *205*
Land, P. W., 272, *277*
Landau, V. I., 54, 55, *59*
Langer, S. K., 37, *45*
Larsen, T., 68, *77*
Latash, L. P., 214, *218*
Latash, M. L., 214, *218*
Lawrence, A. B., 55, *60*
Lazareva, O. F., 113, *122*
Lazaroff, C., 25, *26*

Leavens, D. A., 244, 245, 247, 248, *252, 253*
Leavens, D. L., 245, *253*
Lee, D. N., 211–213, *218*
Lee, R. E., 8, *15*
Lehrman, D. H., 82, *94*
Leighty, K. A., 38, *45*
Lerner, J., 92, *96*
Lerner, R. M., 82, 94, *95, 96*
Lesner, S. A., 145, *160*
LeVay, S., 272, *277*
Levinson, P. K., 162, *172*
Lewin, K., 175, 196, *205*, 290
Lewin, R., 230, *234, 290*
Li, P. W., *28*
Lidz, C. S., 132, *133, 134*
Liebal, K., 264, *267*
Lieberman, D. E., 20, *25*
Lieberman, P., 21, *26*
Lifschutz, W., *278*
Lilienfeld, S. O., 54, *59*
Lima, M. P. D., *59*
Linden, E., 227, 230, *234*
Lindsay, P. H., 211, *218*
Lindsay, R. B., 184, *205*
Livet, M. O., 67, *77*
Loeches, A., 246, *252*
Lofting, H., 224, *233*
Lohmann, A., 162, *171*
Loomis, J. M., 38, *45*
Lord, M. W., 165, *172*
Lorenz, K., 175, 196, 198, 202, *205*
Loreto, V., 197, *204*
Lovejoy, A. O., 176, *205*
Lubinski, D., 127, *132*
Lucas, R. E., 55, *58*
Lukas, K. E., 11, *15*
Lyn, H., 258, 259, 261–264, *267*
Lyons-Weiler, M., 256, *267*

MacGregor, L., 244, *252*
Mach, E., 184, *205*
MacInnes, W. J., 177, *205*
MacKay, D. M., 109, *122*
Mackintosh, N. J., 202, *205*
MacMillan, N. A., 72, *77*
MacNeilage, P. F., 243, *253*
Magnusson, D., 94, *96*
Mancini, J., 67, *77*
Mandler, G., 165, *172*
Mannarelli, T., 49, *58*
Maple, T. L., 11, 13, *15, 16, 36*, 48, *58*
Marchant, L. F., 249, *253*
Margenau, H., 184, *205*
Marino, L., 54, *59*, 70, *78*, 244, *252*
Markowitz, Hal, 12
Marks, J., 19, *26*

Marler, P., 250, *252*
Marr, D., 116, *122*
Marr, M. J., 11, *15*, 99, 101, 103, 105, *107*
Marston, J. R., 185, *204*
Martin, P., 54, *59*
Masterson, R. B., 91, *96*
Matsuzawa, T., 17, *26*, 30, *36*, 39, *44*, *45*, 162–164, *170*, *172*, 202, *204*, 231, *233*
Mattson, M. E., 238, *242*
Maxwell, J. C., 176, 184, *205*
Maynard Smith, J., 83, *96*
Mayr, E., 17, *26*, 100, 101, 105, 107, *108*
Mazzoni, G., 70, *78*
McBeath, M. K., 143, *159*
McCrae, R. R., 51, 52, 56, *58*, *59*
McGee, K., *78*
McGonigle, B. O., 146, 148, *159*, *160*
McGrew, W. C., 17, 21, *26–28*, 35, *36*, 249, *253*, *267*
McKinney, M. L., 17, *27*
Mechner, F., 161, *172*
Medin, D. L., 145, *160*
Meehl, P. E., 53, *58*
Mellon, J., 11, *16*
Mendl, M. T., 55, *60*
Mendoza, S. P., 53, *58*
Menotti-Raymond, M., *27*
Menzel, C. R., 176, 187, 197, 202, 203, *205*, 215–217, *219*
Menzel, E. W., 176, 187, 202, 203, *205*
Merzenich, M., 272, *277*
Michel, G. F., 82, *96*
Miglino, O., 177, 198, *206*
Miles, H. L., 227, 230, *233*, *234*
Miller, D. J., 162, 163, *171*
Miller, G., 207, 208, 211, 213, 218, *219*
Miller, N. E., 137, *142*
Miller, R., 130, *133*
Milligan, N. B., 29, *36*
Mills, S. T., 227, 232, *233*
Minda, J. P., 119, *122*
Mishkin, M., 272, *277*
Mitchell, R. W., 270, *277*
Mohlberg, H., *251*
Molenaar, P. C. M., 82, *96*
Moll, H., 263, *267*
Moore, C. L., 82, *96*
Moore, J., 35, *36*
Morcillo, A., 246, *252*
Morford, M., 256, *266*
Morgan, C. L., 63, 66, 71, *78*, 82, *96*, 213, *219*
Morimura, N., 29, *36*
Morowitz, H., 162, *172*
Morris, D., 223, *234*
Morris, M. E., 49, *58*
Morris, R., 223, *234*
Mukobi, K. L., 162, *171*

Muller, R. U., 187, 198, *205*
Mural, R. J., *28*
Murphy, J., 97, 267, *277*
Murphy, W. J., *27*
Myers, D. G., 54, *59*
Myers, E. W., *28*
Mylander, C., 256, 262, *266*

Nagel, T., 48, 56, *59*
Nakano, Y., 21, *27*
Nakatsukasa, M., 21, *27*
Narens, L., 64, *78*
Nash, W. G., *27*
Neisser, U., 116, 118, *122*, 209, 212, *219*
Nelson, T. O., 64, 70, *78*
Neuringer, A., 103, 104, *108*
Neville, H. J., 272, *277*
Newell, A., 235, *242*
Newell, K. M., 82, *96*
Newton, I., 87, *96*
Nicolis, G., 81, 86, *96*
Niemann, J., 162, *171*
Nisbett, R. E., 49, *59*
Nishida, T., 17, *27*, *28*, *267*
Nishimura, T., 21, *27*
Noble, W., 288, *290*
Noll, J., 128, *134*
Norman, D. A., 211, *218*
Norman, W. T., 50, *59*
North, S. C., 187, *204*
Norton, B., *36*
Novak, M. A., 55, *59*
Núñez, R. E., 185, *205*

O'Brien, S. J., 19, *27*
Odbert, H. S., 50, *57*
Ogawa, H., 35, *36*
Ogihara, N., 21, *27*
Ohio State Research, 232, *234*
Ommaya, A. K., *277*
Ondrejko, M., 212, 218, *219*
O'Reilly, J., 103, *108*
Orlov, T., 151, *159*
*Oxford English Dictionary*, 101, *108*

Pach, J., 187, *205*
Page, S. L., *26*
Pajor, A. E., 29, *36*
Palkovich, A. M., 162, *172*
Palmer, A. R., 249, *253*
Paour, J. -L., 125, *133*
Park, H. S., *26*
Parker, S. T., 17, *27*, 93, *96*, 270, 272, *277*
Parkinson, J. K., 145, *160*

Partridge, T., 83, 92, *95*, *96*
Pascual-Leone, A., *277*
Pate, J. L., 63, 76, *78*, 91, 92, *96*, 139, 140, 142, *142*, 166, *172*, 197, *204*
Patterson, F. G., 227, 230, *234*
Pearson, K., 249, *253*
Pederson, A. K., 54, *59*
Pennington, B. F., 218, *219*
Pepperberg, I. M., 106, *108*, 144, *158*, 162–164, *172*, 227, 230, 231, *234*
Pepys, S., 226, *234*
Perusse, R., 162, 166, 168, *171*, *172*
Peterson, D., 17, *28*
Petros, T. V., 145, *160*
Pfungst, O., 225, *234*
Phelps, M. T., 158, *160*
Phillips, J. B., 49, *58*
Piaget, J., 129, 130, *134*, 146, *160*
Pilcher, D., 244, 251, *252*, *253*
Pinker, S., 224, 227, *234*
Platek, S. M., 70, *79*
Plooij, F. X., 266, *267*
Plutchik, R., 48, *58*
Pons, T. P., 272, *277*
Porter, C. A., *26*
Posner, M. I., 210, *219*
Poss, S. R., 244, 249, *253*
Poucet, B., 197, *205*
Povinelli, D., 245, *253*
Premack, D., 207, 209, *219*, 227, 229, *234*
Preuss, T. M., 20, *27*
Pribram, K., 207, *219*
Prigogine, I., *96*
Pringle, J. W. S., 83, 91, *96*
Pronko, N. H., 82, *96*
Pryor, K., 11, *15*, 103, *108*
Pubols, B. H., 275, *277*, *278*
Pubols, L. M., 275, *277*
Putney, R. T., 207, 212, *219*
Pylyshyn, Z. W., 165, *172*, 236, *242*

Quigley, K. S., 148, *159*

Rachlin, H., 104, *108*
Raghanti, M. A., 152, 153, *160*
Raleigh, M. J., 243, 251, *253*
Rand, Y., 129, 130, *133*
Rapp, P. R., 146, 148, *160*
Rashotte, M. E., 202, *205*
Razran, G., 91, *96*
Reder, L. M., 64, 76, *78*
Redfield, R., 87, *96*
Redford, J. S., 64, 77, *78*
Reese, T. W., 165, *172*
Reynolds, V., *28*, *267*

Rice, C. P., 10, *15*
Richardson, W. K., 39, *45*, 66, *78*, 197, *204*
Riepe, M. W., 41, *44*
Riesen, A. H., 8, *15*
Rijksen, H. D., 31, *36*
Riley, D. A., 113, *122*
Rilling, J., 244, *252*
Rilling, M., 161, *172*
Ristau, C. A., 144, *160*
Riva, G., 38, *45*
Roberts, R. J., 212, 218, *219*
Roberts, W. A., 68, *78*, 158, *160*
Roche, B., 104, *107*
Rogers, C. M., 202, *205*
Rogers, L. J., 246, 247, 250, *251*–*253*, 271, *276*, *277*
Roitman, J., 162, *171*
Romanes, G. J., 37, *45*
Rorty, R., 37, *45*
Rosch, E., 210, *219*
Rosenblatt, J. S., 82, 84, *94*, *96*
Ross, L., 49, *59*
Rowan, A., *36*
Rumbaugh, D. M., 8, 10, 11, 13, *15*, 20, 21, *26*, *27*, 30, *36*, 38, 39, *45*, *46*, *59*, 63, 66, 76, *78*, *79*, 91, 92, *96*, *97*, 100, 104–106, *108*, 110–115, 121, *122*, *123*, 140, 142, *142*, 143, 158, *160*, 162–167, *170*, *172*, 177, 197, *204*, *205*, 207, *219*, 227, *234*, 257, *267*, 270, 271, 273, 275, *277*, 280, 290, *290*, *291*
Russell, T. M., *95*
Russon, A. E., 93, *96*, 270, 272, *276*
Ryder, O. A., 256, *267*
Rylands, A., 25, *27*

Sadato, N., 272, *277*
Sahlins, M., 23, *27*
Sameroff, A. J., 94, *96*
Sanchez, I. C., 197, *205*
Sanderson, C. A., 55, *58*
Sands, S. F., 68, *79*
Sandstrom, N. J., 41, *45*
Santiago, H. C., 68, *79*
Saucier, G., 51, *59*
Saunders, P. T., 83, *96*, *97*
Savage, C., 106, *108*
Savage-Rumbaugh, E. S., 13, *15*, 21, *27*, 30, *36*, 39, *45*, 63, 66, *78*, 92, *97*, 162, 164, 166, *172*, 176, 177, 187, 197, 203, *204*–*205*, 207, 209, 215, *219*, 224, 227, 229, 230, *234*, 257–259, 261, 266, *266*–*267*, 270, 272, *277*, 290, *290*–*291*
Schank, R., 210, *219*
Schapiro, S. J., 249, *252*
Schick, K. D., 270, *277*, 279, 290, *291*

Schilhab, T. S. S., 48, *59*
Schleicher, A., *251*
Schneider, D. M., 23, *27*
Schneider, H., *26*
Schneider, W., 74, 75, *78*
Schneirla, T. C., 83–85, *96*, *97*
Schull, J., 64, *78*
Schunn, C. D., 64, 76, *78*
Schwartz, B. L., 64, 70, *78*, *79*
Searle, J. R., 260, 261, 265, *267*
Sebeok, T. A., 233, *234*
Seidenstein, S., 275, *278*
Sevcik, R. A., 92, *97*, 257, *267*, 290, *277*, *291*
Seyfarth, R. M., 144, *159*
Shanker, S. G., 257, *267*
Shebo, B. J., 165, *172*
Sheets-Johnstone, M., 18, *27*
Sheldrake, R., 81, *97*
Shepherdson, D. J., 11, *16*, 29, *36*
Sherman, W. R., 40, *45*
Sherwood, C. C., 244, *253*
Shettleworth, S. J., 68, 70, 77, *78*, 105, *108*
Shields, W. E., 64, *78–79*
Shiffrin, R. M., 74, 75, *78*
Shore, B., 22, *27*
Shore, D. I., 177, 201, *205*
Shoshani, J., *26*
Shouse, B., 19, *27*
Showalter, K., 187, *206*
Shreyer, T. A., 148, *159*
Shumaker, R. W., 30, *36*, 162, *172*
Sidman, M., 100, 106, *108*
Siemann, M., 162, 164, *173*
Silk, J. B., 256, *266*
Simões, A., *59*
Simon, H. A., 235, *242*
Simon, T., 125, *132*
Simons, D. J., 272, *277*
Singer, P., 17, *25*
Singleton, I., *28*
Skeen, L. C., 91, *96*
Skiena, S. S., 186, *205*
Skinner, B. F., 93, *97*, 103, 104, *108*, 109, 121, *122*, 140, *142*, 227, *234*
Smith, H. L., 55, *58*
Smith, J. D., 64, 66, 68, 69, 71, 73, 76, 77–79, 119, *122*
Smith, J. H., 260, 261, *267*
Son, L. K., 70, *79*, 151, *160*
Spelke, E. S., 162, 164, *170*, *171*
Spence, K. W., 113, *122*
Sperry, R. M., 86, *97*
Spitzer, M., 41, *44*
Staddon, J. E. R., 99, *108*, 146, 147, *160*
Stanford, L., 177, *205*
Stanley, J. C., 127, *134*
Stanyon, R., *27*

Stauffer, R. L., 256, *267*
Stead, M., 187, *205*
Stebbins, G. L., 83, *97*
Stein, G. J., 18, *27*
Stein, J. L., 49, *58*
Steinbock, O., 187, *206*
Steklis, H. D., 243, 251, *253*
Stendorf, F., *59*
Stengers, I., *96*
Sternberg, R. J., 128, 132, *134*
Stevenson-Hinde, J., 48, 53, *59*
Stewart, I., 94, *97*, 101, *108*
Stillwell-Barnes, R., 48, *59*
Stoinski, T. S., 11, *16*, *36*, 244, *253*
Stokoe, W., 250, *251*
Stones, M. J., 56, *59*
Stroop, J. R., 167, *172*
Strote, J., *78*
Sugiyama, Y., 28, *267*
Suh, E. M., 55, *58*
Sulis, W., 82, *97*
Sulkowski, G. M., 162, *172*
Sullivan, B. T., 41, *45*
Sutton, G. G., *28*
Sutton, J., 70, *78*
Swartz, K. B., 146, 148, 151, 154, *159*, *160*
Switzky, H. N., 130, *133*, *134*
Szathmáry, E., 83, *96*

Taglialatela, J. P., 21, *27*, 224, *234*
Tailby, W., 100, *108*
Tarr, M. J., 41, *44*, *45*
Taub, E., *277*
Taylor, T. D., *26*, *36*
Taylor, T. J., 257, *267*
Terrace, H. S., 146, 148, 151, 154, 157, *159*, *160*, 162, *171*, 227, *234*
Terrell, D. F., 162, 165, *172*
Theall, L. A., 245, *253*
Thinus-Blanc, C., 197, *205*
Thomas, R. K., 110–112, *122*, 162, 165, *172*
Thompson, C. R., 240, *242*
Thompson, R., 145, *159*
Thompson, R. K. R., 212, 213, 218, *218*
Thorndike, E. L., 38, *45*, 63, *79*
Tobach, E., 82–85, *94*, *97*
Tolman, E. C., 63, *79*, 138, 140, 142, *142*, 175, 197, 203, *206*
Tomasello, M., 22, 23, *25*, *27*, 244, 252, 263, 266, *267*
Tomczak, R., 41, *44*
Tomonaga, M., 39, *45*, 162, 163, *172*
Tóth, A., 187, *206*
Toth, N., 277, 279, 290, *291*
Toyoda, A., *36*
Trabasso, T., 146, *159*

Treichler, F. R., 145, 149–152, 153, 155, 157, 158, *160*
Treisman, A., 116, *123*
Trevarthen, C., 262, *267*
Trick, L. M., 165, *172*
Triesch, J., 41, *45*
Trofimova, I., 82, *97*
Tsao, F., 162, *171*
Turner, J. H., 23, *27*
Turvey, M. T., 211, 212, 214, *218, 219*
Tutin, C. E. G., 35, *36*
Tuttle, R. H., 18–23, 25, *27, 28*
Tversky, A., 121, *122*
Tyler, E. B., 23, *28*
Tzuriel, D., 130, 132, *133, 134*

Umiker-Sebeok, D. J., 233, *234*
Underwood, B. J., 139, *142*
Uyeyama, R. K., 228, *233*
Uylings, H. B., *251*
Uzgiris, I. C., 130, *134*

Vallortigara, G., 250, *251*
Van Cantfort, T. E., 224, *233*, 259, *266*
van Hoof, J. A. R. A. M., 24n, 25, *28*
van Lawick-Goodall, J., 17, *28*
van Schaik, C. P., 17, 21, *28*
Van Tilburg, D., 149–152, 153, 155, 157, 158, *160*
Vauclair, J., 39, *44*, 272, *276*
Vaught, S., 130, *133*
Venter, J. C., 19, *28*
Vernier, J., 19, *26*
Vickery, J. D., 162, *172*
Volkman, J., 165, *172*
von Fersen, L., 146, 147, *160*
Von Hofsten, C., 212, *219*
von Noorden, G. K., 272, *278*
Vygotsky, L. S., 129, *134*

Wachs, T. D., 130, *133*
Waddington, C. H., 271, *278*
Walker, R., 177, 198, *206*
Wallman, J., 227, *234*
Walter, A., 256, *267*
Warnick, J. E., *95*
Warnick, R., *95*
Warren, J. M., 249, *253*
Warren, W. H., 41, *44, 45*
Washburn, D. A., 13, *15*, 21, 27, 30, *36*, 39, 42, *44–46*, 59, 63, 64, 66, 77–79, 92, *96*, 100, *108*, 110, 112–115, 121, *122, 123*, 143, *160*, 162, 164, 166, 167, *172*, 177, *205, 206*, 280, *290*
Wasserman, E. A., 39, *46*, 113, *122*, 144, *160*, 211–213, *219*
Watanabe, H., 26, *36*
Weary, M. D., 29, *36*
Weaver, S. J., 130, *133*
*Webster's Encyclopedic Unabridged Dictionary of the English Language*, 237, *242*
Weir, A., 106, *108*
Weiss, A., 51, 57, *59, 60*
Weiss, E., 83, *95*
Welker, W. I., 275, *278*
Wemelsfelder, F., 55, *60*
Wesley, M. J., 245, 248–249, *252–253*
Wetsel, W., 145, *160*
Whalen, J., 164, *173*
White, L. A., 23, *28*
White, O., 19, *26*
Whiten, A., 17, 21, 22, *25, 28*, 38, *46*, 110, *123*, 250, *253*, 256, 260, 266, *267*, 270, *278*
Wienberg, J., *27*
Wiener, N., 214, *219*
Wiesel, T. N., 272, *277*
Wilcox, S. E., 250, *251*
Wilkins, V. M., 70, *79*
Williams, S. L., *97*, 257, 266, *267, 277*
Wind, J., 21, *28*
Wise, S. M., 13, *16*
Wolfram, S., 89, *97*
Woodcock, R. W., 129, *134*
Woodruff, G., 207, 209, *219*
Wrangham, R. W., 17, *28*
Wright, A. A., 68, *79*
Wunderlich, A. P., 41, *44*
Wynne, C. D. L., 105, *108*, 146, 147, *160*

Xia, L., 162, 164, *173*

Yada, T., *26*
Yakovlev, V., 151, *159*
Yarbrough, G. L., *95*
Yaspo, M. L., *36*
Yerkes, A. W., 9, *16*
Yerkes, R. M., 9, *16*, 47, *60*, 226, *234*
Young, M. E., 39, *46*, 113, *122*

Zajonc, R. B., 210, *219*
Zentall, T. R., 70, *79*, 106, *107*
Zihlman, A., 256, *267*
Zilles, K., *251*
Zimmerman, T., 40, *46*
Zohary, E., 151, *159*
Zunz, M., 48, *59*

# Subject Index

Abreu, Rosalia, 9
Accumulator, 164
Achievement tests, 129
Action theory, 208–212
Aesop, 224
Affirmative–goal relationship, 261–262
Affirmative use, 262, 265
Africa, 35
African grey parrot (Alex), 106, 227, 230
Age effects, 52
Agreeableness, 50–52, 56
Ambiguity, 75, 236–237
American Psychological Association (APA), 232, 294
American Sign Language, 230, 270
Anagenesis, 82–83
Analysis of variance (ANOVA), 192–193
*Animal Cognition* (Clive Wynne), 105
*Animal Cognition and Sequential Behavior* (S. B. Fountain, M. D. Bunsey, J. H. Danks, & M. K. McBeath), 143–144
*Animal Cognition in Nature* (R. P. Balda, I. M. Pepperberg, & A. C. Kamil), 144
Animal language, human vs., 239–242
Animal language research, 223–233
　beginnings-of-empiricism stage of, 224–226
　experimentation stage of, 227–228
　financial support for, 231–233
　future stage of, 228–233
　myths/fables stage of, 223–224
*Animal Mind* (C. L. Morgan), 213
Animal training, 11
Animal working memory, 212
ANOVA. *See* Analysis of variance
"Anthropomorphic halo," 52–53
Anthropomorphism, 48
Anticipation, 214–215
Anticipatory memory, 213
APA. *See* American Psychological Association
Apes
　capabilities of, 13
　conservation of, 25
Arabic numeral system, 91, 166–168
Arbitrariness, 263–264
Aronson, L. R., 82
Artificial language, 229
Ask Jeeves Web site, 240
Assessment, 125–132
　of human ability, 126–127
　of intelligence, 125
　and prediction of success, 125
　transactional perspective on, 127–132

Association, 138, 141, 147–149
Association learning, 203
Associative drift, 114
Associative learning, 141
Atomicity, 236, 239
Attractors, 87

Baboons, 162, 223–224
Balking, 280
Bartlett, Frederick, 209–210
Behavior, 63–77
　cognition vs., 202–203
　and cross-task analyses of cognitive processing, 68–72
　and grammar, 73–74
　integrative comparative perspective on, 75–77
　levels of, 74–75, 83–85
　problem of interpreting, 63
　and signal detection theory, 72–73
Behavior analysis, 106–107
Behavior dynamics, 102–105
Behaviorism, 100
Behavior prediction, personality and, 52–54
Beran, Michael, 275
Bernstein, Nickolai, 212–215
Bioko red colobus, 24
*BioScience* (journal), 8
Biosocial level, 83–84
Biotaxis level, 83
Black-faced lion tamarin, 24
Body schema, 210
Boesch, C., 23
Bonobo(s)
　Kanzi, 13, 178, 183, 191, 229, 230, 257–259, 262, 264, 279–290
　Panbanisha, 229, 230, 257–259, 261–265
　path selection by, 178, 191
　protolanguage acquisition by, 93
　speech of, 21
Book of Genesis, 223
Boredom, 55
Boule, Pierre, 23
Bourne, Geoffrey, 7
Boysen, Sarah, 232
Brachiating gibbons, 212
Brain
　as rational agent, 105–107
　and sensory arrays, 120
　size of, 275
Broca's area, 243, 244
Brown, R., 259–260

309

Brown spider monkey, 24
Bush, George W., 238, 241
Byrne, D., 88, 92–93

Capuchin, 24
Categorical lists, 145
Categorization, 118
Causality, 88
Cave Automatic Virtual Environment (CAVE) system, 40, 43
Central tendency effects, 210
Change blindness, 41
Chickens, 271
Chimpanzee(s), 9
  Ai, 163
  Austin, 164, 183, 191, 215–216
  Carl, 191
  Chim, 47
  Clint, 191
  counting behavior in, 162–165
  as cultural beings, 22
  DNA of humans and, 30
  GARI study of, 29–35
  genomic closeness of humans and, 19–21
  Gua, 226
  husbandry of, 30–31
  Joni, 226
  Lana, 13, 38–39, 141, 163–165, 168, 183, 191, 227, 240
  Loi, 32
  manual gestural/intentional vocalization in, 244–245
  Mercury, 165, 178, 183, 191
  Misaki, 32
  Mizuki, 32
  nonvisible-set numerousness judgments by, 168–170
  Panzee (Panpanzee), 47, 178, 188, 191, 216–217, 257–265
  path selection by, 178, 191
  Patrick, 191
  prospective control in, 215–217
  Sarah, 232
  Sheba, 163
  Sherman, 165, 168, 183, 191
  speech of, 21
  Tsubaki, 32
  Viki, 226
  visible-set numerousness judgments by, 165–166
  Washoe, 227, 232, 259, 270
  Winston, 191
  Zamba, 32
ChimpanZoo, 54, 56
Clade, 256–257
Cladistic analysis, 256–257

Classical conditioning, 117
Clever Hans (horse), 163, 224–225
Climbing towers, 33
Cognition
  animal potential for, 11
  behavior vs., 202–203
  comparative, 143–144
  cross-task analyses of, 68–72
  intelligence vs., 127–130
  and Morgan's canon, 66–68
*Cognition, Evolution, and Behavior* (Sara Shettleworth), 105
Cognitive maps, 138, 203
Coherence, 88
Colobus, 24
Comparative cognition, 143–144
Comparative psychology
  enlightenment view vs., 37–38
  and technology, 38–39
Complex behaviors
  as emergents, 92–93
  and learning theory, 138
Complex cognitve capacities, 273–275
Complex expression, 236
Complexity
  and brain as rational agent, 105–107
  levels of, 83–85
  and path selection, 197–199
  of syntax, 240–242
Complex learning, 91
Complex multiproblem learning, 90
Complex systems, stability in, 101–102
Compositionality, 236
Comprehension, 226
Computational irreducibility principle, 89–90
Computerized Test System, 66
Computers, 38, 229–230
Concurrent conditional training, 146–150
Concurrent discrimination, 145
Conditioned responding, 114
Conditioned response (CR), 117
Conditioned stimulus (CS), 117
Confidence, 53
Congo, 35, 51, 52
Conkouati Douli National Park (Congo), 51, 52
Conservation, 8, 23–25
Constitutive hierarchies, 100
Constructive enumeration, 164
Contingency–species interaction, 107
Control
  personal, 55
  prospective. *See* Prospective control
  relational, 106
Controlled observation, 225
Convergent validity, 53

Cook, Robert, 106
*The Co-ordination and Regulation of Movements* (N. Bernstein), 212
Correlation, 88
Counting behavior, 161–170
  in animals, 163–165
  chimpanzee numerousness judgments, 165–166
  nonvisible-set numerousness judgments, 168–170
  rhesus monkey numerousness judgments, 166–168
  visible-set numerousness judgments, 165–168
CR (conditioned response), 117
Creative adaptation, 293
Creative synthesis, 141
Creativity, 13, 271
Cross River gorilla, 24
CS (conditioned stimulus), 117
Cuba, 9
Culture, 21–23, 262

Darwin, C., 18, 269
DataGlove, 40–41
Davis, L. H., 208–211
Deacon, T. W., 91
Delacour's langur, 24
Dependability, 50, 51, 56
Dependence, 265
Descartes, René, 203–204
*The Descent of Man and Selection in Relation to Sex* (C. Darwin), 18
Determinism, 109
Deterministic causal model, 88
Differential brain function, 105
Direct perception theory, 117–118
Discriminant validity, 53
Discrimination reversal, 140
Discriminative cue, 145
Distance, 185
Distress, 55
DNA, 30
Dobkin, D. P., 201
Dolphins, 227–229
Dominance, 50, 52, 57
Douc, 24
"Do Willful Apes Know What They Are Aiming At?" (R. T. Putney), 207
Dynamic assessment, 131–132
Dynamical models, 82, 88

Early mother–infant relationship, 32
Eastern Black crested gibbon, 24
Eastern gorilla, 24
Ecological psychologists, 212

Ecological self, 209, 212
Egypt, ancient, 223–224
Eighth Kent Psychological Forum, 143
Einstein, Albert, 87
Embryonic posture, 271
*Emergence* (journal), 87
*Emergence* (S. Johnson), 175
Emergence/emergents, 85–90, 99–107, 109–122, 141–142, 271
  and behavior dynamics, 102–105
  behaviors not qualifying as, 111–115
  and brain function, 105–107
  case for, 115–119
  characteristics of, 112–113
  complex behaviors as, 92–93
  as concept, 143
  criteria for, 88
  definition of, 87
  dynamics of, 100–102
  of long-range correlations, 102
  and meaningful failures to learn, 113–115
  operants/respondents vs., 112–113
  and organized monkey memory, 158
  perception of, 116–118
  and stimulus control/reinforcer control/brain control, 119–122
  R. K. Thomas's distinctions in, 110–111
Emergent behaviors, 76, 110
Emergent event, 86
Emergent factor, 86
Emergentism, 110–111
Emergent properties, 86
Emotionality, 50, 52, 56
Empathic ability, 48
Endangered species
  of primates, 24, 25
  scientific management of, 7
Enlightenment philosophy, 37
Entity-demonstrative, 264
Entity-location, 264
Environmental enrichment, 12–13
Environmental stimuli, 272
Epigenesis, 271–272
Equability, 53
Equivalence relations, 100
"Equivalence Relations and the Reinforcement Contingency" (M. Sidman), 106
Ethological laboratory, 33
Euclidean distance, 185
Evolution, 103
*The Evolution of Cognition* (C. Heyes & L. Huber), 143
Excitability, 53
Exemplar-based theories, 118
Expectancies, 138
Experience, 38

Experimental methods, 38
Extrapolation to the future, 215
Extraversion, 50–52, 54, 56

Fables of animal language, 224
Face-processing tasks, 67
Facial expressions, 18, 246, 247
Factor constancy, 50–52
Feedback, 214
Feeding behavior, 84–85
Ferster, C. B., 162
Festschrift, xiii, 294
Findley, J. D., 162
*Firing Line* (television show), 109
Five-factor model of personality, 48, 50
Five-item task, 147, 149
Five-term task, 146
Fixed consecutive number, 161
Flexibility, 211–212
Flexible prospection, 216
Flexible subroutines, 217
Flight simulators, 41
Flow of causal information, 88–89
Fouts, Deborah and Roger, 227, 232
Free-form context, 38
Free will, 109
"Functional Analysis of Emergent Verbal Classes" (M. Sidman), 106
Functional stimulus, 137
Fundamental attribution error, 49
Furness, W. H., 225

$g$ (general component of intelligence), 129
Galago, 24
Gannets, 212
Gardner, Beatrix and Allen, 227, 232
GARI. *See* Great Ape Research Institute
Garner, Richard Lynch, 224
General component ($g$) of intelligence, 129
Genesis, Book of, 223
Genomic closeness, 19–21
Georgia State University Language Research Center, 66, 93, 168, 257, 270, 297
Gestalt, 199
Gestures, 226, 230, 231
  chimpanzee, 244–248
  combinations of, 258
  integration of, 265–266
Gibbons, 21, 24
Gibson, James J., 116–118
Global level, 88
Golden-headed langur, 24
Goldstein, J., 90, 94
Gombe Stream Reserve, 270
Goodwin, Brian, 82
Gorilla(s)
  Albert, 13, 47

Congo, 9
  endangered species of, 24
  Gargantua, 226
  Koko, 227, 230, 232
  Toto, 226
Gray-shanked douc, 24
Great Ape Research Institute (GARI), 29–35
  activities of, 30
  aims of, 29
  basic concept/policy of, 30–31
  daily routine at, 34–35
  environment of, 29
  facility of, 32–34
  future plans for, 35–36
  individual chimpanzees at, 32
  research/applications at, 31–32
Great Ape Trust, 297
Greater bamboo lemur, 24
Group life, 34
Guenon, 24
Guizhou snub-nosed monkey, 24
Gymnasiums, 33

Habit, 138
Habitat Ecologique et Liberté des Primates (HELP), 51
Habitats, 35
Haeckel, Ernst, 18
Hainan black-crested gibbon, 24
Handedness, 243, 246–251
Happiness, 54–57
Harbor seals, 106
Hardware, husbandry, 30
Harlow, Harry, 10, 63, 140, 293
Hayashibara Biochemical Laboratories, Inc., 29
Hayes, Keith and Cathy, 226
Haywood, Carl, 297
Head, Henry, 210
Head-mounted VR display, 40
Hebb–Williams intelligence test, 182, 201
Hediger, Heini, 11, 14, 185
HELP (Habitat Ecologique et Liberté des Primates), 51
Heritability, 57
Herman, Louis, 227
Hershberger, W. A., 120, 121
Hierarchies, 82–83, 100
Holistic approach, 88–89
Home rearing, 225
*Homo sapiens. See* Humans
Horton Plains slender loris, 24
Hoyt, Maria, 226
Hull, C. L., 138
Hull's Theorem 61, 202
Hulse, S. H., 144
Human language, 239–242

SUBJECT INDEX 313

Human memory research, 139
Human mental ability
　individual differences in, 127
　intelligence vs. cognitive processes, 127
　and motivation, 127–128
　multidetermined, 127
　multifaceted, 127
　transactional perspective on, 127–132
Humans
　and apes, 17–19
　behavioral variation in, 103–104
　DNA of chimpanzees and, 30
　genomic closeness of chimpanzees and, 19–21
　gripping ability of, 20
　path selection by, 191
　speech of, 20–21
　uniqueness of, 37
　vocal capacity of, 274
Human state of mind, 262–263
Human subjective judgments, 48–49
Human verbal behavior, 107
Husbandry, 30–32

Ideation, 9
Immune response, 54
Inauguration addresses, 241
Information pickup, 118
Informativeness, 261
Insight, 203
Instinctual drift, 114
*Integrative Activity of the Brain* (J. A. Konorski), 213
Integrative comparative perspective, 75–77
Intelligence
　cognitive processes vs., 127–130
　models of, 125
Intelligence testing, 125
Intention, 208–211
Intentional action, 260–261
Intention-in-action, 261, 265
Intentions, 207–208
Invariants, 117
IQ testing, 125

James, William, 176, 186, 204
Japan, 39
*Journal of the Experimental Analysis of Behavior*, 106
Joysticks, 39, 64–65, 178
Jump metric, 185–186

Kahneman, Daniel, 121
Kantor, J. R., 82
Kauffman, Stuart, 82, 85
Kellogg, Winthrop and Luella, 226
King, J. E., 71

Kluckhohn, C., 22–23
Knapping stone, 279–290
　experimental design for, 280–281
　Kanzi's mastery of, 288–290
　overcoming initial obstacles in, 282–283
　redirecting Kanzi to, 284–288
　throwing technique for, 284
Koehler, O., 161
Köhler, Wolfgang, 47, 175
Kohts, Nadesh, 226
Konorski, J. A., 213
Kroeber, A. L., 22–23

Language, 235–242
　cognitive components of, 230
　as complex behavior, 93
　connecting natural animal communication to human-designed, 230
　construction of, 274
　context independence of, 236–237
　as emergent property, 86, 87
　human/animal, 239–242
　neurological perspective on. *See* Neurological perspective on language
　production of new, 238–239
　systematicity/atomicity of, 238–239
Language learning studies, 140, 141. *See also* Animal language research
Language Research Center. *See* Georgia State University Language Research Center
Langurs, 24
Latent learning, 106
Laterality, 245–248, 250
Lateralization, 272
Learned behavior, 109–110
*Learning as Self-Organization* (K. Pribram and J. King), 143
Learning sets, 99, 105–106, 140, 297
Learning theory, 137–139
Lee, David, 212, 213
Lemur, 24
*Levels of Integration in Biological and Social Systems* (R. Redfield), 87
Levinson, P. K., 162
Lexigrams, 38–39, 141, 227, 229, 231, 257, 258
Lindsay, R. B., 184
Linking lists, 142–158
List positions, 151–152
Living standards for zoo animals, 12–13
Locke, John, 138, 141
Logic, 236
Logical positivists, 238
Logic modes, 129
Long-term memory, 145
Long-term reward contingencies, 155
Lorenz, K., 198

Loris, 24
Luria, Alexander, 213–214

Macaques, 145, 149, 269. *See also* Rhesus monkeys
Machine translation, 237
MacKay, Donald M., 109
Mackintosh, N. J., 202–203
Macro level, 88
*Man and Animal in the Zoo* (Heini Hediger), 11
Mangabey, 24
Margenau, H., 184
Markowitz, Hal, 12
Mason, William A., 297
Material reductionism, 110–111
Matsuzawa, Tetsuro, 39, 231
Maxwell, J. C., 176
Mayr, Ernst, 100
McGrew, W. C., 17
Meaningful failures to learn, 113–115
Mechner, F., 161
Mediational learning, 141–142
Memory
 and control, 212–213
 monkey, 144–145
 for number values, 165
 for serial information, 146–147
Memory monitoring, 68–72
Mental constructional skills, 273–275
*The Mentality of Apes* (Wolfgang Köhler), 47, 175
Menzel, C. R., 216–217
Menzel, Emil, 178
Menzel, Emilie, 200
Metacognition, 129, 130
 and Morgan's canon, 66–68
 perspectives in comparative studies of, 64–66
Micro level, 88
Miles, Lyn, 227, 232
Mill, James, 138, 141
Mill, John Stuart, 141
Miller, N. E., 137
Miller's grizzled surili, 24
Mirror self-recognition, 270, 271
Miss Waldron's red colobus, 24
Model-rival approach, 231
Models, 19
Monkey memory, 144–145
 and emergents, 158
 and knowledge of list positions, 151–152
 and linking lists, 149–150, 152–158
 and organization vs. association, 147–149
 organizing factors in, 144–145
 for serial information, 146–147

Monkeys
 cognitive processing in, 68–70, 72–73
 endangered species of, 24
 list-making by, 148
Morgan, Lloyd, 71, 82, 161, 213
Morgan's canon, 63, 66, 71
Motivation, 127–128, 130–132
Motor constructional ability, 274
Motor program, 214
Mt. Rungwe galago, 24
Mountain gorilla, 24
Multimodal ape, 231
Muriqui, 24
Myths and fables of animal language, 223–224

Nagel, Thomas, 48, 56
National Aeronautics and Space Administration (NASA), 41, 297
National Science Foundation, 297
Natuna banded leaf monkey, 24
Nature, nurture vs., 201–202
*Natürlich Schöpfungsgeschichte* (Ernst Haeckel), 18
Neanderthals, 21
Negative feedback, 101–102
Negative reinforcement, 103
Negative reinforcers, 211
Nelson, T. O., 70
Nervous system complexity, 91
Neuringer, Allen, 103, 104
Neurological perspective on language, 243–251
 asymmetries, 244
 chimpanzee manual gestures/intentional vocalization, 244–245
 and handedness in apes, 247, 249
 laterality/gesture/facial expression, 245–248
 theoretical implications/future directions of, 249–251
*The New Behaviorism* (John Staddon), 99
New Caledonian crow, 106
Newton, Isaac, 87, 184, 203, 204
Newtonian space, 184
Newton's laws of gravity, 89
Nilous, Horapollo, 223
Nim Chimsky (ape), 232
Nonlinear dynamical complexity, 101
Nonlinear dynamic systems, 82
Northern muriqui, 24
"Nothing but-tery," 109
Numerical symbols, 114–115
Numerosity judgments, 39, 162
 of nonvisible sets, 168–170
 of visible sets, 165–168
Nurture, nature vs., 201–202
Nut cracking, 34

Observation, controlled, 225
Observational learning, 231
Observation booths, 33
Ohio State University, 232
*On Dexterity and Its Development* (N. Bernstein), 214
Open enclosure, 33
Openness, 50, 52
Operants, 109, 112–113, 227
Orangutan(s)
 Chantek, 227, 230, 232
 endangered species of, 24
 gripping ability of, 20
 Julius, 9
 language in, 225
 Madu, 178, 183, 191
 path selection by, 178, 191
 sexual maturity in, 271
 speech of, 21
Ordinal position, 151–153
Organization, association vs., 147–149
Osten, Herr von, 224
Ostensive, 88

Pagai pig-tailed snub-nosed monkey, 24
*Pan troglodytes*. See Chimpanzee(s)
Parrot (Alex), 106, 227, 230
Parsimony principle, 66
Parsing, 259–260
Path selection, 41–43
Path selection research, 175–204
 analysis of test patterns/travel paths in, 183–188
 apparatus used in, 178
 and behavior vs. cognition, 202–203
 discussion about, 195–203
 and gestalts vs. stimuli, 199–200
 and nature vs. nurture, 201–202
 and Newton vs. Descartes, 203–204
 and physics/aesthetics/intelligence, 200–201
 preliminary training used in, 182
 procedure used in, 182–183
 results of, 188–195
 and simplicity/complexity, 197–199
 stimuli used in, 178–180
 subjects in, 178
 test program used in, 180–182
Pattern recognition, 118
Patterson, Penny, 227, 232
Pepperberg, Irene, 106, 227, 231
Perrier's sifaka, 24
Personal control, 55
Personality, 47–57
 animal, 13
 and anthropomorphism, 48
 assessment difficulties with, 48–49
 and behavior prediction, 52–54
 in chimpanzees, 49–50
 and factor constancy, 50–52
 five-factor model of, 48, 50
 future studies of, 57
 and happiness, 54–57
 interrater reliabilities of, 49
Pfungst, Oskar, 225
Phase transition, 86, 87
Phonemes, 229
Phylogenetic continuity, 37
Piaget, Jean, 130
Piagetian framework, 273
Pigeons, 106, 147, 148, 158, 161, 162
*Planet of the Apes* (Pierre Boule), 23
Plans, intentions as, 207–208
Plants, 33
Planum temporale (PT), 244
Plastic symbols, 229
Porpoises, 103
Positive feedback, 101–102
Positive reinforcers, 211
Predictability, 90
Preferred orders, 261, 263–264
Premack, Ann and David, 227, 229, 232
Premise pairs, 146
"The Primary Role of Primates in Comparative Psychology" (William A. Mason), 297
Primate field studies, 269
Primate Laboratory at Kent State University, 144–145
Primate Research Institute (Inuyama, Japan), 39
*Principia* (Isaac Newton), 184
Principle of computational irreducibility, 89–90
Pringle, J. W. S., 83
Prior intent, 260–261, 265
Probabilistic causal model, 88
Probing tools, 270
Prospective control, 207–218
 action theory of, 208–212
 in chimpanzees, 215–217
 and intentions as plans, 207–208
 levels of, 212–213
 Nickolai Bernstein on, 213–215
Prostheses, vocal, 229
Protoculture, 270
Protolanguage acquisition, 93
Prototype-based theory, 119
Prototypes, 210
Pryor, Karen, 103
Psychological space, 184
"Psychologist's fallacy," 186
Psychology laboratory, 34
Psychomotor development, 130
Psychopathy, 54
Psychosocial level, 84

Psychotaxis level, 84
PT (planum temporale), 244
Publicity, 232–233
Punishment, 103
Pylyshyn, Z. W., 236
Pythagorean formula, 184–185

Quality of life, 35
Quinta Palatino, 9

Rachlin, H., 104
Racism, 18
Radical behaviorism, 121, 144
Radical novelty, 88
Rational behaviorism, 99, 121–122, 271, 273, 275, 293
Rats, 103, 105–106, 161, 163
"Recognition–categorization by specifics" theory, 118
Reconstructions, 19
Reduction, 111–112
Reductionistic science, 89
Reinforcement, 102, 103, 293
Reinforcers, 119–122, 211, 294
Relational control, 106
*Relational Frame Theory* (S. Hayes, D. Barnes-Holmes, and B. Roche), 104, 106
Relational learning, 113, 114, 293
Relational thinking, 273, 275
Relative numerousness, 165
Republic of the Congo, 51, 52
Respondents, 109, 112–113
"Respondents, Operants, and Emergents" (D. Rumbaugh, D. Washburn, and W. Hillix), 100
Response alternators, 139
Response class, 104
Response differentiation, 103
Responsive enumeration, 164
Restrospective working memory, 212–213
Retained information, 145
Reward contingencies, 155
Reynolds, Richard, 7
Rhesus monkeys, 66, 148
    meaningful failures to learn in, 113–115
    numerousness judgments by, 166–168
    path selection by, 178, 191
    personality studies of, 53
Right-arm bias, 286–287
Ringling Brothers Circus, 9
Robotics, 177
Roloway guenon, 24
Romanes, George, 161
Roosevelt, Theodore, 240, 241
Rosenblatt, J. S., 84
Rumbaugh, Duane M., xiii, 7, 14
    and complexity, 90
    and cranial capacity, 275
    and emergent behaviors, 63
    on emergents, 110
    grammar use demonstrated by, 270
    on learning sets, 99
    lexigram research by, 227, 232
    photo of, 9, 10
    symbol research by, 257
    symbol use pioneered by, 38–39
    and transfer index, 92, 140

San Diego State College/University, 9, 90, 295–297
San Diego Zoo, 7–14, 47, 90, 296, 297
Sanje mangabey, 24
Savage-Rumbaugh, E. Sue, 227, 229, 231, 232, 257, 270, 297
*Scala Natura*, 176, 177
Scalar variability, 164–165
Scenarios, 19
Schema (term), 210
Schick, Kathy, 279
Schneirla, T. C., 82–84
"The Science of Animal Cognition" (E. A. Wasserman), 144
Science of self control, 104
Scientific zoos, 11–15
Scripts, 210
SDT. *See* Signal-detection theory
Selective rewards, 155
Self, 209
Self-concept studies, 106
Self-directed inputs, 120–121
Self-organization, 81, 87
Self-regulatory systems, 214
Semantic illusions, 238, 239
Semantic information, 250
Semantic transparency, 236
Sensory arrays, 120
Sequences of units, 141
Serial gap distance, 153, 154
Serial information, 146–147
Serial list integration, 152
Serial-probe recognition (SPR), 68–70
Setonaikai National Park (Japan), 32
Sexual selection, 57
Shakespeare, William, 238–240
Shaping, 140
Sherlock Holmes (fictional character), 117
Shettleworth, Sara, 105
Short-term memory, 213
Sibling species, 17
Sidman, M., 100, 106–107
Sifaka, 24
Signal-detection theory (SDT), 72–73
Sign language, 34, 226, 227–230, 270
Sign significate, 138

# SUBJECT INDEX

Silky sifaka, 24
Simian immunodeficiency virus, 54
Simplicity principle, 66, 197–199
Simultaneous chaining technique, 146–151
Skiena, Steve, 186
Skinner, B. F., 81, 93, 103, 104, 109, 121, 140, 227
Sleeping arrangements, 34
Smith, Maynard, 83
Sociability, 53, 54
Social learning, 269–270
Social sharing, 263
Software, husbandry, 30–31
Space (concept), 176, 184, 195
Speech/spoken language
  as criterion for culture, 22
  human, 20–21
Spontaneous alternation behavior, 139
SPR. *See* Serial-probe recognition
Squirrel monkeys, 148
S-R. *See* Stimulus–response
Stability, 101–102
Staddon, John, 99
Stebbins, G. L., 83
Stewart, I., 94
Stimuli, 137
  environmental, 272
  gestalts vs., 199–200
Stimulus alternators, 139
Stimulus control, 120–121
Stimulus equivalence, 106
Stimulus–response (S-R), 91, 104, 207, 293
Stone knapping. *See* Knapping stone
Stroop-like interference, 167
Subitization, 165
Subjective well-being (SWB), 54–57
Successive inclusion, 149
Sumatran orangutan, 24
Summation, 165, 166
Surili, 24
*Survey of the Primates* (film), 8, 297
SWB. *See* Subjective well-being
Symbol combination in apes, 255–266
  awareness of human state of mind, 262–263
  cladistic analysis of, 256–257
  cultural relations, 262
  data collection/recording of, 257–258
  integration of communication modalities, 264
  intentional action, 260–261
  nature/size/evolutionary significance of, 258
  ordering convention, 263–264
  parsing, 259–260
  participants in study of, 257
  preferred orders, 261–262
Symbolic distance effect, 148–149, 167
Symbolic learning, 91

Symbols/symbology, 21–23, 38–39, 141
Symbol system, 235–236
Syntax, 228, 236, 240–242, 270
Systematicity, 236, 238, 239

Tamarin, 24
Tana River mangabey, 24
Tana River red colobus, 24
Tanzania, 35
Task-intrinsic motivation, 127–128, 130
Taxis level, 83
Technology, 38–39
10-item serial list, 149, 150
Terrace, Herbert, 232
Test, operate, test, exit (TOTE) unit, 207–208
Theoretical behaviorism, 99
Theory of mind, 263
Thomas, R. K., 110–111
Three-body problem, 101
Throwing technique, 284
Thumbs, 20
Thumb-to-index fingertip pinch, 20
TI. *See* Transfer index
Tobach, Ethel, 83, 84
To-be-learned response, 137
Tolman, Edward, 126, 138, 142, 197, 203
Tomasello, M., 23
Tonkin snub-nosed monkey, 24
Toolmaking, 270
  ape vs. human, 274
  by Kanzi, 279–290
Tool use, 106
TOTE unit. *See* Test, operate, test, exit unit
Toth, Nicholas, 279–281
Touch-screen testing systems, 39
Training, 231
Transactional perspective on human mental ability, 127–132
  individual differences, 127
  intelligence vs. cognitive processes, 127–130
  motivation, 127–128
  multidetermined ability, 127
  multifaceted ability, 127
Transfer index (TI), 13, 91–92, 140–141
Transient memory, 213
Treat, Wolcott, 295–297
Trigrams, 139
Typicality ratings, 210

*Umwege*, 175
Uncertainty monitoring, 64–76
U.S. Naval Medical Research Institute, 296

Value transfer theory, 147
Variability effects, 210
Viewing public, excessive interest in the, 54

Virtual reality (VR), 40–44
　challenges with, 43
　and imagination, 40
　immersive environment of, 40
　interactive nature of, 40
　promise of, 42–43
Visual-density discrimination task, 64–66
Voas, Robert B., 296
Vocal production, 226–228
　in chimpanzees, 244–245
　and gestures, 246
VR. *See* Virtual reality

Wasserman, E. A., 213
Wernicke's area, 243, 244
Western purple-faced langur, 24
"What Is It Like to Be a Bat?"
　(Thomas Nagel), 48
White-cheeked gibbon, 9
White-collared lemur, 24
White-headed langur, 24
White-naped mangabey, 24
William's syndrome, 67

Wisconsin General Test Apparatus, 12, 90, 145, 152, 165
Wolfram, S., 89
Working memory, 212, 213
Workshop on Gestural Communication in Nonhuman and Human Primates, 264
World Heritage Sites, 25
Wundt, Wilhelm, 141
Wynne, Clive, 105

Yellow-breasted capuchin, 24
Yerkes, Robert, 8–10, 47–48
Yerkes Primate Research Center, 10, 297
Yunnan snub-nosed langur, 24

Zoo Atlanta, 7–8, 11
Zoological Society of Atlanta, 7
Zoos
　as classrooms, 8
　levels of U.S., 11
　need for, 7–8
　pragmatic model for, 8
　as scientific resources, 11–15

# About the Editor

**David A. Washburn, PhD,** is professor of psychology and director of the Language Research Center at Georgia State University, Atlanta, from which he received his PhD in 1991. His research is broadly focused on the emergence of cognitive competence. More specifically, he and his collaborators investigate the significance of individual and group (including species) similarities and differences in attention, executive functions, and learning. He has authored or coauthored about 100 journal articles, book chapters, and other publications, including *Intelligence of Apes and Other Rational Beings*, a book coauthored with Duane M. Rumbaugh. His research enjoys current and recent support from the National Institute of Child Health and Human Development, the U.S. Army Medical Corps and Materiel Command, the Federal Aviation Administration, and other agencies. Additional information about this research and the Language Research Center is available at http://www.gsu.edu/lrc.

**DATE DUE**

| | | | |
|---|---|---|---|
| OhioLi | | | |
| JAN 1 5 REC'D | | | |
| NOV 2 1 2008 | | | |
| OCT 3 1 REC'D | | | |

GAYLORD PRINTED IN U.S.A.

SCI QL 737 .P96 P738 2007

Primate perspectives on behavior and cognition